T5-CVD-894
R0172066502

COMBUSTION AND FLAMES

Chemical and physical principles

COVER PHOTOGRAPH

A qualitative study of the structure of turbulent flames using the laser tomography technique, conduced at the *Office National d'Études et de Recherches Aérospatiales* (ONERA) (French National Institute for Aerospace Studies and Research)

The photograph shows a laminar diffusion flame created by a jet of methane in air at ambient temperature. In accordance with the method proposed by L. Boyer of University of Provence, France, the methane and the air are seeded with very fine particles which disappear above a certain temperature. When the flame is illuminated by a plane of laser light, the low-temperature regions of the flame zone are revealed, in green on the photograph. The luminosity created by the flame, seen in yellow, is due to the soot formed in the high-temperature zone, and predominates over the green light of the laser. Photograph taken using ONERA facilities during J.P. Dumont's doctoral thesis.

FROM THE SAME PUBLISHER

- Direct Numerical Simulation for Turbulent Reacting Flows
  Edited by T. BARITAUD, T. POINSOT, M. BAUM
- Principles of Turbulent Fired Heat
  G. MONNOT
- Fuels and Engines
  Technology. Energy. Environment (2 vol.)
  J.-C. GUIBET
- Automobiles and Pollution
  P. DEGOBERT

Roland BORGHI
Professor at the University of Aix-Marseille II

Michel DESTRIAU
Professor at the University of Bordeaux I

with the collaboration of

Gérard DE SOETE
Senior Engineer at the Institut Français du Pétrole

COMBUSTION AND FLAMES

Chemical and physical principles

Translated from the French
by Richard Turner

1998

ÉDITIONS TECHNIP 27 RUE GINOUX 75737 PARIS CEDEX 15

The figures and tables listed below have been printed in this book with the permission of the publishers.

For each figure and table :
– the author, year and title of the publication are given in the caption of the figures and/or in the title of the tables.

- *Académie des Sciences, Paris* : figure 6.11b.
- *American Chemical Society* : figures 2.4, 2.5a.
- *American Institute of Aeronautics and Astronautics – AIAA* : figures 5.13, 6.14, 8.7.
- *American Society of Mechanical Engineers* : figure 9.12.
- *Chapman & Hall* : table 7.1.
- *The Combustion Institute* : figures 2.3, 4.1, 5.7, 6.12b, 6.13a, 6.25, 6.26b, 6.27, 6.31, 6.32, 6.33, 6.36, 8.8, 9.3, 9.9a, 10.4.
- *Éditions Européennes Thermique et Industrie* : figures 10.5, 10.6, 10.7.
- *Elsevier Science Publishing Co, Inc.* : figures 6.10, 6.37, 8.1a, 8.2, 8.4, 8.5, 8.6.
- *Gordon & Breach Science Publishers* : figures 6.17, 6.18, 6.19, 6.35, 6.38.
- *Louvain University Press* : figure 5.11.
- *Masson* : table 7.1.
- *Pergamon Press, Inc.* : figures 6.28, 6.34, 9.10b.
- *The Royal Society* : figure 6.15.
- *Éditions Scientifiques Elsevier* : figure 4.3, 4.4.
- *Société de Chimie Physique* : figure 2.5b.
- *Stanford University Press* : figure 6.30.
- *University of California Press* : tables 7.1, 7.3.
- *Williams and Wilkins* : figure 6.2.

Translation of
La combustion et les flammes,
R. BORGHI, M. DESTRIAU

Published with the Financial Aid of the French Ministry of Culture.

ISBN 2-7108-0740-8

FOREWORD

The aim of this book is to provide a balanced introduction to a scientific study of combustion and flame phenomena for new entrants to the field. It can be understood by readers with a basic scientific background in the various relevant aspects of physics at about second year undergraduate level.

Modern society utilizes combustion and flame phenomena in many ways including transportation, power generation and chemical process engineering. On the other hand society also demands that all the resulting combustion processes be clean, safe and efficient.

Combustion science involves the study of complex interactions between processes of chemical change and physical phenomena in the field of fluid mechanics and heat and mass transfer. Although it has a long history, until recently, combustion science has had relatively little impact on engineering practice, which has relied on empirical "cut and try" design methods. However the situation has now clearly changed. Progress to more scientific design methods was encouraged in part by growing environmental concerns because the traditional empirical techniques proved inadequate to optimize performance while satisfying increasingly severe emissions legislation. There have been great advances in our understanding of all aspects of the subject while the dramatic and continued growth in computing power has made possible the application of this new understanding in engineering practice.

Because this book is addressed to new entrants to the field it does not set out to provide exhaustive coverage of all the most recent developments. A unified theoretical approach is adopted leading to a clear and well balanced description of the most important chemical, fluid mechanical and heat and mass transfer aspects of the subject.

Borghi and Destriau have written a book providing a firm foundation for more detailed scientific study of a topic which has always fascinated mankind. It also helps readers to understand how combustion processes can be more efficiently controlled and utilised for the benefit of society.

K.N.C. Bray
Emeritus Professor
University of Cambridge

PREFACE

Combustion and flames are very common features of nature that we accept in everyday life much like we do gravity. However, unlike the latter, they are highly complex phenomena insofar as they are produced by the interplay of several elementary phenomena, each associated with a different aspect of the physical world: thermodynamics, heat and mass transfer, the mechanics of fluids and chemistry all play a part.

The purpose of this book is to provide an initial introduction to combustion and flames through studying the various interactions of the physico-chemical phenomena involved. This book is aimed at anyone wishing to further their knowledge in this field, and can be understood by readers with a basic scientific background, of about second-year undergraduate level, in the various aspect of physics involved.

With this purpose in mind, it has not been possible to explore fully the richness of the physical and chemical behaviour associated with flames, nor to describe details of the theoretical methods that have been created for their study. Only the main aspects have been presented, simplified such that only their essential elements have been retained. This decision was based on the authors' desire to explain clearly the phenomena involved in each case in physical terms, and to discuss the existing theoretical models. For this reason, no chapter truly exhausts its subject matter, even the more traditional chapters on thermodynamics and transport phenomena. Moreover, certain chapters have been left out altogether since their usefulness was judged to be secondary (such as a chapter on numerical calculation methods for flames).

This book has been written simply to offer an intelligent introduction to the study of combustion and flames. Other works, more advanced, but possibly less complete and certainly presenting greater initial difficulty, have been or will be written. The authors hope that this book will provide the reader with a more complete and more rapid understanding of this fascinating field of such wide practical utility.

CONTENTS

Foreword V

Preface VII

Notation and usual units XVII

Introduction
DISCOVERING COMBUSTION AND FLAMES

1 **Fire and combustion: multi-aspect phenomena** 1

2 **Combustion and flames in industry and nature** 6

A. Engines 7

B. Burners 18

C. Fires 23

3 **The spirit and contents of this book** 26

Chapter 1
COMBUSTION THERMODYNAMICS

1.1 **Phenomena and definitions** 31

1.2 **A macrophysical perspective of the First and Second Laws of thermodynamics: adiabatic temperature produced by combustion in a closed system** 33

1.2.1 Consequences of the First Law: heat evolved, enthalpies of formation 34

1.2.2 Consequences of the Second Law in combustion chemistry 38

1.2.3 Calculating the adiabatic combustion temperature 40

1.3 **A microphysical perspective of the concept of temperature** 44

1.4 **Industrial calorific value** 46

1.5 **Equivalence ratio and Ostwald diagrams** 47

1.5.1 Equivalence ratio 47

1.5.2 Ostwald diagrams 48

Worked examples 51

Chapter 2
CHEMICAL KINETICS APPLIED TO COMBUSTION

2.1 General points and definitions ... 61

2.2 Rates of elementary reactions: definition and measurements ... 63

2.2.1 Definition ... 63
2.2.1.1 In a closed system ... 63
2.2.1.2 Open systems: relationship between flow of matter and rate, or extent of reaction ... 64

2.2.2 Measuring ... 66
2.2.2.1 Measuring in a closed system ... 66
2.2.2.2 Measuring in an open system ... 66

2.3 Experimental rate laws for elementary reactions ... 72

2.4 Rates of reaction for balance reactions: the steady state approximation ... 72

2.5 Experimental rate laws for balance reactions ... 75

2.5.1 Concentration dependence ... 75

2.5.2 Temperature dependence ... 77

2.5.3 Total-pressure dependence ... 77

2.6 Theoretical laws for elementary reaction rates ... 79

2.6.1. Molecular collision theory ... 79

2.6.2 Activated complex theory ... 79

2.7 Chain reactions ... 84

2.8 Chemical kinetics of combustion ... 85

2.8.1 General points ... 85

2.8.2 The H_2—O_2 reaction system ... 86

2.8.3 The CO—O_2 reaction system ... 86

2.8.4 Combustion of hydrocarbons ... 87

Worked examples ... 88

Chapter 3
MASS AND ENERGY TRANSPORT BY CONVECTION AND DIFFUSION

3.1 Relevance of transport phenomena to flames ... 93

3.2 Molecular diffusion of mass, energy and momentum ... 95

3.2.1 A simplistic description of molecular diffusion 95
3.2.2 Heat conduction 100
3.2.3 The diffusion of momentum 102

3.3 The balance equations for aerothermochemistry 103
3.3.1 Balance equations in a one-dimensional medium 104
3.3.2 Equations for a three-dimensional medium 106
3.3.3 Specific simple forms of the balance equation for energy 107

3.4 Simplified calculation of a laminar diffusion flame 109
3.4.1 Principal equations 109
3.4.2 Specific relationships 112
3.4.3 Very fast chemical reactions 113
3.4.4 The final step 114

3.5 Turbulent transport 116
3.5.1 What is turbulent flow? 116
3.5.2 Balance equations describing "mean" turbulent flow 117
3.5.3 Turbulent diffusion fluxes 121

Worked examples 123

Chapter 4
SELF-IGNITIONS IN CLOSED SYSTEMS

4.1 General points 125

4.2 Temperature distribution in a closed chamber containing a combustible mixture 127

4.3 Explosion limits 129

4.4 Octane number 135

4.5 Explosion theories 136
4.5.1 Thermal theory of explosions 137
4.5.1.1 Balance equations and critical conditions for a non-convective system 137
4.5.1.2 Balance equations and critical conditions when the cell temperature is equalised by convection (for diathermal walls) 141
4.5.1.3 Temperature and concentration profiles in the case of a closed system in the absence of convection 144
4.5.2 Theory of isothermal chain branching 146

Worked examples 149

Chapter 5
LAMINAR FLAMES AND DEFLAGRATIONS

5.1 General points 153

5.2 The stability and propagation of a premixed flame front in laminar flow 154

5.3 Quenching 161

5.4 Premixed flames: the Hugoniot-Rankine equation 162

5.5 Premixed flames: deflagration propagation mechanism 164

5.5.1 Thermal balance 164

5.5.2 Mass balances 168

5.6 Chain-branching theory 170

5.7 Flammability limits 171

5.8 Laminar diffusion flames 173

5.9 Theoretical results for laminar diffusion flames 174

5.10 Diffusion flames in more realistic chemical applications 176

5.11 Stabilisation of laminar flames 178

5.11.1 Problem definition 178

5.11.2 Stabilisation of a premixed, Bunsen burner flame 181

5.11.3 Stabilisation of diffusion jet flames 183

5.11.4 Stabilisation in a recirculation zone 185

Worked examples 190

Chapter 6
TURBULENT FLAMES AND DEFLAGRATIONS

6.1 What is a turbulent flame? 195

6.2 Fundamental concepts and characteristics of turbulence 198

6.2.1 The kinetic energy of turbulence 199

6.2.2 The integral turbulent length scale 200

6.2.3 The "spectrum" of turbulent length scales 201

6.2.4 General overview of the "$k - \varepsilon$" turbulence model 204

6.3 The structure of turbulent premixed flames 206
6.3.1 The three types of turbulent premixed flame 206
6.3.2 Experimental evidence 210
6.3.3 Burning velocity and thickness of a premixed turbulent flame 216
6.4 "Modelling" premixed turbulent flames 221
6.4.1 The context of the modelling problem 221
6.4.2 The "Eddy Break-Up" model 223
6.4.3 The case where the chemical reactions are not infinitely rapid 226
6.5 Detailed example: a turbulent gas burner 228
6.5.1 Velocity field 228
6.5.2 The average rate of reaction field 229
6.6 The structure of turbulent diffusion flames 232
6.6.1 Description of the structure of turbulent diffusion flames 233
6.6.2 The various types of turbulent diffusion flames 236
6.6.3 Experimental evidence 240
6.7 Modelling non-premixed, turbulent flames 245
6.7.1 What should be calculated and how? 245
6.7.2 Calculating non-premixed turbulent flames, assuming rapid chemical reactions 247
6.7.3 Calculating an inert species in a non-premixed turbulent flame 252
6.7.4 Comparison between calculation and experiment 256
6.8 The influence of combustion on turbulence 260
Worked examples 267

Chapter 7
DETONATION AND SUPERSONIC COMBUSTION

7.1 Shock and detonation: phenomena and definitions 269
7.2 One-dimensional or "planar" shock waves in gases 273
7.3 One-dimensional (or planar) detonations in a gas: the Chapman-Jouguet condition 275
7.4 One-dimensional shocks in a gas: properties as a function of Mach number 280
7.5 One-dimensional detonations in a gas: properties as a function of Mach number 281

7.6 Limits of detonability 284

7.7 Appearance of complex three-dimensional structures 284

7.8 Detonations in condensed media 287

7.9 Supersonic flow with premixed combustion 289

Worked examples 293

Chapter 8
FLAME IGNITION

8.1 The ignition phenomenon 299

8.2 Hot spot ignition 302

- 8.2.1 Results of a typical calculation 302
- 8.2.2 Physical explanation 304

8.3 Ignition through energy deposition 308

8.4 Ignition with consecutive pressure perturbations 312

Chapter 9
COMBUSTION OF LIQUIDS AND SPRAYS

9.1 General points 319

9.2 Combustion of an isolated drop 321

- 9.2.1 Quasi-stationary theory of the combustion of a spherical drop 322
- 9.2.2 The d^2 rule 327
- 9.2.3 Combustion of a drop in motion 329

9.3 Combustion of a spray of droplets 331

- 9.3.1 General points 331
- 9.3.2 Combustion in premixed sprays: the various types of flames 332
- 9.3.3 Rate of combustion and flame velocity in premixed sprays 338

9.4 Diffusion jet flames with droplets 342

- 9.4.1 Description 342
- 9.4.2 "Modelling" droplet-spray diffusion flames 345

Chapter 10
POLLUTANT EMISSIONS IN COMBUSTION REACTIONS

10.1 General points 349

10.2 Pollution by CO_2, CO, aldehydes, small hydrocarbons, alkenes, peroxides, tetraethyl lead and PAN 351

10.2.1 CO_2 351

10.2.2 CO 352

10.2.3 Aldehydes, alkanes, alkenes, peroxides, tetraethyl lead and PAN 353

10.3 Pollution by soots and polyaromatics 355

10.4 Pollution by nitrogen oxides (NO_x) 357

10.4.1 General points 357

10.4.2 Gas-phase mechanisms leading to NO_x emission 358

10.4.3 Heterogeneous production and reduction of NO and N_2O 362

10.4.4 Controlling NO_x emissions 362

10.4.4.1 In the gas phase 362

10.4.4.2 Heterogeneous control 363

10.5 Pollution by sulphur oxides (SO_x) 365

10.5.1 General points 365

10.5.2 Mechanisms leading to SO_2 and SO_3 in flames 365

10.5.3 Controlling SO_x emission 366

INDEX 367

NOTATION AND USUAL UNITS

- a m/s, velocity of sound
- a_T m^2/s, thermal diffusivity
- A_T m^2/s, turbulent thermal diffusivity
- C progress variable
- C, c m/s, shock velocity
- c_p, c_v kJ/kg K, specific heat (at constant p, constant v)
- D_i m^2/s, diffusion coefficient of a species i
- d common value of all the D_i, if equal
- d_T turbulent diffusion coefficient
- E activation energy
- G, g Gibbs function (free enthalpy)
- H, h enthalpy
- ΔH° standard enthalpy of reaction
- k turbulence kinetic energy
- K_p (bar units), dissociation constant
- l m, mean free path
- l_t integral turbulent length scale
- p bar, pressure
- R J/mol K or J/kg K, gas constant
- S, s entropy
- S_L burning velocity of a laminar flame
- S_T burning velocity of a turbulent flame
- U, u internal energy
- V volume
- v m^3/kg, specific volume
- V_j, v_j m/s, speed
- w reaction rate
- X_i molar fraction
- Y_i mass fraction
- Z mixture fraction
- $\mathrm{Z_i}$ Zeldovitch variable
- z extent of reaction
- γ ratio of specific heats, c_p/c_v
- η Kolmogorov length scale
- λ W/m K, thermal conductivity
- μ dynamic viscosity
- μ_i chemical potential
- ν m^2/s, kinematic viscosity = μ/ρ
- ν_i stoichiometric coefficient
- ρ kg/m^3, density (or volumic mass)
- τ_c chemical time
- τ_K Kolmogorov time
- τ_L Lagrangian integral time scale
- τ_t integral turbulence time scale
- $\tau_{\alpha\beta}$ friction tensor

Dimensionless numbers

- Damköhler, $Da = \tau_t/\tau_c$
- Lewis, $Le = a_T/D_i$
- Mach, $Ma = V/a$
- Peclet, $Pe = ve/d$ (or $S_L d/a_T$ in some combustion problems)
- Prandtl, $Pr = \mu c_p/\lambda$
- Reynolds, $Re = v_{\mathrm{ref}} l_{\mathrm{ref}}/\nu$
- Turbulent Reynolds, $Re_T = k^{1/2} l_t/\nu$
- Schmidt, $Sc_i = \mu/\rho D_i = \nu/D_i$

(v is a speed, ν the kinematic viscosity, ν_i the stoichiometric coefficient of the i-species)

INTRODUCTION

DISCOVERING COMBUSTION AND FLAMES

1 FIRE AND COMBUSTION: MULTI-ASPECT PHENOMENA

Combustion is an extremely widespread phenomenon in nature, and it is mainly due to man's ability to use it to his advantage that human activity has developed, and continues to develop; although sometimes with undesirable consequences! From ancient times up until the middle ages, **fire** was considered to be one of the four basic elements in the universe. The study and understanding of all aspects of combustion and fire is, therefore, of paramount importance.

But what is fire, what precisely is combustion, and indeed what is a flame? Scientists have striven for centuries to determine answers to these questions. In today's industrialised world innumerable applications exploit the phenomenon of combustion, although far from optimally in many cases. This is because difficulties arise from the fact that combustion is the result of the interplay of various, quite different, physical and chemical phenomena.

The fundamental phenomena are based first on **chemical aspects**. In any **combustion** process there is rarely simply one chemical reaction involved, but numerous, all occurring simultaneously. There are various types of combustion reaction, dependent mainly on the reactants involved. The first aspect of fire is therefore a chemical aspect; and from now on will be referred to more specifically as combustion.

The second aspect is that of **heat transfer.** The chemical reactions mentioned above raise the temperature of the medium, and are themselves highly sensitive to temperature. Heat transfer will occur whenever there is a difference in temperature from one point in a medium to another. In a flame produced by combustion processes, heat is transferred by conduction, sometimes by radiation, and often by diffusion resulting from the vortex motion known as turbulence. Whilst heat transfer may sometimes only be implicated on a secondary level in terms of the path followed by the chemical reactions of combustion, they are occasionally essential, for example in the case of flame propagation.

Most flames burn in gaseous media, which explains the involvement of the third aspect: **mass transfer**. There are two types of this form of transfer: the convective motion of some or all of the gases which make up the flame, or the transfer by diffusion (molecular or turbulent) of certain species in relation to others in the medium. Gas motion is generated either by the flow supplying the flame, or is due to buoyancy effects (in the gravitational field) which cause hot gases to rise and cooler gases to be drawn in to replace them. The phenomena related to the diffusion of certain species, on the other hand, result from significant differences in the gas composition from one point in the flame to another. In certain cases the motion caused by convection and diffusion plays an important role, and can produce remarkable specific features. One striking example of this is a candle burning in zero-gravity, which produces large volumes of smoke, and often goes out after a short while: this example clearly shows that the function of supplying unburnt gas and removing burnt products, particularly soot, normally provided by natural convection is an essential prerequisite in enabling the flame to self-sustain. This third aspect is related to the **mechanics** and physics of fluids.

To illustrate more fully how these three aspects interplay and combine in a common flame, let us first consider the flame produced by a candle. This apparently simple flame is in fact regulated by a complex series of phenomena; indeed it is practically the most complex of all flames.

• The flame is located just above the wick, in a medium which is clearly gaseous. The candle is made of a solid material, the top of which melts to produce a pool of liquid at the base of the flame which soaks into the wick. Let us first consider the gaseous medium, the flame itself, and describe the interactions between the three related aspects, chemistry, heat transfer and the mechanics of the system.

• The chemical aspect, the combustion itself, does not immediately appear to be easy to understand. However, it is the very chemistry of the system which governs the combustion processes occurring in the heart of the flame. The system can be described as being a set of chemical reactions whose principal reactants are oxygen in the air, and gaseous vapours evaporating off the wick. The candle is made up mainly of stearin, a product containing carbon, hydrogen and a few atoms of oxygen. Although solid at room temperature, stearin liquefies when heated gently (at about 150°C) to become gaseous; a process which is associated with a decomposition into lighter species.

The chemical reactions which occur involve not only oxygen and these species, but also the resulting atoms and radicals: O, OH, CH_3, HCHO, etc. The presence of these radicals can be confirmed by taking precise measurements inside the flame, as well as simply by observing the slightly blue colour at the base of the flame, produced by the emission of visible light by the CH radical. The various chemical reactions form CO_2 and water vapour, and increase the temperature of the medium. However, in certain cases, the combustion process also forms soots, which are particles of very slightly hydrogenated carbon. Moreover, there is more soot in the flame than above it, indeed the particles of hot soot are only visible due to the yellow light that they emit.

- The physical aspect of heat transfer is clearly apparent, since it is partly produced by the yellow radiation which is the most visible feature of the flame. This radiation is responsible for the candle's primary function, but it is also essential insofar as it is the high temperature produced by the flame which heats and liquefies the top of the candle and then vaporises the liquid wax to create the gaseous fuel which then reacts within the flame.

- The role of the mechanical aspects of the system, in combination with the above two aspects, is to bring together the reactants. In simple terms, the buoyancy of the hot gases and burnt products (consisting mainly of atmospheric nitrogen, CO_2 and water vapour) enables them to rise and be replaced by entrained cooler, unburnt air, in addition to fuel vapours from the wick. Furthermore, these two groups of different chemical species mix intimately by molecular diffusion in the flame. The molecules approach sufficiently closely to other molecules and to the atoms or radicals in the flame to enable chemical reactions to occur. The molecular diffusion of the various species is as important as the diffusion of heat

In addition to these three primordial aspects, a candle's flame also brings into play secondary aspects. The physical phenomena of state changes: liquefaction and vaporisation, have already been mentioned. In addition, the phenomena associated with the nucleation of soot particles, followed by their growth and coagulation are very important factors especially when studying the generation of smoke by flames. Heat conduction in the porous wick, and the capillary attraction of liquid stearin are further physical phenomena associated with the production of the flame (and which may extinguish it in certain cases...).

All these elementary phenomena in the flame are closely linked, and each play their part in the overall process. The flame can only continue to burn if each element functions correctly. Moreover, a well-defined structure is produced since each element has its own characteristic position in relation to the flame, although clearly its geometric form can vary slightly depending on the individual conditions. Figure 1 shows a simplified representation of the structure of a candle's flame. The zone where the chemical reactions occur separates a gaseous medium containing oxidising gases (outside the flame) from a gaseous mixture containing reducing gases (within the flame, around the wick). This is the case for numerous other flames, all of which will be referred to as diffusion flames, or "non-premixed flames".

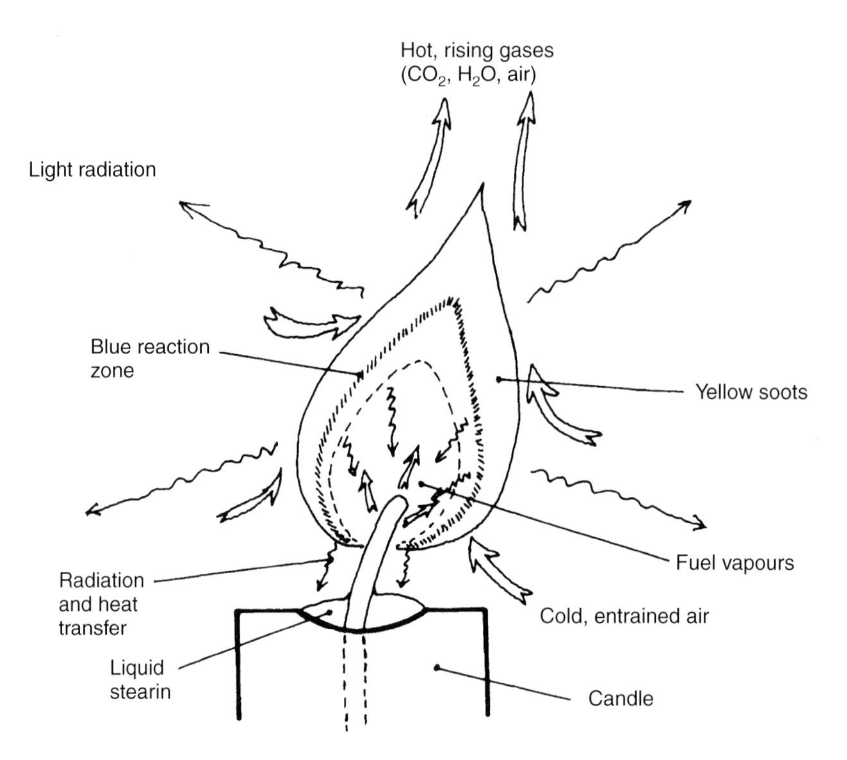

Figure 1 The various physical phenomena involved, at different positions, in the candle's flame.

However, the manner in which the various physico-chemical phenomena interact is not always the same, and consequently not all flames resemble a candle's flame; indeed they are not all diffusion flames. Fortunately, there are only two types of flame. If a flame is not of the type described above, i.e. not premixed, it must therefore be premixed. In fact, in the case of candles, the chemical reactions require the presence of two different types of species (air (the oxidant) and the vapours produced by the candle (reductants)) which approach the combustion site from different spatial positions. In the case of premixed flames, the chemical reactions require either a single chemical substance (in which case it is a potentially dangerous "explosive"), or two different substances which are mixed before they reach the combustion site. Practical examples of premixed flames are gas burners or a spark-ignition engine. A flame of this type can be produced very easily (and safely!) to show its principal property, which is to diffuse, in a simple experiment that anyone can carry out, so long as a few basic safety procedures are respected. This experiment is described below (taken from Pedro Garcia Ybarra, of the *Universidad Nacional de Educación a Distancia*, Madrid).

Take an empty glass bottle, and fill it with gas from an unlit gas lighter. The best way to do this is to hold the lighter in the mouth of the bottle for about 30 seconds (for a one-litre bottle). Then mix the contents of the bottle with a long rod, blocking the mouth with your hand to limit any escape of gas. Finally light the lighter and hold the flame in front of the mouth of the bottle.

So long as you do not put your fingers or your face too close to the opening the experiment will not be dangerous, since even though the temperature of the burning gas is very high (about 1500°C) the quantity of energy produced is very small and only causes a slight increase in the temperature of the bottle itself, and even less that of your hand. A flame, seen as a thin blue surface, forms in the neck of the bottle and propagates towards the base of the bottle, whilst remaining more or less perpendicular to its direction of propagation (see Figure 2).

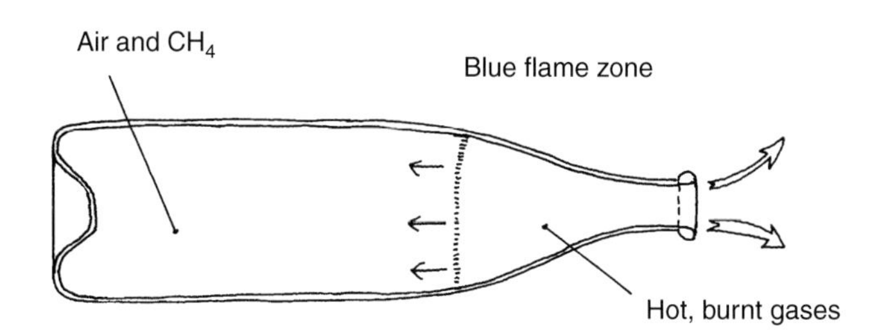

Figure 2 Flame propagation in a premixed blend of gases; an experiment that may be performed in a bottle, so long as basic safety precautions are respected.

In this experiment, the flame, a fine, bluish propagating zone, is very different from the flame produced by a candle. However, the same (or at least very similar) phenomena are involved based on the three principal aspects mentioned above.

Firstly, the chemical reactions associated with combustion still occur in the flame: i.e. reactions between oxygen and the gas from the cigarette lighter (mainly methane) and are very similar to those taking place in the candle's flame. Moreover, the blue colour of the flame is also due to the CH radical, and if the quantity of methane had been greater (in which case it would have been more difficult to ignite the blend of gases) then the flame would also have been characterised by soots radiating a yellowish light.

Heat transfer again plays an important role; not to illuminate the surroundings or vaporise the reactants but to heat the gaseous medium containing the premixed air and methane. This blend would not otherwise react spontaneously; a section of the mixture can only react and "ignite" if its temperature is first raised by heat transfer from a nearby flame (which is initially the case immediately surrounding the lighter's flame). The "flame front", as it travels down the bottle, separates the unburnt mixture (between the bottom of the bottle and the front) from the hot, burnt gases (between the front and the neck). The hot gases and the flame itself increase the temperature of a small portion of unburnt mixture ahead of the front by heat conduction which ignites, to heat the next portion of unburnt gas, which in turn ignites, and so on. This is the mechanism involved in the propagation of the flame towards the unburnt gas, as so clearly observed in the bottle.

Finally, mass transfer phenomena also participate in flame propagation, in two ways. Firstly, the burnt gases (CO_2, H_2O) tend to diffuse, at the same time as the heat, from the zone behind the flame into the unburnt gases ahead of the front, and secondly, the greater volume occupied by the heated gases cause these gases to escape from the bottle at high speed (a few metres per second), a result which it is not advisable to check with your finger...

The structure of premixed flames is quite fundamentally different from that of diffusion flames. Initially, one might assume that the basic nature of the flames is different, but this is not the case. The apparent contrasts can be explained by the differences in the interactions between the same three aspects : chemical, physical and mechanical.

For centuries, combustion theory has presented scientists with major difficulties. However, recognising that its behaviour may be explained by the combination of these three aspects, or principal elementary phenomena (in addition to other secondary phenomena) is an approach which simplifies all analysis, and opens the door to a more complete understanding.

2 COMBUSTION AND FLAMES IN INDUSTRY AND NATURE

Industry uses combustion and flames in numerous applications. The above descriptions of the two types of flame, the non premixed candle and the flame propagating in a premixed blend of fuel and air (in a bottle), are nonetheless already sufficient to enable us to understand qualitatively how the various industrial devices operate.

A. Engines

The first main category of applications is engines, which harness the heat energy released by combustion to provide the motive power for various vehicles (aircraft, rockets, boats, cars, etc.). Most of the general types will be described below, classified in order of increasing complexity.

The spark-ignition engine exploits combustion in a quite uncomplicated manner. The gas enclosed in the cylinder after the inlet valve(s) has closed is a blend of air and petrol vapours. This mixture is compressed, thus heated, and then ignited by a spark, whose timing ensures that the mixture burns over a specific time period. As the piston is pushed down by the expanding gases, the mixture (whose chemical nature has been changed by the reaction, and which had been heated to approximately 2500 or 2700 K) cools before finally being expelled into the exhaust pipe when the exhaust valves open. The piston then rises again. A highly idealised approach considers that combustion occurs over an infinitely short time when the piston is at its highest point (Top Dead Centre, TDC). This is not the case in practice since the spark is normally fired when the crank angle is at approximately 20 degrees before TDC. The reason for this is that complete combustion of the gases contained in the cylinder head requires the time equivalent to the crank advancing from 20° before TDC to 20° after TDC. If for example, we assume that the engine is rotating at 3000 r.p.m., this would mean that the "combustion phase" lasts for approximately $(40/360) \times (60/3000)$ seconds, i.e. about 2 milliseconds–not very long in human terms, but relatively slow compared to some chemical reactions.

What actually happens during these 2 milliseconds? In essence, exactly the same process as we observed with a premixed flame propagating along a bottle. The mixture enclosed in the cylinder is a substantially homogenous (so long as the droplets of petrol have been uniformly dispersed and correctly vaporised in the inlet manifold) pre-mixture of air and gaseous fuel. The electric spark, fired near one wall of the cylinder head, triggers the propagation of the premixed flame through the gaseous mixture. If the gas in the cylinder was at rest, the premixed flame would propagate equally in all directions, and would adopt a hemispherical form so long as the side walls did not disturb the propagation, as shown in Figure 3a. This is more or less what happens in engines designed to operate at low speeds, however, for most engines, the supply of gas via the valve, and the rise of the piston create motion and eddies in the mixture which do not have time to dissipate before the spark ignites the gas. The premixed flame thus propagates in a highly turbulent medium, and since in general the scale of the turbulent eddies is greater than the thickness of the blue flame (observed in the bottle experiment) then the flame front wrinkles, as shown schematically in Figure 3b. This phenomenon whereby the flame and turbulence are coupled to produce what is known as a "turbulent premixed flame" is very important; indeed an entire chapter is devoted to this phenomenon later in this book. In fact, this effect is highly beneficial, since it significantly increases the speed of propagation (more correctly termed "burning velocity") of the flame zone. It is now accepted as a fundamental design element implicit in the normal operation of modern engines. The reason for this is that a laminar flame front (i.e. in the absence of all turbulence) propagates at approximately 1 m/s inside the cylinder head. Assuming a piston travel of 4 cm, the flame has approximately 40 ms to propagate from the top to the bottom of the cylinder; if the engine runs faster than 750 rpm (80 ms per rotation) there is insufficient time for this to occur. Clearly, engines greatly exceed this figure (certain racing cars operate at over 12000 rpm...) and are able to do so due to turbulent propagation.

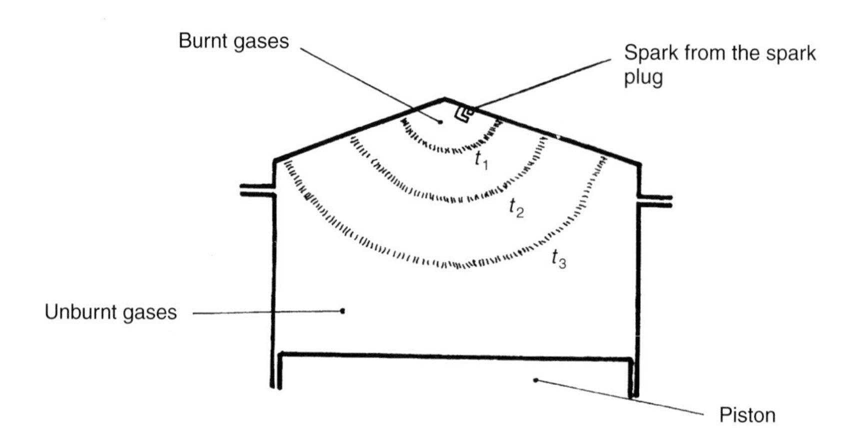

Figure 3a Flame front propagation in an idealised spark-ignition engine, without turbulence at various times t_1, t_2, t_3.

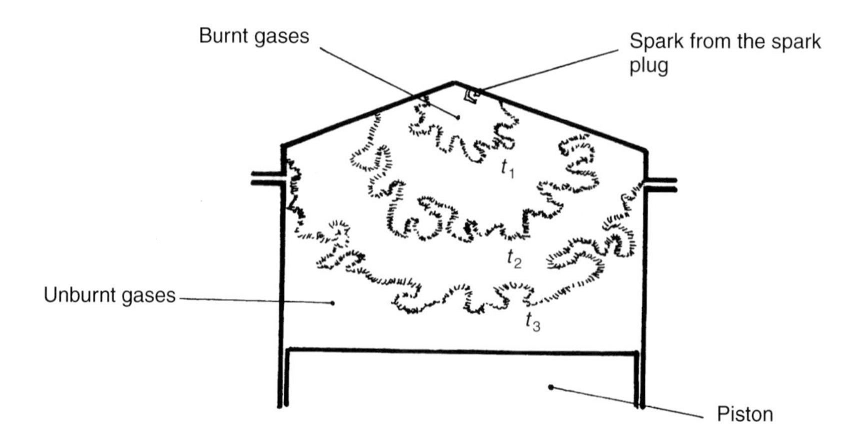

Figure 3b Turbulent propagation of the flame front in a spark-ignition engine at various times t_1, t_2, t_3.

The quantitative calculation of the times required for the "combustion phase", as a function of engine geometry and adjustment, the prediction of the temperatures reached on the cylinder walls and by the engine and the composition of the exhaust gases are all problems whose solution is of great importance when designing engines. Progress has been made in this field, although much remains to be achieved, especially in terms of the composition of the gases produced which may include pollutants.

In contrast to the spark-ignition engine which uses a premixed flame, **"Diesel engines"** are powered by the combustion processes resulting from a diffusion flame similar to that produced by a candle. The gas enclosed in the cylinder and compressed by the piston is simply air. At approximately 20° before TDC a jet of liquid diesel fuel is injected into the cylinder (in the case of direct injection, which we shall consider here). This jet of liquid develops into a "spray" of very fine droplets which disperse, mix with the air and burn as a diffusion flame with an elongated plume. Since the air is hot, the droplets are heated and quickly vaporise to produce the fuel vapours which spontaneously burn with the air. For the duration of the injection, a diffusion flame is obtained similar to that of the candle, but with the wick replaced by a very dense core of liquid droplets (see Figure 4). In fact, other than for the scale factor, the flame resembles closely that produced by a flame thrower in a circus. The liquid jet ignites a few fractions of a millisecond after being injected. In general, the injection continues for one or two milliseconds, and the combustion phase finishes a few fractions of a millisecond later. The piston is forced down and the exhaust valves open, in the same way as a spark-ignition engine.

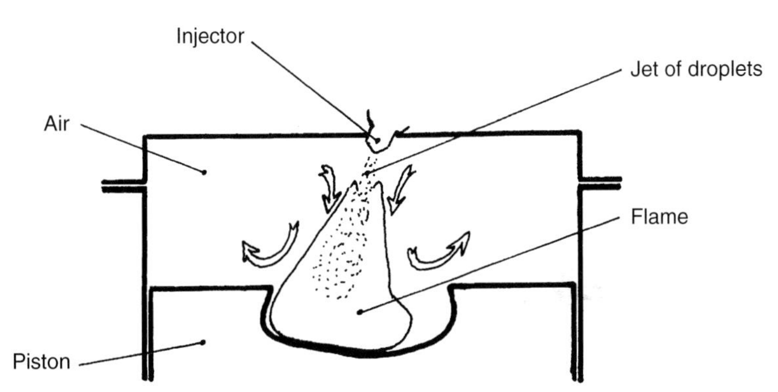

Figure 4 Schematic representation of a flame in a direct injection Diesel engine.

Again the combustion phase is not instantaneous, and lasts for the equivalent of about 40 degrees of crankshaft rotation. However, the major difference is in the structure of the flame, which is a diffusion flame with a fuel-rich core and an air-rich peripheral zone. The result is that a Diesel engine generally produces more smoke than a spark-ignition engine, i.e. it releases more soot into the environment via the exhaust fumes. This is offset to a certain extent by the greater overall thermodynamic efficiency of Diesel engines, due to the higher temperature of the compressed gases, which also enables more unrefined fuel to be used.

The reasons for studying the combustion phase in a Diesel engine are the same as those for a spark-ignition engine. However, the analysis for Diesel engines is more complex, mainly due to the presence of the jet of liquid which creates the spray of fine droplets. Understanding this aspect of combustion requires studying firstly how droplets are generated from a jet of liquid, and secondly the mechanisms involved in the combustion of a droplet/air mixture. This book will not cover phenomena related to the "spraying" or "atomising" of jets of liquid, which are in themselves fairly complex phenomena in the field of fluid mechanics, however, an entire chapter is devoted to the combustion of droplets and sprays of droplet.

Combustion is also responsible for providing the thrust produced by aircraft engines, and the last 50 years have seen tremendous advances in the development of engines other than piston engines. The simplest example of these is the **ramjet**, which can be used by an aircraft which has already reached, by other means, a speed of approximately twice the speed of sound. Its simplicity has led to it being nick-named "the flying flue pipe" and it is indeed no more than a cylindrical tube containing a few basic devices for injecting the fuel and for holding and stabilising the flame.

As in the case of a piston engine, the gases are compressed, burnt and then allowed to expand, although the flow is continuous through the engine and every small portion of gas undergoes these processes in succession as it follows its path through the engine. Compression occurs in the air intake, at the front of the engine, and expansion in the nozzle at the rear. Combustion occurs in the central part of the engine, in the combustion chamber, whose structure is shown schematically in Figure 5a. For subsonic-combustion ramjets, the air intake supplies a current of air at a speed much less than the speed of sound (at approximately Mach 0.1 or 0.2, i.e. 0.1 or 0.2 times the speed of sound). A spray of liquid fuel is injected into this current via a perforated pipe. The fuel used by aircraft engines is called kerosene, and there are some varieties, each differing slightly. At this stage, it is sufficient to know that it is a mixture of hydrocarbons with an approximate chemical formula of $C_{10}H_{20}$. At the air temperatures present (about 600 K) the sprayed fuel can disperse in the air and vaporise in exactly the same way as the spray injected into a Diesel engine. However, unlike the Diesel engine this mixture cannot spontaneously ignite under these conditions. A mixture of air and kerosene vapours is thus prepared, prior to combustion, in the first part of the chamber. A flame, or rather several flames are then lit and held in the second part of the chamber. The device which fixes and stabilises the flames is a ring-shaped gutter, called a "flame-holder". From each point on this device two premixed flames de-

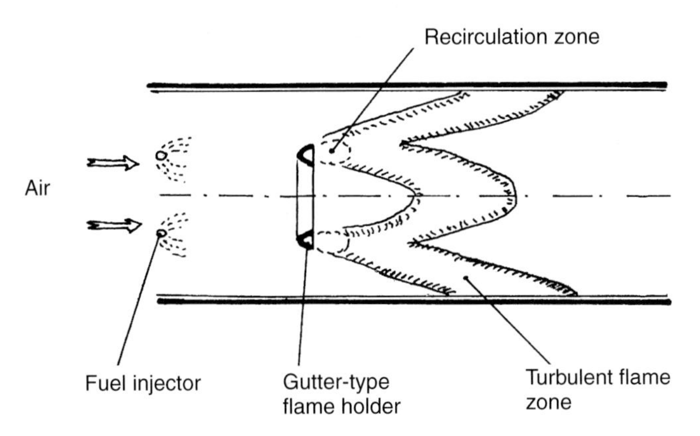

Figure 5a Schematic sketch of the combustion chamber of a cylindrical ramjet, with a single flame holder in the form of a toric trough.

velop, at an angle to the gutter, one towards the axis of the tube and the other towards its periphery. Once these two flames have reached the axis and wall respectively, then the entire flow of gases which passes through the combustion chamber must pass through the flame zone, and only burnt gases will enter the nozzle.

How does the flame-holder enable the flame to self-sustain, and why is the flame produced conical in shape? The answer to these questions is provided by a characteristic that was first introduced above, namely the burning velocity. A premixed flame propagates in a premixed medium at a velocity of the order of one metre per second. It is therefore possible to maintain a premixed flame stationary (i.e. relative to the chamber) simply by allowing a premixed flow to travel through the combustion chamber at this velocity. However, the flow velocity in the combustion zone is much greater than the burning velocity (approximately 50 m/s) which is why a flame-holder is required.

The gutter of the flame-holder generates in its wake what is known as a paired vortex system, and in which the gaseous flow velocity, as shown schematically in Figure 5b. This recirculation zone continuously exchanges mass with the surrounding flow, although the gas flow velocity entering and leaving this zone is much slower than 50 m/s, such that the (average) time that the gas spends in this zone (known as the residence time) is several milliseconds. Once lit, the combustion in this recirculation zone is self-sustaining, so long as the residence time of the gas in the zone is sufficiently long. The flame produced in the combustion zone does not have the same structure as the diffusion flame typified by a candle, or even the premixed flame propagating in a bottle, and it should not, strictly speaking, be called a flame, but rather a combustion zone. We shall study this zone in greater detail in a section devoted to flame stabilisation.

This recirculation zone thus forms a stable zone containing hot gases, which continuously releases a certain flow rate of these gases into the main flow. As a consequence of this exchange of hot gases, combustion is transmitted to the main flow and, since this flow is travelling towards the rear of the combustion chamber at high speed, the flame zone develops obliquely away from the recirculation zone. Away from this attachment zone, combustion again occurs in the form of a premixed flame, similar to that which propagates in a bottle. However, there are two main differences, one of which is minor, the other more important. The first is that the flame is oblique in relation to the flow, and thus that its so-called fundamental burning velocity is the component of the flow velocity normal to the flame. The more important difference involves the fact that the heat and mass exchange processes occur due to turbulent eddies, as in the case for piston engines. This turbulence is beneficial since it greatly increases the fundamental burning velocity and thus the overall angle that the oblique flame makes to the axis of the chamber. Without this effect, the combustion chamber of the ramjet would be so long that its weight would make the engine impractical.

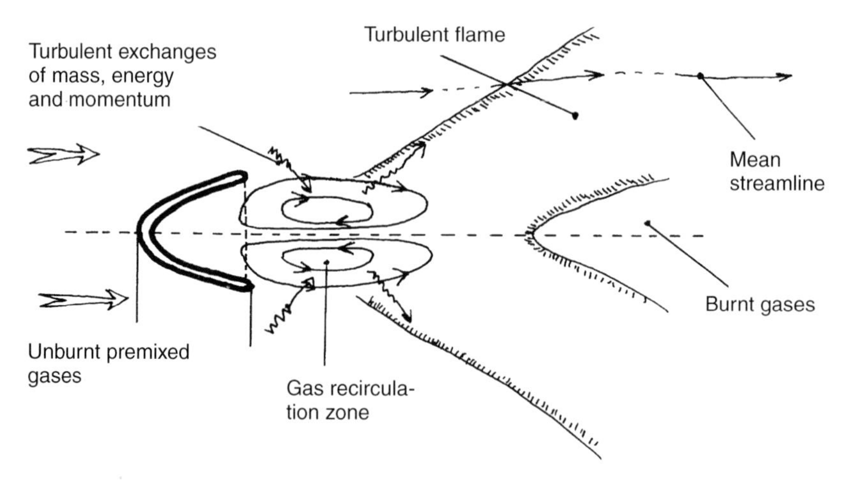

Figure 5b Detail of the stabilisation of a flame by a recirculation zone, behind the flame-holder.

Since the invention of the first ramjet by Leduc shortly after the Second World War, numerous improvements have been made to the basic design shown in Figure 5a; principally relating to the injection and flame-holding systems. The development of these systems requires a detailed study of turbulent premixed flames; a subject to which an entire chapter of this book is devoted. The study and understanding of the phenomena occurring around the flame-holders is especially important in improving the design of the latter.

Since they are unable to provide power at aircraft take off speeds, the potential applications for ramjets are relatively limited, and they are mainly used for certain military missiles. However, the principles developed for the combustion chamber of ramjets have been applied to "post-combustion" or afterburning in turbojets.

Turbojets are now used widely by virtually all types of aircraft. Turboprops, which power propeller-driven aircraft and helicopters, and gas turbines used by boats, trains and cars, are all derivations of turbojets. In these engines, air is drawn in and compressed by a rotating compressor (to pressures of approximately 20 or 30 atmospheres at take-off, although 4 or 5 times less than these figures when at altitude) and is then guided into a combustion chamber. A fine spray of kerosene is injected into the combustion chamber where it mixes with the air and burns. The mixture of burnt gases then gives up some of its energy in a first turbine stage to power the rotation of the compressor. At this point there are two possibilities depending on the type of turbomachine. In a pure turbojet the gases are exhausted directly at high speed via the nozzle, and it is the thrust provided by the jet of gas which propels the aircraft. Alternatively, the energy in the gases is extracted by further turbine stages to power the propeller (in **turboprops**) or the fan blades (in **turbofan** engines) and it is the thrust provided by the propeller or fan which moves the aircraft through the air. The combustion chamber for a turbojet is very different from that of a ramjet, for reasons which will be described below. A diagram is given in Figure 6a.

The combustion chamber of modern turbojets has an annular geometry, which is why only a half-section is shown in Figure 6a. Within the case (see Fig. 6b) is a "flame tube" inside which combustion is stabilised, and to which are attached the kerosene injectors in its upstream section. About fifteen injectors are arranged around the periphery of the flame tube, as indicated on the axial view shown in Figure 6b. The inside of the flame tube may be divided in the direction of flow into three zones: firstly the primary zone, which is fed with kerosene via the injector (which sprays the fuel in the form of fine drops) and also receives air via the injector itself as well as via the first row of lateral openings in the flame tube (called primary orifices). The secondary zone precedes the dilution zone, which is located downstream of the second row of flame tube openings (known as dilution openings). The flame tube is designed in this way so that the temperature of the hot gases at the turbine inlet is not allowed to exceed 1700 or 1800 K (since the turbine blades are unable to resist such high temperatures). A consequence of this restriction is that only 1/3 or 4/10 of the passing air can be used. However, the combustion zone stabilises more easily if the temperature is high, which is achieved if all the air is burnt. Thus, the primary zone is designed such that only 1/3 (or 4/10) of the air enters. The temperatures reached are therefore maximal (approximately 2200 to 2500 K, depending on the conditions), and combustion attachment is optimal. A further 1/3 of the air can then be mixed in the dilution zone with the gases from this primary zone, in order to reduce the temperature of the hot gases, and the remaining third (approximately) is injected along the walls of the flame tube in order to prevent the walls from overheating.

The way in which combustion occurs in the primary zone of the flame tube is very similar to the situation in a recirculation zone behind a flame-holder. The degree of turbulence is very high, and a recirculation pattern is created, often as a consequence of turbulent motion along the longitudinal axis caused by the kerosene injector itself. Although the air and kerosene are not mixed before entering this zone, the strong turbulence present is able to mix the gases very effectively under normal operating conditions.

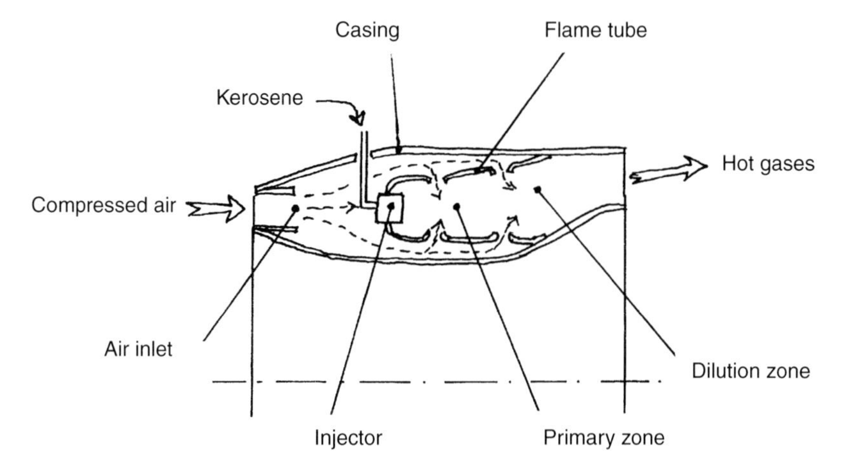

Figure 6a Axial section of the annular combustion chamber of a turbojet.

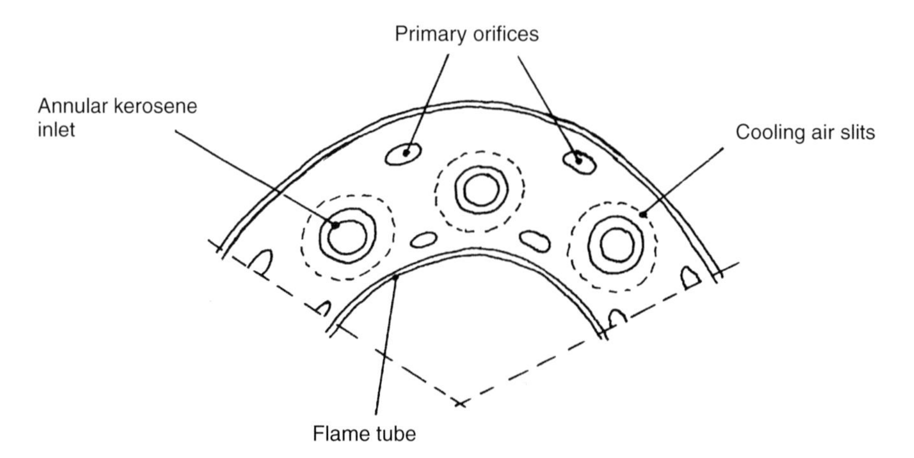

Figure 6b View of the injector heads in a portion of flame tube, seen from the rear of the engine.

The most important characteristic of a turbojet's combustion chamber is combustion efficiency (roughly equal to the percentage of fuel efficiently burnt), which must be greater than 99% over a broad operating range (at take-off, taxiing on the runway, climb, cruise, at altitude) within which the engine's flow rate and pressure characteristics vary greatly. Combustion stability must be guaranteed across this entire operating range, particularly in terms of being able to reignite the primary zone when at altitude (at low pressure), which sets a high minimum size for the volume of this zone. Moreover, recent restrictions aimed at limiting the emissions of pollutant products such as nitrogen oxides, carbon monoxide and soot often require difficult compromises to be found. The design of turbojet combustion chambers was for many years a question of expertise, based on tests and successive correction. However, over the last ten years or so, (and mainly due to restrictions related to pollution about which no experience had previously been gained), detailed calculations of all the chemical, thermal and flow phenomena associated with chambers are being increasingly used in industry. Understanding all the aspects of flame-related phenomena is thus required, and combining existing knowledge with the numerical calculation tools available has led to a great deal of progress in this field.

Another type of engine used in aeronautics is the **rocket engine**. The principal advantage offered by these engines is that they do not consume air, which means that they can power vehicles in space where there is no air, or achieve very high performance levels. Although air is freely available, approximately 3/4 of its volume is made up of a product which is virtually inert from a combustion point of view: nitrogen. This is why (and we shall analyse this in more detail later) the presence of nitrogen in combustion processes is a hindrance in terms of the temperatures reached. Moreover, other chemical products are more powerful oxidisers than oxygen itself, and thus offer more efficient alternatives.

The first rocket engines were invented around the turn of the twentieth century in Russia, France, the USA and in Germany (Tsiokolski, Esnault Pelterie, Goddard, Obert) with interest generated by the availability of products capable of releasing more energy than air plus a hydrocarbon. Various combinations of an oxidant and reductant are possible: oxygen-kerosene, N_2O_4-UDMH (*unsymmetrical dimethylhydrazine*), or simply oxygen-hydrogen. Solid products may also be used: ammonium or potassium perchlorate, and polyurethane-type plastic materials. Certain products containing both oxidising and reducing functions on the same molecule may also be used, with or without the addition of a reductant: this is the case for hydrazine N_2H_4, nitroglycerine or nitrocellulose. UDMH and ammonium perchlorate are also of such a type.

The principal factor in combustion chamber design for rocket engines is thus the physical state of the reactant(s) (known as propellants): a differentiation is made between bi-liquid rockets, solid (or powder) propellant rockets and hybrid rockets, where the oxidant is a liquid and the reductant a solid.

A simplified diagram of the combustion chamber of a bi-liquid rocket engine is given in Figure 7. The bottom of the chamber is covered with a large number of injectors which spray jets of liquid propellants. In certain cases, one of the propellants (hydrogen for example) is first used to cool the walls of the combustion chamber before being injected in gaseous form. The droplets of liquid fuel and oxidant are then dispersed, mixed and partially vaporised to form a spray. The flame burning in this spray raises the temperature, thus ensuring the vaporisation of the droplets inside the flame over the first ten or twenty centimetres of the chamber. The downstream part of the combustion chamber consists of a convergent-divergent nozzle. The gases finally approach chemical equilibrium as they flow through the convergent nozzle, then expand in the nozzle where they provide the engine's thrust.

For certain pairs of substances, such as N_2O_4 and UDMH, there is no need to ignite the engine since these propellants (called hypergolic propellants) ignite spontaneously on contact under normal conditions of temperature and pressure. For others, such as hydrogen and oxygen for example, the propellant must be ignited by a jet of hot gas before the flame zone can establish.

Although the propellants are liquid when injected into the chamber (although in some rockets only one of them is), in most cases combustion occurs in the gaseous phase. The flame zone consists of a spray of droplets in

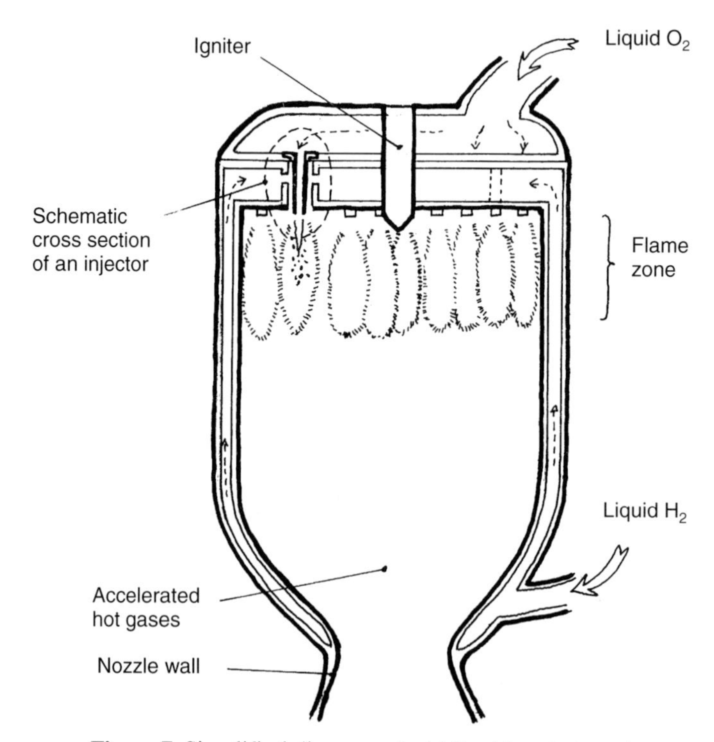

Figure 7 Simplified diagram of a bi-liquid rocket engine.

gases, produced either from the vaporisation of the two types of droplet, or from combustion. This flame may be considered as being premixed insofar as the injectors mix the droplets of both propellants, although this is only true at the macroscopic scale of about one hundred droplets. On an individual droplet scale, certain zones may be clearly identified as being fuel-rich and others rich in oxidant. The reaction zones are located in the spray between the droplets; in fact small diffusion flamelets surround drops or groups of drops. These flamelets may be considered as being similar to a candle's flame, except that in this case the wick is replaced by a droplet of fuel (or, on the contrary by a droplet of oxidant, where the surrounding air is replaced by a reducing gas) and the droplet has a diameter of approximately 10 to 50 μm, compared with the flame which is from 500 μm to one millimetre long. An entire chapter is devoted to the combustion of droplets and sprays of droplets. It explains in greater detail the structure of such flames, and also describes the methods behind the quantitative calculations.

The design of solid propellant rocket engines is very different. The combustible substance in this case takes the form of a solid block, which looks very much like plastic. When one of its surfaces is ignited the substance is consumed by the combustion processes and retreats, releasing large quantities of hot gases which are the products of decomposition and combustion, the majority of which are gaseous. The surface retreats at a rate of a few millimetres per second (depending on the product and the pressure conditions), and the gas flow rate of the gases produced is proportional to the area of burnt surface. This explains why the combustion chamber of an engine of this type has the basic design shown schematically in Figure 8. The chamber is similar in shape to a gas cylinder, closed on one side and elongated on the other by a convergent-divergent nozzle. The block of propellant is cast inside this container, firmly attached to the side walls and base, and with a space left along the axis, connecting the igniter and the throat of the nozzle. Once ignited, (initiated by a jet of hot gases, itself produced by a spark-ignited mini jet unit), the inside surface around the central space burns and retreats. The shape of this space is designed such that the surface of the propellant is consumed in a controlled manner. This provides a means of adjusting the gas flow rate, and

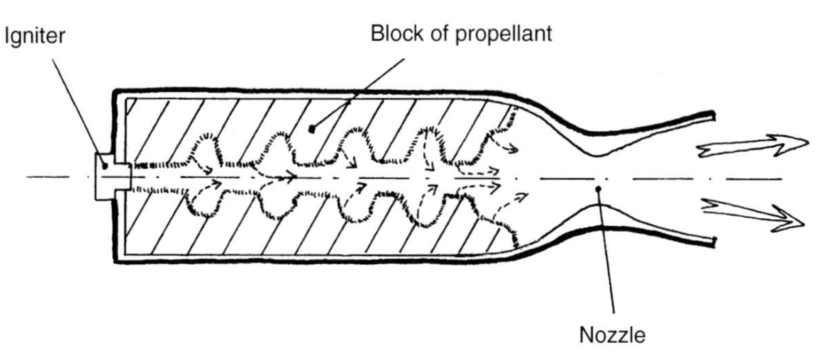

Figure 8 Sectional view of a solid propellant rocket engine.

since the size of the throat is fixed, it also regulates the pressure in the chamber and thus the thrust provided by the motor.

The flame produced in an engine such as this is no more than a very thin zone located a few tens of microns above the surface of the block of propellant. Usually, the solid decomposes and vaporises, and the flame burns mainly in the gaseous phase. It is therefore a premixed flame when viewed on a macroscopic scale, which propagates by consuming the propellants. On a smaller scale, its structure is more complex when the solid propellant is a composite (i.e. formed from a mixture of two substances, one an oxidant, the other a reductant), compared with when it is a homogenous propellant (i.e. when the same molecule is both a reducer and oxidant, such as nitrocellulose, in which case the product is extremely dangerous). No further reference will be made in this book to this type of flame, due to the highly specialised nature of its applications.

Other types of rocket engine, known as hybrids, were manufactured and used for a certain time. The fuel was a solid, annular block, on the inside surface of which a liquid oxidant was sprayed. The advantage offered by these devices was that propellant pairs could be used which released extremely large amounts of energy, but which could still be handled relatively safely. One such high-energy couple is ClF_5 (oxidant)/HLi (reductant). Due to the difficulties associated with the use of these propellants, as well as the noxious nature of the products of combustion (ClH and FH in particular), engines of this type have not been developed, to our knowledge. Again, no further mention will be made of this type of combustion.

B. Burners

Burners are the other principal industrial devices which exploits combustion, and their applications range from domestic cookers to industrial furnaces and boilers. They do not propel vehicles of any description, but instead provide a source of heat, or destroy or transform certain products. They are used in industry and in power stations. The engines described above which generate mechanical energy must first produce heat energy (i.e. hot gases) which is transformed (in the nozzle, turbine, or piston) into useful "work". The combustion chambers of engines are therefore simply burners, and the clear similarities between them and industrial burners should come as no great surprise.

A distinction may be made between gas burners, fuel oil burners and solid fuel burners which use coal. A description of each type will be given. There are numerous designs, slightly or significantly different, although at this point we shall simply give a few examples of typical burners. Firstly, however, let us briefly consider two extremely simple and common gas burners: the gas cigarette lighter and the simple gas burner.

The gas lighter is the simplest possible burner: it produces a jet of natural gas (i.e. containing approximately 95% methane), which entrains the surrounding air. Once ignited (by a spark from the flint) a diffusion flame, such as that shown schematically in Figure 9a, is obtained which is very similar to a candle's flame, except that the supply of combustible gas comes directly from the lighter's reservoir, instead of vaporising from the wick. The diffusion flame obtained is quite long, especially when compared with the diameter of the gas supply orifice. The calculation of the flame length is a problem which will be addressed in chapter 3.

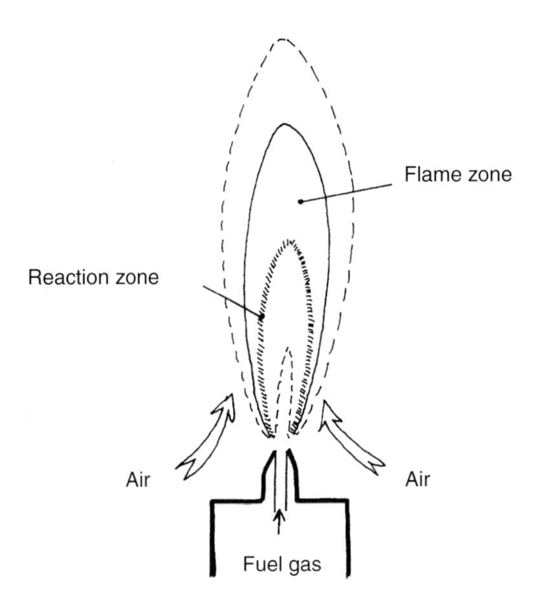

Figure 9a Jet flame from a gas cigarette lighter.

The gas burner such as the "Bunsen" burner popular in schools or the rings on domestic gas cookers, produces a premixed flame. Consider the design of the traditional gas burner that we have all probably used at some time in the school chemistry laboratory. It consists of a tube, into which gas is supplied via a central injector. Air is entrained past the collar by the high velocity of the jet of fuel gas and the two gases mix as they rise up the tube (see Figure 9b). Closing the collar reduces the quantity of entrained air. The air and gas are very well mixed by the time they reach the tube outlet, and, after ignition, a conical premixed flame attaches to the rim and stabilises. Clearly the flame length produced is much less (for an equivalent tube diameter) than for a diffusion flame, although the device is more complex. The flame is conical for the following reason: the velocity of the gases leaving the tube is greater than the burning velocity of a premixed flame, however, the gas velocity is much lower near to the walls and the flame may attach to the rim (if the burner is correctly designed) and develop obliquely from this point. There are certain similarities here with the way in which the flame zone in a ramjet is held by the flame-holder. Under certain conditions, the flame can travel back against the flow in the tube to attach to the gas injector. This malfunctioning quickly damages the burner if it is not corrected. The gas burner is used frequently in the laboratory, and allows the structure and properties of laminar flames to be studied. This type of burner will be discussed again in detail later in this book.

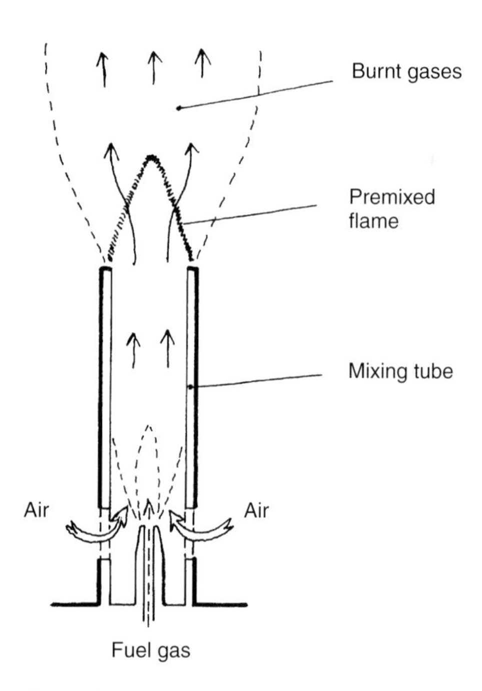

Figure 9b Flame over a classical gas burner.

Industrial burners are more complicated than these two domestic burners. The reason for this is firstly that they must supply the highest possible flow rate of gas for the smallest possible overall size, and secondly that they must be able to operate at different settings without requiring too many modifications:

- **An industrial gas burner** is shown in Figure 10. The air is introduced in a well-controlled manner, and not simply entrained by the jet of gas. Moreover, a highly intense swirl is produced in the burner, which extends the time during which the fluid remains inside the burner's nozzle, and also produces intense turbulence. The result is that combustion occurs in the burner (once ignited) in a similar way to how it occurs in the primary zone of a turbojet's combustion chamber. The phenomenon can best be described as taking place in a combustion zone rather than in a flame. In this design of burner, a colder, methane-rich zone is created around the gas inlet, and the exhaust gases are hotter at the centre than at the peripheries. Both these factors are beneficial in terms of the life of the materials used to manufacture the burner.

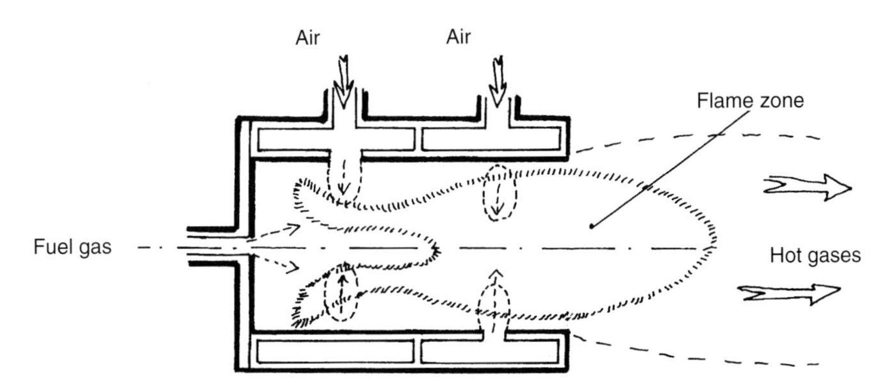

Figure 10 "Counter-rotating" gas burner. Air is injected through two series of holes; the first series creates swirl in one direction, and the second swirl in the other.

• **The flame of a fuel oil burner** is more similar to a diffusion flame. The liquid fuel oil is injected in the form of droplets whose diameter is in general fairly large (about one hundred microns) and it is difficult to devise a way of mixing these droplets as effectively as those devised for gases only. In fact, the fuel oil used is an extremely viscous liquid at room temperature, and must be preheated to approximately 150°C before it can be injected under satisfactory conditions without the need for excessively high pressures. A diagram of a fuel oil burner is given in Figure 11. The fuel oil is injected, in droplet form, into the centre of the burner and the air jet (which is also often preheated) enters via an annular peripheral channel. In general, to attach the flame without requiring an obstacle, the air jet is rotated (or swirled) in the axis of the burner, to create a slight pressure drop which produces a recirculation zone on the central axis, very close to the fuel-oil injector. Once ignited, this recirculation zone, which is fed both with air and by some of the droplets produced by the injector (the smallest droplets), stabilises a core zone of hot burning gases, performing the same role as a flame-holder. The diffusion flame which develops from this zone has a core rich in droplets and fuel oil vapours, whilst its periphery is richer in air. In general, burners such as these operate in furnaces or boilers, and a consideration of flame length is an essential factor in their design. The flames tend to be longer for this type of burner compared with the gas burner described previously, and radiate to a greater extent due to the presence of soot within the flame (just like a candle). However, these features are not necessarily shortcomings and can be used to advantage if their characteristics are known quantitatively: many applications require flames which heat an object through radiation rather than by direct contact with the hot gases. Moreover, to a certain extent it is possible to shorten the flame by swirling the air more. The proportion of fuel burning in the recirculation zone increases at the expense of the diffusion flame, which shortens. Like the other devices which use liquid fuels, the flame burns in a medium in which the droplets and gases are mixed, and thus its structure is not homogenous when viewed on a small scale: the reaction zones are effectively located around groups of droplets.

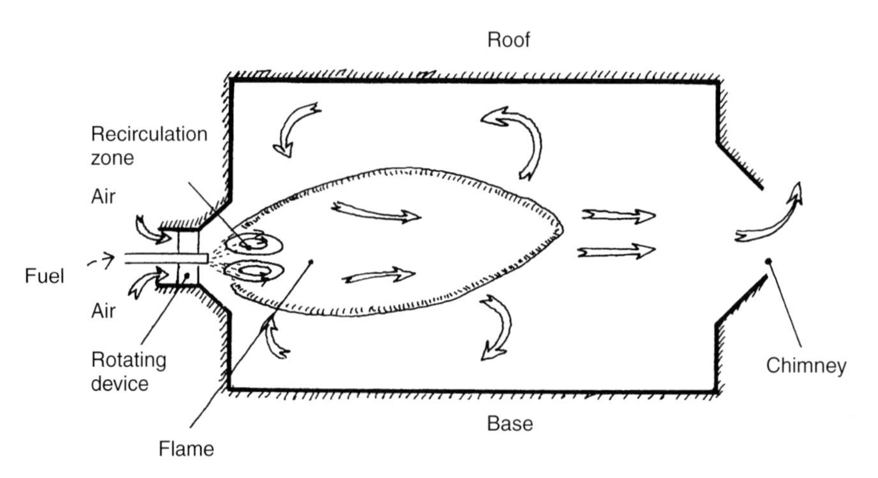

Figure 11 Diagram of a fuel oil burner in an industrial furnace.

• **Coal-fired burners** are used widely across the world, but now more rarely in developed countries. Moreover, their design has been fairly radically changed. The first coal-fired burners or furnaces used the same (or at least very similar) principle to the sort of cooking stove or range popular in the days before gas cookers. The coal, broken into large lumps (several centimetres across) were spread on a grill, and burnt with the air drawn up through this grill and around the lumps. Furnaces of this type are no longer used today in industry, except when waste or other products need to be burnt instead of coal. Modern coal-fired burners use pulverised or powdered coal, i.e. coal ground down into particles whose diameter is smaller than one millimetre. These particles may be carried by a current of air through the supply circuit and injected into the burner in more or less the same way as a liquid would be. Pulverised-coal burners are thus not dissimilar now to fuel oil burners.

Coal is increasingly burnt in industry in what are known as "fluidised beds", which are more like chemical reactors than burners. Their operation does not strictly involve flames, and for this reason they will not be considered in this book. Occasionally, a combustible gas is extracted from the coal, using these fluidised beds, and this gas is then burnt in a conventional burner.

The combustion of coal is thus fairly different, from a chemical perspective, from that of liquid fuels, or even from that of other solid fuels. A significant fraction, which may be as much as 50%, of chemical reactions occur between the solid phase and the surrounding air: these are therefore heterogeneous reactions (the remaining reactions occurs between gases derived from the coal and air). Clearly, a critical role is played by radiation phenomena either between particles, or between the gas and the particles, or involving soot. However, the heat transfer phenomena (between the coal particles and the gaseous medium or between the different constituents of the gaseous medium) and the mass transfer phenomena (due to diffusion or convection of gases and particles) are the same for pulverised-coal flames, as they are for fuel-oil flames, candle's flames or indeed for the flames produced in the combustion chambers of turbojets.

C. Fires

Fires and explosions are the last category of situation encountered in industry or in nature in which combustion plays a dominant role. Once again we can describe and analyse the various situations of this type in terms of premixed or diffusion flames in the gaseous phase or involving liquids or solids. The information gained about the behaviour of engines or industrial furnaces can be applied to attempt to control such incidents, predict the mechanisms and possible outcomes and thus estimate the risks and associated solutions. One notable specific feature of fires is, however, related to the important role played by radiation and natural convection: the flame size in a fire is much greater than in an engine, and for this reason buoyancy effects can cause large-scale movements. Moreover, the absorption and emission of light energy occurs much more effectively.

Four different situations will be used as typical examples. Let us first consider what is known as a "deflagration". This type of situation is caused by, for example, a gas leak in a room which produces an air-fuel mixture. Evidently, a spark with the right characteristics (which may be, and have been, studied in detail) is sufficient to trigger a premixed flame which will subsequently propagate through the house: this event is known as a deflagration. The speed at which the deflagration propagates, the state and temperature of the gases behind the flame and the flow field of the induced gas, may all be studied quantitatively with the same tools as those used to study these phenomena occurring in the cylinder head of a piston engine. Such deflagrations have been studied in detail for many years, in fact the first significant results obtained on laminar flames were due to the work of Mallard and Le Chatelier, mining engineers who studied firedamp explosions in 1881.

The propagation of a deflagration in a medium at rest is preceded and accompanied by pressure variations. Under certain conditions, they may be coupled with the deflagration itself to produce what is known as a detonation, which then propagates at an incredibly high speed (at about 1000 m/s or more instead of 1 m/s!). This phenomenon is also sufficiently important to have been studied in depth for many years, and is related to the use of solid and liquid explosives. A chapter of this book is devoted to this phenomenon, which discusses the connexion of detonation with the phenomena implied in combustion at supersonic flows, information which is required when studying the propulsion of hypersonic vehicles by ramjets. Combustion in supersonic flow has many fundamental aspects in common with the propagation of what are known as "weak" detonations.

Another situation which merits study is typified by the case of a burning, vertical combustible wall, such as a curtain or a polymer wallcovering in a room. The flame produced is a diffusion flame which locates at about one centimetre from the wall, and which separates the air in the room and the vapours released by the burning material (see Figure 12a). The hot gases are entrained upwards, and cooler air is drawn in at the bottom of the flame by natural convection. The heat conduction and radiation from the flame vaporise and possibly also decompose the wall, which then sustains the flame by supplying the reaction zone with fuel vapours.

A diffusion flame may also stabilise on a layer of liquid fuel spread across a container, or floating on the sea. Once again the heat produced by the flame vaporises the combustible product by radiation, more rapidly if the latter is volatile, and if natural convection provides the lateral supply of unburnt air. This situation is shown schematically in Figure 12b.

One final situation, unfortunately all too common, is a **forest fire**. Considered on a large scale, it most closely resembles the propagation of a premixed flame over a surface (and not in three-dimension space). A simplified analysis considers that a forest is a surface on which the fuel (trees) and the oxidant (air) are premixed. As soon as a fire breaks out at a specific point, a premixed flame front is created which is linear rather than a surface and which propagates from this point. Over flat land, with no wind and homogenous vegetation, the propagation pattern would be circular, although in reality

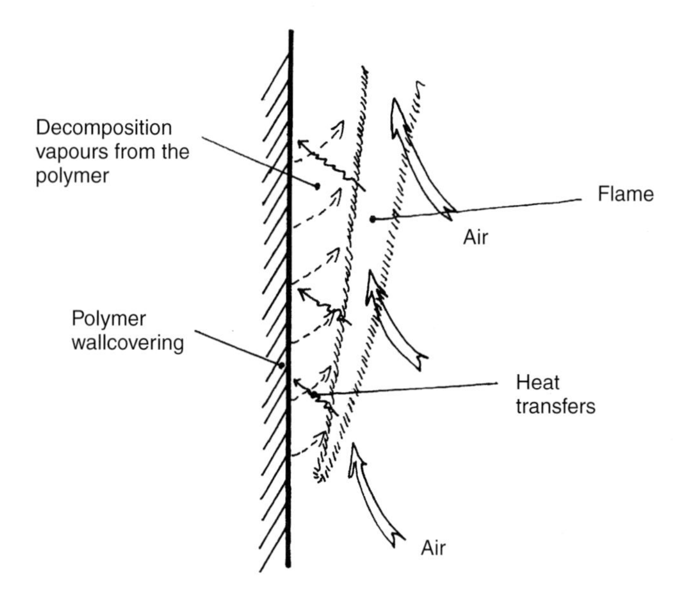

Figure 12a Flame spreading near to a vertical wall covered with a polymer.

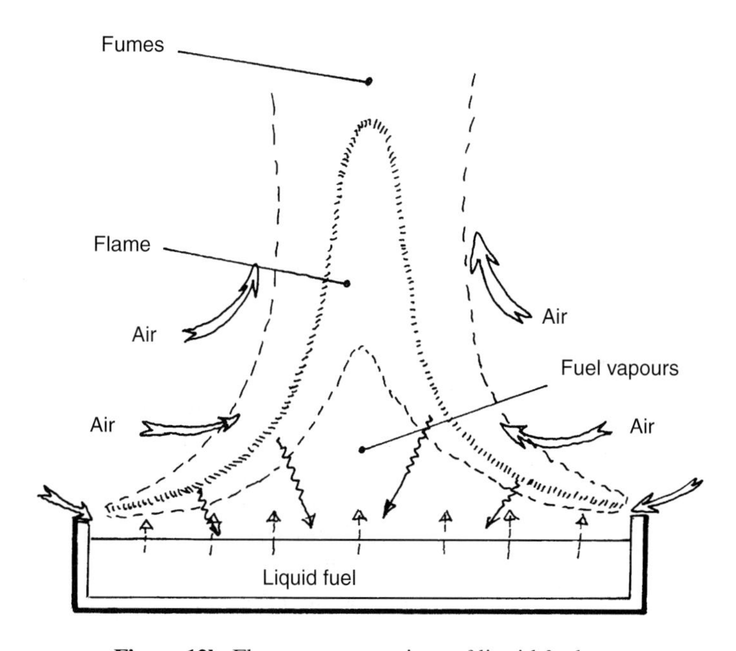

Figure 12b Flame on a container of liquid fuel.

every slight contour in the ground, the effect of wind, and differences in vegetation will influence the behaviour of the premixed flame. Clearly, if the flame is observed on a smaller scale, for each individual tree, then the situation is more like diffusion flamelets surrounding each tree or group of trees and which may be compared with the diffusions flamelets surrounding drops or groups of drops in a premixed spray. Once again, and perhaps slightly paradoxically, knowledge gained in studying combustion in rocket engines can be applied to deal with a forest fire. Of course, only the basic knowledge is of value, and must be suitably adjusted to take into consideration radiation and natural convection phenomena specific to a forest fire.

3 THE SPIRIT AND CONTENTS OF THIS BOOK

The above descriptions have demonstrated the diversity and importance of combustion and flame phenomena both for industry and for society; proof of which is the large number of applications. Is a mainly theoretical rather than simply empirical study possible? Our reply is yes.

The explanations for the various combustion applications described above were structured around a consideration of the interplay of the three basic phenomena associated with flames: chemical combustion reactions, physical phenomena of heat and mass transfer and mechanical gas flow phenomena. Breaking down a phenomenon may appear, at first sight, simply to complicate the analysis; but this is not the case here. On the contrary, knowing how to recognise the respective roles of these three fundamental aspects of a flame enables a unified description to be made of practically all known flames. Moreover, it then becomes possible to apply the understanding obtained about one type of device to the design or control of another device or another situation. This is, we hope, the impression that the reader will have gained from the above discussions, and it is on this basis that the rest of the book has been structured and written.

This book is designed to provide an introduction to the study of combustion and flames. Our intention is not to provide answers to problems not yet resolved, but to spread a general understanding of combustion and flames to a wider group than simply those specialists working in this field. As we have just indicated, our aim is to present a unifying perspective of the various types and configurations of flames. For this reason, our discussions will favour the theory, and will endeavour to explain the phenomena observed based on certain principles, rather than presenting an empirical panorama of these phenomena. However, our objective is not to provide long and tedious mathematical proofs, but to prefer explanations of a physical nature. This introduction has probably already provided an example of our approach. Having said this, it would be impossible to avoid completely the use of equations, since they are essential for a quantitative understanding of the phenomena involved.

This study of combustion and flames will begin with three chapters devoted to the fundamental aspects of combustion. The first will be concerned with thermodynamics, an aspect that has not been covered up until now but which is implicitly the basis of all the theoretical analyses that will subsequently be developed in relation to flames. This first chapter will have two objectives; the first will simply be to detail the concepts and definitions, particularly temperature, which will play a key role in subsequent chapters, and secondly to show how to calculate the maximum temperature reached inside flames; a criterion which can be used to classify, as a first approximation, the propellants as a function of the potential amounts of energy that they can supply.

The second chapter will consider the chemical aspects of combustion reactions. In this respect, the analysis will concentrate mainly on reaction kinetics, i.e. on the rates at which they occur, and which depend greatly on the temperature and concentration conditions of the mixture. The concept of elementary reaction will be discussed, and we will see that the combustion reactions when considered as a whole virtually always constitute a chain-reaction process.

The third chapter will be devoted to the physical and mechanical phenomena of heat and mass transfer by diffusion and convection of gases. Starting with a demonstration of the effects of molecular diffusion phenomena in simple cases, the analysis will continue by explaining how to formulate equations to describe convection and diffusion phenomena. From a physical perspective, the equations analyse the system by balancing what enters with what leaves and what remains in a specific, pre-defined system; much like how an accountant analyses receipts, expenditure and cash reserves. Heat-exchange phenomena due to radiation will not be covered in this book. They are often important in studying flames, but almost never essential. Our decision to disregard them stems also from the fact that the formulation of the equations required to explain them (to a degree which would enable them to tie in with the other phenomena described in this book) presents significant theoretical complexities. Differential balance equations are applied to the simple case of a diffusion flame similar to that produced by a cigarette lighter, which should allay any fears the reader may have had relating to the complexity of these differential equations. At the end of this chapter, similar equations are used to show how it is possible to consider heat exchange due to turbulent diffusion which is much more effective than molecular exchanges in most industrial devices.

These first three chapters create a solid foundation for the subsequent theoretical explanations of combustion and flames. The following chapters build on these foundations to consider in greater detail the various situations in which combustion is encountered.

Chapter 4 is entitled "self-ignitions", and looks into the ignition of a fuel/oxidant mixture in a closed container when the mixture is virtually homogenous, both from a composition and temperature perspective. In such a situation, the existence of chain reactions is demonstrated by a characteristic abrupt increase in the rate of reaction (so long as the correct conditions are provided), a phenomenon which was the reason why the term "explosion" was coined for this type of reaction.

Chapter 5 considers flames in a laminar medium, i.e. in a medium in which exchanges due to turbulent diffusion are virtually non existent. The applications are almost exclusively laboratory burners which do not involve turbulence, but which are very useful for the in-depth study of flame structure. In this chapter, the first part treats premixed flames, and the second part diffusion or non premixed flames. Finally, the stabilisation of these flames will be studied, particularly in recirculation zones.

Chapter 6 will consider turbulent flames at length, both for premixed and diffusion flames. The first part of this chapter will include a short introduction to the concept of turbulence and its principal characteristics. Since turbulence affects the flame, and conversely the flame can affect turbulence, the end of this chapter will introduce this problem. In this chapter, possibly more so than in the rest of the book, the ideas developed should only be considered as an introduction. This field is the subject of a great deal of research work and many properties are still not fully understood or quantified. However, in general terms, turbulent combustion is sufficiently well understood qualitatively to justify its inclusion here, even as an overall description.

Chapter 7 is devoted to the specific phenomenon of detonations, resulting from the coupling of a deflagration with a shock wave. Unlike deflagrations, the very high pressure differences associated with detonations are highly destructive. Only a general study of this complex phenomena can be presented here, described in relation to combustion in supersonic flows, since detonations propagate at a speed much greater than the speed of sound

Aspects related to the ignition of flames are considered in Chapter 8. Once again, the aim is to provide a comprehensive introduction to the subject. A book about flames, stabilised or propagating, would have been incomplete without a discussion of the exact moment when they are ignited.

Chapter 9 looks into the combustion of liquids in the form of sprays. When reviewing the various jets and burners we noted that many of them use a liquid fuel, and occasionally even a liquid oxidant. An understanding of flames in media made up of sprays of droplets is therefore at least as important as that of flames in gaseous media. However, much less research has been conducted in this field. After discussing the simplified theory of the combustion of a single droplet with the surrounding air, the structure of spray flames and the associated calculation methods will be reviewed. This chapter is probably that which is least well supported by experimental data and proven theoretical approaches, and includes certain conjectures which we hope to see confirmed sometime in the future.

The last chapter discusses a new aspect of flames: their ability to form pollutant products in the burnt gases. Over the last twenty years, this subject has been studied in depth, and a certain amount of knowledge has been accumulated. Once again, simply an introduction to this field will be given, to prepare the way for a more detailed subsequent study.

CHAPTER 1

COMBUSTION THERMODYNAMICS

1.1 PHENOMENA AND DEFINITIONS

Combustion is a **chemical reaction,** exothermic when considered as a whole, and which accelerates after a slow start to become rapid and even violent. Radiation is emitted and there is an associated rise in temperature. The reaction is globally exothermic insofar as the overall result is a release of heat, even though processes may be involved which actually absorb heat. Combustion can occur in an "open" or "closed" system (i.e. with or without the exchange of matter between the system and its surroundings) or in an "isolated" or "non-isolated" system (i.e. without or with the exchange of matter and energy). An enclosed combustion chamber with diabatic walls is a closed system, whereas rigid, adiabatic walls produce an isolated system. An example of an open system is a flame produced by a gas burner.

Chemical reactions can only occur when the chemical species (atoms, ions, radicals and molecules) either come into contact or are sufficiently close to each other to be mutually modified. All other considerations apart, reactions must be slow if the speed of motion of the species is slow, since the encounters will be relatively infrequent. This is the basic reason why combustion reactions mainly occur in the gas phase, and why in general liquids burn only after vaporization. One exception is carbon, which burns primarily in the solid state even though it releases volatile products on heating.

In general, combustion processes are oxidation-reduction reactions, with the oxidising agent known as the **oxidant** (pure or dilute oxygen, ozone, chlorine, nitrates, etc.) and the reducing agent the **fuel** (H_2, CO, hydrocarbon, etc.). A **complex mechanism** is involved, consisting of a large number of so-called "elementary" reactions, some of which consume, and others which release heat. The amount of heat released soon overtakes the amount being absorbed. If the fuel is completely oxidised, the general equation can be written:

$$a \text{ oxidant} + b \text{ fuel} \rightarrow c\ CO_2 + d\ H_2O + ...$$

This equation simply indicates the initial and final products of the process, and no more, and is known as the **balanced stoichiometric reaction**. It does not provide any information whatsoever about what happens during the combustion process itself, nor necessarily does it indicate the actual nature of the products at the end of combustion, since, as will be seen later, these products are not always completely oxidised. The balanced reaction does however provide a starting point: if the initial mixture is "stoichiometric", i.e. if the ratio of the number of moles of fuel to the number of moles of oxidant is exactly *b/a* (where *a* and *b* are the so-called "stoichiometric" coefficients of the above reaction) then the composition of the mixture is optimal and the reaction may continue to completion; meaning that when the reaction ends, no oxidant or fuel remains. Whilst the products in a stoichiometric mixture will not necessarily reach complete combustion, it is true to say that if the stoichiometric ratio is not respected, then combustion will be incomplete.

Occasionally the same molecule contains both the oxidising and reducing functions. For example, there is sufficient oxygen contained within nitrocellulose molecules to combine, under the right conditions, with other atoms in the molecule. This is also the case for nitroglycerine. Compounds such as acetylene, C_2H_2, may also be thought to behave in the same way, since for the exothermic reaction:

$$C_2H_2 \rightarrow 2\ C + H_2,$$

a formal analysis would show that C has been oxidised by H (its oxidation number increasing from –1 to 0) and H reduced by C (its oxidation number dropping from +1 to 0). For this reason, C_2H_2 can burn alone, with no contribution from its surroundings, to produce what is known as a decomposition flame. Furthermore, acetylene may also burn in oxygen, releasing heat due to the oxidation of C to CO_2 and H_2 to H_2O in addition to the decomposition of C_2H_2. The ability of such compounds to burn by consuming only the oxidant contained in its molecule has been clearly demonstrated in numerous accidents. The reaction can be extremely violent, with the reactants literally exploding; which is why these compounds are known as "explosives". Local heating or even a slight impact or jolt can trigger the phenomenon: with the massive amounts of energy locked up in the molecule then being released violently and often unexpectedly. In 1886, the Swedish scientist Alfred Nobel found a solution to this problem, at least for nitroglycerine. By mixing it with a porous substance, diatomaceous earth, he produced dynamite, a product which can be handled relatively safely since it requires a violent shock (produced by a detonator) to trigger its explosive decomposition.

During the Second World War, and then with the space race, rocket engines were developed and with them a second generation of products, **propellants,** pure or mixed products containing both oxidant and fuel which could burn without requiring a supply of air. In addition to offering stability during handling and storage, such products, unlike explosives, burn progressively, providing the rocket with thrust for a predefined duration. These products must be ignited with sufficient violence to ensure that combustion is initiated reliably, and can then continue in a controlled manner, without causing an explosion.

In practice, combustion may occur haphazardly, in fires and accidental explosions, or in a controlled fashion, on burners, in rocket engines or in internal combustion engines. In the latter case, combustion is controlled either electrically, in a "spark-ignition" engine or through adiabatic compression in a "Diesel" engine. Depending on the conditions, the phenomenon may continue independently or slow down gradually and stop. In order to study the two main phases; the initial ignition or lighting phase and the development phase, thermodynamics and chemical kinetics must be applied to the chemical phenomena, whereas mechanics and the science of heat are required for the physical phenomena.

The first question to consider when analysing combustible mixtures is what temperature could be reached if the combustion occurs without heat exchange with the surroundings, i.e. adiabatically. This may be calculated, and is purely a question of thermodynamics, since it is totally unaffected by the chemical mechanisms of the reactions which actually occur. This chapter is primarily concerned with answering this question. The solution can be found by applying the First Law of Thermodynamics and by establishing an energy balance. However this balance can only be produced if the final composition of the burnt gases is known, which in turn requires the application of the Second Law in order to calculate the chemical equilibria of these gases. The way in which the adiabatic temperature at the end of combustion can be calculated will be explained in detail, but will be given for a few simple examples only, and using a general calculation method. A description of the calculations involved in determining these values for any mixture and when several chemical reactions occur simultaneously, is beyond the scope of this book (although very accurate numerical calculation programmes are available to do this).

1.2 A MACROPHYSICAL PERSPECTIVE OF THE FIRST AND SECOND LAWS OF THERMODYNAMICS: ADIABATIC TEMPERATURE PRODUCED BY COMBUSTION IN A CLOSED SYSTEM

If a fuel/oxidant mixture burns adiabatically, its final temperature is generally the maximum temperature reached during the reaction. In certain cases, however, for example mixtures of air with an excess of hydrocarbon, the temperature varies during combustion, passing through a maximum. This can be explained by the complexity of the chemical processes involved which may include endothermic reactions. In practice, heat is almost always transferred away from the combustion system, and thus the temperature rises to a maximum, (which is nonetheless lower than the adiabatic case), before falling more or less rapidly. The phenomena are governed by the First and Second Laws of Thermodynamics; and thus before analysing the problem, we shall consider the consequences of the First and Second Laws for chemical reactions. Without going into detail about the laws and their implications, it is worth emphasising that, in general, these laws introduce "state functions" whose properties are such that any changes between a given initial and final state depend only on these states, and are independent of the processes which occurred in passing between the two states.

1.2.1 Consequences of the First Law: heat evolved, enthalpies of formation

A chemical reaction may evolve or consume heat. According to an agreed **convention**, this heat is considered as being negative in the first case and positive in the second. It is therefore negative for a combustion reaction.

Assume that the combustion reaction as a whole is represented, for a closed system, by the balanced stoichiometric equation:

$$a \text{ fuel} + b \text{ oxidant} \rightarrow c\, CO_2 + d\, H_2O + ...$$

Then assume that a sufficient quantity of the heat generated is transferred away, such that the final temperature of the products (CO_2, H_2O, etc.) is the same as the temperature of the initial fuel/oxidant system. Under these conditions, the energy balance does not include a contribution from a change in temperature; its sources are purely chemical. This therefore reveals that the chemical nature of the final system is not the same as that of the initial system, i.e. that the molecules are different.

For this case, it may be deduced from the First Law that:

- For constant volume, the heat evolved is equal in absolute terms to the change in internal energy ΔU between the beginning and end of combustion.

- For constant pressure, the heat evolved is equal in absolute terms to the change in enthalpy ΔH.

The terms constant volume or constant pressure are not strictly correct in this context since they do not necessarily mean that these parameters do not vary during the reaction, but simply that their values at the end of the process are the same as at the start. This follows on from the use of state functions such as U and H, as is demonstrated by thermodynamic theory. Only changes in the functions U and H can be measured; "absolute" values of U or H cannot be determined. However, this is not a problem since we are only interested in expressing difference between these values. Upper case letters U and H (and later, S, F, G, C_p, C_v, etc.) are used for systems containing either a certain number of moles, or a particular mass, and lower case letters u and h (and later s, f, g, c_p, c_v, etc.) either for one mole, or for unit mass. Lower case letters are also used, in conjunction with the symbol Δ, for "reaction" energies, enthalpies, etc. for any reaction which occurs under the conditions described above. These are the enthalpies, energies, etc. of the products minus those of the reactants.

Enthalpy of formation, Δh_f, is defined, for a given substance, as the changes in the reaction enthalpies involved in the formation (i.e. synthesis) of this substance at a given pressure and temperature. These are the same for the initial system as for the final system, and depend to a greater extent on the temperature than on the pressure. The reactants are, according to a **convention**, elemental substances in their most stable state under given conditions (C graphite, H_2, O_2, N_2 gas...). For example, at ambient temperature and 1 bar, the reaction for water is:

$$H_2 + 0.5\ O_2 \rightarrow H_2O \text{ liquid}$$

and for CO_2:

$$C \text{ (graphite)} + O_2 \rightarrow CO_2,$$

since graphite is more stable than diamond. The same is also true for CH_4 in the reaction:

$$C \text{ (graphite)} + 2\ H_2 \rightarrow CH_4.$$

All these reactions are considered as being complete, regardless of whether they are or not, and even whether in practice they actually occur, which excludes any consideration of the attainment or not of equilibrium. However, reactions are rarely complete at the high temperatures produced by combustion phenomena, and considering them to be so can lead to an overestimation of the heat actually generated. The factor of incomplete combustion must be taken into account when actual values are calculated. In accordance with the logic of the convention mentioned above relating to the use of lower case letters, Δh_f indicates, depending on the context, the enthalpy of formation for one mole or per unit mass. Obviously, any change in the physical state of the species, of both reactants or products, contributes to the change in enthalpy. Take the example of the synthesis of water; if the water produced is assumed to be steam and not water, the enthalpy (latent heat) of vaporisation must be added to the enthalpy of formation of one mole of water.

Let us now consider a general reaction, symbolised by:

$$\nu_A A + \nu_B B + \ldots \rightarrow \nu_{A'} A' + \nu_{B'} B' + \ldots,$$

where ν_A, ν_B, $\nu_{A'}$, $\nu_{B'}$ etc. are the stoichiometric coefficients. The First Law still implies that the Δh of the reaction is related to the Δh_f of A, B, A′, B′, etc., by the equation:

$$\Delta h = \nu_{A'}\Delta h_f(A') + \nu_{B'}\Delta h_f(B') + \ldots - \nu_A \Delta h_f(A) - \nu_B \Delta h_f(B) - \ldots \qquad (1.1)$$

(and the same is true for Δu and Δu_f). The Δh_f for many compounds may be found in tables, usually at 298 K, and under the "standard" conditions indicated in these tables. This value is written $\Delta h^\circ_{f,\,298}$ (which indicates that the standard conditions are 1 bar + a second condition. For gases, this second condition is that A, B, etc. are pure, i.e. prior to mixing, in their initial state, and equally that A′, B′, etc. are pure, unmixed, in their final state).

The state functions h and u, depend on the temperature T and, to a lesser extent, on the pressure (as indeed do heat capacities). For a perfect gas, when no change of state or chemical reaction takes place, u and h depend only on T, thus:

$$\mathrm{d}h = c_p(T)\,\mathrm{d}T \quad \text{and} \quad \mathrm{d}u = c_v(T)\,\mathrm{d}T$$

which, by integrating the first equation between T_1 and T_2 gives:

$$h \text{ at } T_2 - h \text{ at } T_1 = \int_{T_1}^{T_2} c_p\,\mathrm{d}T \qquad (1.2)$$

or between 0 K and T, and writing h at 0 K = h_0:

$$h - h_0 = \int_0^T c_p\,\mathrm{d}T \qquad (1.2')$$

or a relationship of the same type between 298 K and T. Numeric values of c_p° and (h° at T minus h° at 298 K) for a large number of substances are given, in addition to other values, in the *Janaf Thermochemical Tables*[1].

By introducing molar magnitudes and using n_i to designate the number of moles of each species i, the following relationship may be written for any mixture with a set composition:

$$\Sigma n_i h_i \text{ at } T - \Sigma n_i h_i \text{ at } T_0 = H \text{ at } T - H \text{ at } T_0$$

and thus:

$$H \text{ at } T - H \text{ at } T_0 = \int_{T_0}^{T} \Sigma n_i c_{ip,\,i}\,\mathrm{d}T = \int_{T_0}^{T} C_p\,\mathrm{d}T \qquad (1.3)$$

(with m_i instead of n_i if specific (i.e. per unit mass) magnitudes instead of molar magnitudes are considered) where C_p is the heat capacity of the mixture. The internal energy can be treated in the same way.

---

1. *Janaf Thermochemical Tables*, 3rd edition, published by the American Chemical Society and the American Institute for the National Bureau of Standards, 1985.

In the same way, Δh_f and Δu_f are temperature-dependent and the following equation can be written for the enthalpies of formation at a given pressure:

$$\Delta h_f \text{ at } T - \Delta h_f \text{ at } T_0 = \int_{T_0}^{T} \Delta c_p \, dT \tag{1.4}$$

where Δc_p is the difference in c_p between the compound produced, e.g. H_2O, minus the c_p of the elemental substances making up the reactants, for example the c_p of one mole of H_2 and half a mole of O_2. At the standard conditions:

$$\Delta h_f^\circ \text{ at } T - \Delta h_f^\circ \text{ at } T_0 = \int_{T_0}^{T} \Delta c_p^\circ \, dT \tag{1.5}$$

and equally for Δu:

$$\Delta u_f^\circ \text{ at } T - \Delta u_f^\circ \text{ at } T_0 = \int_{T_0}^{T} \Delta c_v^\circ \, dT \tag{1.5'}$$

taking T_0 to be equal to 298 K, 0 K or any other value. Similar laws can be deduced for all the Δu and Δh involved in the reaction. Finally, any change in state of a reactant or product, must be considered, which may occur between the temperatures T_0 and T.

Starting with the chemical reaction:

$$\nu_A A + \nu_B B + \ldots \rightarrow \nu_{A'} A' + \nu_{B'} B' + \ldots$$

the "differential of the advancement of the reaction" (or "reaction degree"), dz, can be defined by the equalities:

$$-\left(\frac{1}{\nu_A}\right) d[A] = -\left(\frac{1}{\nu_B}\right) d[B] = \ldots = \left(\frac{1}{\nu_{A'}}\right) d[A'] = \ldots dz$$

where [A], [B], [A'], etc. are the molar concentrations of A, B, A', etc. The advancement z is determined by integration, assuming that in the initial system (when $z = 0$): $[A'] = [B'] = \ldots = 0$, and $[A] = [A]_0$, $[B] = [B]_0$, etc. A linear relationship can thus be obtained between each concentration and z, at any instant:

$$-\left(\frac{1}{\nu_A}\right)([A] - [A]_0) = -\left(\frac{1}{\nu_B}\right)([B] - [B]_0) = \ldots = \left(\frac{1}{\nu_{A'}}\right)[A'] = \ldots = z \tag{1.6}$$

At the start of the reaction $z = 0$, and at the end $z = 1$ (for a unit volume).

In general, for combustion processes, reactions are not complete due to the presence of reverse reactions which lead to chemical equilibrium. Under these conditions, the reaction occurs not once but z times, where z is less than 1; and the heat involved, (evolved or absorbed), is $z\Delta u$ for constant volume and $z\Delta h$ for constant pressure. This can be easily demonstrated, again in relation to the

First Law. Assuming that the previous reaction, occurring at constant pressure, was not complete, and thus that its extent of advancement is z, then the difference between the enthalpy of the final mixture under these conditions and that of the initial mixture, is now, instead of (1.1):

$$[A']\Delta h_f(A') + [B']\Delta h_f(B') + \ldots + [A]\Delta h_f(A) + [B]\Delta h_f(B) + \ldots$$
$$- [A]_0\Delta h_f(A) - [B]_0\Delta h_f(B) - \ldots$$

and since the molar concentrations are all related to z by (1.6), we finally obtain:

$$(\nu_{A'}\Delta h_f(A') + \nu_{B'}\Delta h_f(B') + \ldots - \nu_A\Delta h_f(A) - \nu_B\Delta h_f(B) - \ldots)z = z\Delta h,$$

and similarly for the constant volume case.

1.2.2 Consequences of the Second Law in combustion chemistry

From the Second Law, it can be deduced that a system, and thus also a chemical reaction, proceeds from left to right, at a given temperature:

- At constant volume, if the variation ΔF in the free energy function, $F = U - TS$, is negative (where S is the entropy)
- At constant pressure, if the variation ΔG in the free enthalpy function, $G = H - TS$, is negative

and in the opposite direction if ΔF or $\Delta G > 0$. Like U and H, F and G are state functions, i.e. only functions of the state of the system. Hence, it is possible to determine their variation, and thus the difference in their values between two states, but not the value in each one separately. The **reaction** Δf and Δg quantities are the **differences** in the F or G of the products of the stoichiometric reaction minus those of the reactants. Taking μ_i to be the chemical potentials, ν_i the stoichiometric coefficients, n_i the number of moles of each constituent, a_i the activities (corrected concentrations or pressures) and G the total free enthalpy of the system, we have:

$$\mu_i = \frac{\partial G}{\partial n_i} \quad \text{where} \quad \mu_i = \mu_i^\circ + RT \ln a_i$$

with $a_i = p_i$ (partial pressure) for an ideal gas

$$G = \Sigma n_i \mu_i$$

and:

$$\Delta g = \Sigma \nu_i \mu_i \text{ of the products} - \Sigma \nu_i \mu_i \text{ of the reactants},$$
$$\Delta g^\circ = \Sigma \nu_i \mu_i^\circ \text{ of the products} - \Sigma\, \nu_i \mu_i^\circ \text{ of the reactants}.$$

Thus for the reaction:

$$A + 2\,B \rightarrow 3\,C$$

$$G = n_C\,\mu_C + n_A\,\mu_A + n_B\,\mu_B \quad \text{and} \quad \Delta g = 3\mu_c - \mu_A - 2\mu_B$$

(see also the first worked example at the end of this chapter).

Based on the second law, it can also be shown that the state of **chemical equilibrium** at a given pressure and temperature is such that the Δg of the reaction is zero. This quantity should not be confused with the change in G produced by a variation in the advancement from an initial value z_1, which may be zero, to a final value z_2, which may be the equilibrium value. Moreover, it can be shown that the total G of the system, as shown above, is the sum:

$$\Sigma n_i\,\mu_i \text{ (and not } \Sigma \nu_i\,\mu_i)$$

and that G reaches a minimum value for a system in equilibrium. Variations in Δg and G can thus be represented as a function of the advancement z of the reaction as in Figure 1.1. Since the profile of G is known, but is subject to an unknown constant of integration, then the origin of G can be chosen freely. As

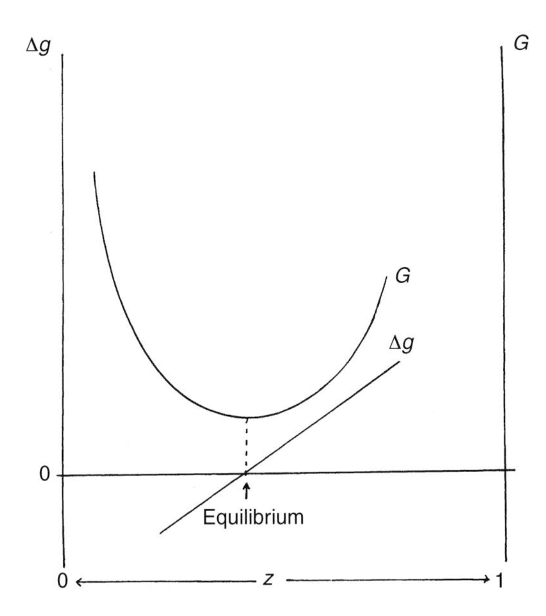

Figure 1.1 Variation with z, advancement of the reaction, of the Δg of reaction ($\Sigma \nu_i \mu_i$ of the products – $\Sigma \nu_i\,\mu_i$ of the reactants) and of the total G for the system ($\Sigma n_i \mu_i$). The z value at which chemical equilibrium for the system is reached is where Δg is zero and where G passes through a minimum. G is only known with an additive constant of integration. The origin for the right-hand scale, relating to G, has deliberately not been indicated. Depending on the choice of this origin, the curve $G = f(z)$ is translated up or down. G being a state function, its absolute value is not known, as explained in the text.

a consequence, depending on this choice, the curve showing the variation in G can be translated up or down. This is true for each of the several reactions occurring simultaneously.

It can then be shown that this condition enables the relationship known as the **law of mass action** to be determined for each reaction at equilibrium. This law states that in the classical case of a perfect mixture of ideal gases:

"at the equilibrium point in a reaction, the product of the partial pressures of the products, each increased by a power equal to its stoichiometric coefficient, divided by the corresponding product of the partial pressures of the reactants, is a function of temperature only, and is called the **equilibrium constant**".

Theoretically at least, and with the use of thermodynamic tables, equilibrium constant values can be calculated as a function of temperature. Indeed, for the formation reactions mentioned above, these values may be found from the tables directly, and the equilibrium constants for all the other reactions can be determined by multiplying and/or dividing these equilibrium constants.

When combustion ceases, chemical equilibrium is reached, but not necessarily the final state predicted by the balanced stoichiometric reaction. Using the law of mass action, the extent of advancement at equilibrium can be deduced, as a function of temperature and pressure. Based on this result the adiabatic combustion temperature at constant pressure or volume can be determined. This procedure is the subject of the following section.

1.2.3 Calculating the adiabatic combustion temperature

The adiabatic combustion temperature, T_{ad}, is the final temperature of combustion under adiabatic conditions.

Let us begin by considering the simplest combustion reaction, starting with a stoichiometric mixture, and which will reach completion in the direction of combustion ($z = 1$), i.e.:

$$H_2(\text{gas}) + 0.5\ O_2(\text{gas}) \rightarrow H_2O(\text{gas})$$

and which takes place in a fixed-volume chamber with adiabatic walls (known as a bomb calorimeter). Assume that the heat released when the reaction occurs at T_0 simply increases the temperature of the burnt gases (in this case water vapour) from an initial temperature T_0 to the final temperature T_{ad}. In making this assumption, note that the reaction is not considered to have actually occurred at this temperature T_0 (which is not the case in practice). Since the internal energy (or enthalpy) is a state function, only the initial and final states of the evolution are of interest, and the way in which the states are actually reached is of no importance. The heat released is equal to $(-\Delta u)$ in the above reaction, i.e. approximately 60.5 kcal·mol^{-1}. If we take c_v to designate a mean value between T_0 and T_{ad} of the molar heat capacity of water vapour, at constant volume, i.e. approximately 6 cal·mol^{-1}·K^{-1}, then under these conditions:

$$c_v(T_{ad} - T_0) + \Delta u = 0 \tag{1.7}$$

which gives $T_{ad} - T_0 = 10\,083$ K, a value which is much larger than the experimental final value. Note that the same equation (1.7) and thus the same balance would have been obtained if instead of 1 mole of H_2 and 0.5 mole of O_2, k times more or k times less had been taken. If air rather than pure oxygen had been considered, then since air consists of approximately 0.5 O_2 + 2 N_2, then the burnt gases would contain 2 moles of N_2, and thus twice 5 cal·mol^{-1}·K^{-1} would need to be added to the heat capacity of the burnt gases, giving a total of 16 cal·K^{-1}. In this case $T_{ad} - T_0 = 3781$ K, which although smaller than the previous value, is nonetheless still greater than the experimental value, about 2400 K.

T_{ad} clearly depends on the composition of the initial mixture, and is much higher for the stoichiometric mixture H_2/F_2, for example, than the stoichiometric mixture H_2/O_2. Thus, still based on equation (1.7), T_{ad} is found to be approximately 11 000 K for the stoichiometric mixture CH_4 + 2 O_2 and 7000 K for the CH_4 + 4 O_2 mixture.

The temperature increases do not reach values anywhere near these figures. The reason for this is not so much due to variations in c_v with T (which were ignored above) nor even to non-adiabatic considerations, but to the fact that chemical equilibrium must be assumed to have been attained. In reality, combustion reactions are restrained, when they take place at high temperatures, by reactions which consume the products found on the right-hand side of the balanced stoichiometric reaction. The global balance reaction is impaired, and thus combustion is **incomplete** ($z < 1$). By way of an illustration, using the above example, the water produced dissociates, initially into H and OH, and can then turn back into H_2 and O_2 and to a proportionally greater extent as temperature increases. Even in the simplest cases, involving a single reaction, the difficulties which can arise when determining the final temperature are clearly apparent. The amount of heat released must be known, which requires knowing the advancement of the reaction, which in turn requires knowing the temperature.

Let us return to the case of hydrogen burning in air, and consider only the reaction involving the combustion of H_2 and O_2 to give H_2O, as written above. Two equations must be solved, both containing the temperature T. The correct final temperature value T is that which is correct for both equations. The first equation is the thermal balance equation derived from the First Law, and analogous to equation (1.7). Let c_1, c_2, c_3 and c_4 refer to the molar heat capacities at constant volume for H_2O, H_2, O_2 and N_2, still assumed to be constant between T_0 and T:

$$\left[c_1 z + c_2(1-z) + c_3\frac{1-z}{2} + 2c_4\right](T - T_0) = z\Delta u \tag{1.8}$$

The second equation, derived from the Second Law, is based on the law of mass action for the reaction under consideration, and takes into account the variation in the equilibrium constant with temperature:

$$\frac{\mathrm{d}(\ln K_p)}{\mathrm{d}T} = \frac{\Delta h^{\mathrm{o}}}{RT^2} \tag{1.9}$$

Simplified to approximate an ideal gas system, (which can be easily verified), the law of mass action gives:

$$K_p(T) = \frac{p_{H_2O}}{p_{H_2}\, p_{O_2}^{\frac{1}{2}}} = z\left[\frac{(7-z)^{\frac{1}{2}}}{(1-z)^{\frac{3}{2}}}\right] p^{-\frac{1}{2}} \tag{1.10}$$

The fact that the total pressure p and the advancement z appear in this equation clearly does not mean that K_p is dependent upon them, which would contradict equation (1.9). The final temperature T and z will be given by the solution to (1.8), (1.9) and (1.10). Since z is a function of both p and T as a result of (1.10), it is dependent upon pressure (for details about the numerical solution, refer to worked example 2 at the end of this chapter).

The above calculation still only gives a very approximate value of T. In fact, in addition to ignoring variations in c_v with temperature, there is no justification, in principle, for the presence of the single chemical equilibrium corresponding to the stated balanced reaction, and products such as OH, H, O etc. may also be formed. Experience has shown that these species are only produced in small quantities if the temperature is not too high, but that large amounts are produced at temperatures above approximately 1700 K. Taking them into account in the calculation requires not only the water formation balance reaction, as written above, but also the balance reactions for the formation of products which need to be taken into consideration, in addition to writing the corresponding chemical equilibria. The calculation of T and the composition of the final mixture is then performed as above, but without explicitly involving the advancement z. This is more complicated and can only be determined accurately by computer. Figure 1.2 illustrates the results of one such calculation for the combustion of mixtures of propane and air, under adiabatic conditions.

For the propane/oxygen stoichiometric mixture, at constant pressure, a temperature increase of 6340 K is calculated if it is assumed that the final mixture only contains CO_2 and H_2O, and a value of 3081 K if all the possible equilibria are taken into account, i.e. a final mixture which additionally contains CO, H_2, O_2, H, O, OH, etc. Moreover, complete thermodynamic calculations have been performed, including all the possible chemical equilibria, for a series of pairs of oxidant/fuel, the bi-propellants (designed for use in rocket engines) which should give high temperatures. The following approximate adiabatic

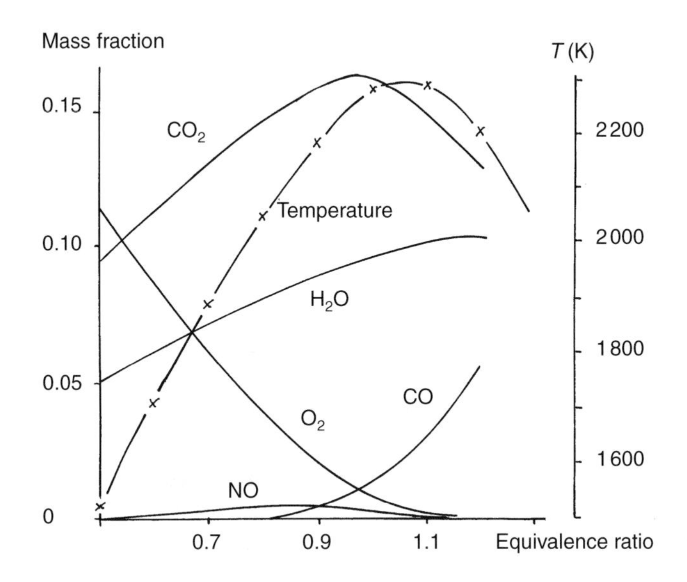

Figure 1.2 Effect of equivalence ratio (*cf.* section 1.5.1) on the composition at equilibrium and on the final temperature reached following adiabatic combustion, at a given volume, of propane-air mixtures. Initial temperature: 298 K. Initial pressure: 1 bar. The maximum final temperature is reached for an equivalence ratio slightly greater than one and the maximum mass fraction of CO_2 for a slightly lower value.

combustion temperatures were determined for the pairs (based on the proportion which gave the highest value):

3400 K for the ClF_3/H_2 pair
5100 K for the ClF_3/Li pair
3900 K for the ClF_3/N_2H_4 pair
3700 K for the ClF_5/H_2 pair
5200 K for the ClF_5/Li pair, etc.

Finally, it should be noted that in a combustion chamber of a finite size in which chemical reactions occur in a closed system, some very rapidly, and others more slowly, which is often the case, then **complete chemical equilibrium** will not necessarily be reached, even at the combustion chamber outlet. In situations such as this, certain reactions may reach equilibrium, but not all. This is why NO, which may be produced at the high temperatures reached in a combustion reaction, may be present in exhaust gases, but is not, in general, present in quantities as high as those found at equilibrium. However, when the gases are cooled to a lower temperature in the exhaust pipe, and even though the NO should dissociate to give $N_2 + O_2$, it does not do so completely. This cooling, which is generally fairly rapid, causes the phenomenon known as **chem-**

ical quenching: below 500 to 1000 K, the rate at which NO dissociates tends towards zero and the $NO/N_2/O_2$ equilibrium, which should give almost entirely N_2 and O_2, is not reached under these conditions.

An even worse situation occurs, as we shall now see, if **thermal equilibrium,** as defined by statistical thermodynamics, is not reached. In this case, the concept of a unique, instantaneous temperature at a point ceases to have any meaning.

1.3 A MICROPHYSICAL PERSPECTIVE OF THE CONCEPT OF TEMPERATURE

In a gas, the chemical species (atoms, ions, molecules and radicals) translate in all directions, resulting in collisions. In a gaseous mixture, certain collisions may result in a redistribution of the atoms in the various molecules, i.e. a chemical reaction. When there is an associated release of energy, i.e. with an additional agitation of the molecules, this energy localises in the created species and is then partially transferred to other molecules by successive impacts. The result is that heat is released (very high energy species may also release this energy in the form of light, without involving any collisions).

A molecule may possess translational energy due to its translational motion, vibrational energy due to the vibration of its constituent atoms, rotational energy due to its rotation and electronic energy due to electron-atomic nuclei interactions. However, it does not possess a "temperature".

The energy of a movement is **quantised:** only being able to assume a series of discrete values, which in increasing order are:

$$E_0 \quad E_1 \quad E_2 \ ... \ E_i \ ... \ E_j \ ...$$

In a pure gas and for a given type of motion, let N_0 be the number of particles with an energy value of E_0, N_1 particles with an energy of E_1, and so on. If for any E_i and E_j, we have the equation containing a parameter T, whose value is the same for all quantum levels:

$$\frac{N_j}{N_i} = \left(\frac{g_j}{g_i}\right) \exp\left[-\left(\frac{E_j - E_i}{k_B T}\right)\right] \tag{1.11}$$

then this parameter is called the temperature of the considered movement.

g_j is the level j degeneracy (number of different levels, in a species, which have the same energy E_j)

g_i the level i degeneracy

k_B the Boltzmann's constant.

Moreover, if the same procedure is carried out for each type of motion, with the same value of T in each case, T can then be said to be the thermodynamic temperature for the gas under consideration (it can be shown that it is the same as that used in classical thermodynamics). Now, in general, this condition is achieved fairly quickly as collisions occur in an isolated pure gas and thus proceeds towards an equilibrium. Such an equilibrium is said to be **thermal**, where T is the thermal equilibrium temperature of the gas under the conditions achieved. In a gas mixture, if all the motions of all the particles occur at the same temperature, this is then the thermodynamic temperature of the gas mixture and this mixture is in thermal equilibrium (but not necessarily in chemical thermodynamic equilibrium).

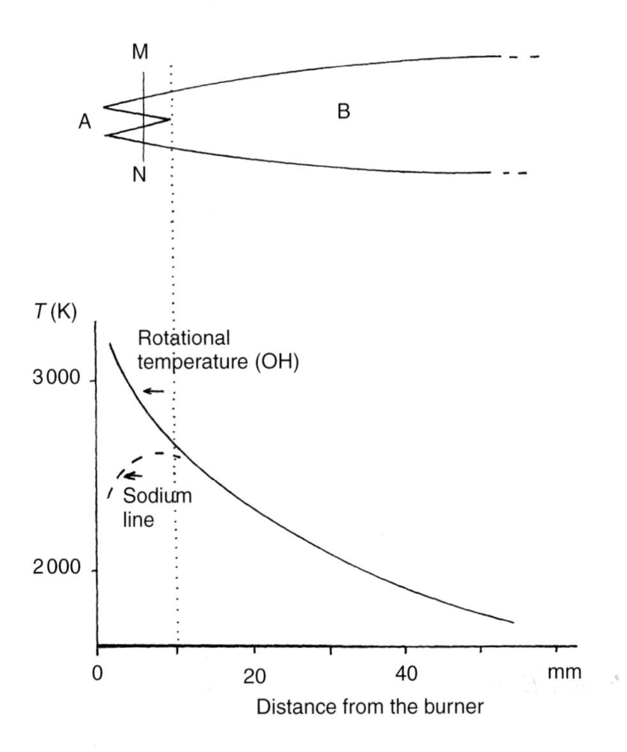

Figure 1.3 An example of thermal imbalance in the combustion of a stoichiometric propane-oxygen mixture. In region A, the rotational temperature of the OH radical, measured along the axis of the Bunsen burner, is greater than that obtained by the reabsorption of the sodium line. Measurements are taken across sections such as MN; and integrate the emissions. Thus, with section MN as shown in the figure: thermal equilibrium is reached in the flame envelope and not in the cone. The difference between the temperature values given by the two methods reduces when MN approaches the summit of the cone from the left, and drops to zero thereafter.

Since thermal equilibrium is generally reached more rapidly than chemical thermodynamic equilibrium, the former is often almost reached and the second not at all. Alternatively, neither may be reached, which is the case for media in which very rapid chemical reactions occur, such as those which take place in the hottest zones of a combustion process. Figure 1.3 shows that for the combustion of a stoichiometric propane-oxygen mixture on a burner, in which oxidant and fuel are premixed, then in the so-called "primary" combustion zone (the conical region A in Fig. 1.3) the temperature measured by the reabsorption of the sodium line (which is an electronic temperature of sodium) differs from the rotational temperature of the OH radical, (which, in this case, is the rotation bands of OH in the state $(A^2\Sigma^+)_{v'=0}$, both being taken from equation (1.11), but for different populations). On the other hand, these temperatures are equal in the so-called "secondary" combustion zone, or "flame envelope", indicated by region B in Figure 1.3. Having made this point, in the rest of this book thermal equilibrium will be assumed to be achieved virtually throughout the reacting medium unless otherwise stated. A consequence of this assumption is that a unique value for the temperature can be defined.

1.4 INDUSTRIAL CALORIFIC VALUE

Industrial calorific value (CV) is defined as the number of kcal of heat produced by the complete combustion in air, **in a calorimeter maintained at 25°C** and at atmospheric pressure, of one kg of a solid fuel, or of one kg, one litre or one cubic metre (depending on the specifications) of a liquid or gas.

The **net** value (NCV) is used when the water produced is considered to be a vapour, and the **gross** value (GCV) when liquid water is produced. The gross value is greater since the GCV is the sum of the NCV and the heat released by the condensation of the water vapour. Clearly this water will automatically exist in vapour form in the combustion zone, and will condense at some distance away from the combustion zone, depending on the cooling conditions, and thus of the recovery of the heat produced. The "useful" calorific value thus depends on the system implemented to recover this heat, and thus on the position and type of any heat exchangers used.

The calorific value can easily be calculated if the composition of the fuel is known. If it is assumed that combustion occurs as a balanced stoichiometric reaction with a final extent of advancement $z = 1$, it is then equal to the opposite of the Δh for this reaction. Neglecting the possibility of incomplete combustion (by assuming $z = 1$) introduces only very minor errors in this case, where the temperature is maintained at 25°C. In fact, the "equilibrium constant" values at this temperature are such that the dissociation of the products of complete combustion are insignificant, and the complete calculation of all possible equilibria would give negligible concentrations for all species other than those indicated in the balanced stoichiometric reaction.

Indeed, empirical relationships are used in many cases, such as that of Goutal for the NCV of solid fuels:

$$\mathrm{NCV} = a\left(\frac{\mathrm{C}}{100}\right) + bV \tag{1.12}$$

where C/100 is the percentage of carbon said to be "fixed" (i.e. combustible carbon minus the volatile matter minus the ash) and V the percentage of volatile matter which is mixed but not combined. The numerical coefficients a and b vary depending on the fuel, and are deduced experimentally.

Calorific value may be used instead of Δh to calculate the maximum combustion temperature at constant pressure, using equation (1.7) (but with c_p). As we have seen above, this presumes that combustion is complete, and thus must overestimate the actual value, although the precision is improved when there is little dissociation of the products and thus a low T.

1.5 EQUIVALENCE RATIO AND OSTWALD DIAGRAMS

1.5.1 Equivalence ratio

This is the ratio:

$$\frac{\left(\dfrac{\text{amount of fuel}}{\text{amount of oxidant}}\right)_{\text{actual mixture}}}{\left(\dfrac{\text{amount of fuel}}{\text{amount of oxidant}}\right)_{\text{stoichiometric mixture}}}$$

The oxidant is usually O_2 or air. The equivalence ratio is the same irrespective of whether the quantity is measured based on mass or on the number of moles, and is in fact a dimensionless number which is equal to 1 when the mixture being studied is stoichiometric.

As has been shown, a "rich" mixture, i.e. where the equivalence ratio is greater than 1, is a mixture rich in fuel. For example, the stoichiometric combustion of methane can be written:

$$CH_4 + 2\,O_2 \rightarrow CO_2 + 2\,H_2O$$

The equimolar mixture: one mole of CH_4 for one mole of O_2, ($CH_4 + O_2$) or ($2\,CH_4 + 2\,O_2$), etc., is a rich mixture, with an equivalence ratio of 2. Other definitions of the "richness" of a mixture are occasionally encountered, particularly in technical documents. In order to avoid any confusion, it is therefore advisable to use the terms fuel rich, or fuel lean, although the term "excess air" also indicates the composition of the mixture: e.g. a mixture with 20% excess air contains 20% more air (by volume) than is required for the stoichiometric ratio.

The combustion of a **rich** mixture automatically results in unburnt reactants. In the case of the equimolar mixture $CH_4 + O_2$, the equivalent of over half the methane is **unburnt**, although this does not necessarily mean that it is unchanged by the process. As discussed above, unburnt products may occur (**incomplete combustion**) even in a stoichiometric mixture. Thus:

$$CH_4 + 2\,O_2 \rightarrow x\,C + y\,CO + z\,CO_2 + n\,H_2O + m\,O_2$$

since the law of conservation of mass must be satisfied:

for C: $x + y + z = 1$
for O: $y + 2z + n + 2m = 4$
for H: $n = 2$

If, on the other hand, we consider the combustion of a **lean** mixture ($CH_4 + 4\,O_2$):

$$CH_4 + 4\,O_2 \rightarrow x\,C + y\,CO + z\,CO_2 + n\,H_2O + m\,O_2$$

$$x + y + z = 1$$
$$y + 2z + n + 2m = 8$$
$$n = 2$$

If combustion is complete, insofar as all the methane is transformed into CO_2 and H_2O:

$$x = y = 0, \text{ hence } z = 1, \text{ and}$$
$$n = 2, \text{ hence } m = 2$$

In this case, there are no unburnt products left after combustion, but rather excess oxygen in the final mixture...

It should be noted that if combustion is complete and produces CO, which makes y different from 0, x at equilibrium is also automatically different from 0, so long as the well-known equilibrium, known as Boudouard's equilibrium, is reached:

$$2\,CO \rightleftharpoons C + CO_2$$

In this case, in addition to unburnt gaseous CO, there is also unburnt solid C.

1.5.2 Ostwald diagrams

In practical industrial applications, the overall combustion process can be represented by diagrams, known as Ostwald diagrams. These are approximate representations, insofar as all the species which could exist in the final mixture are not taken into consideration, particularly atoms and radicals. Thus, in the case of the combustion of carbon in air, the reaction can be written generally as (taking air to be = $O_2 + 4\,N_2$):

$$C + (1+u)\,(O_2 + 4\,N_2) \rightarrow (1-v)\,CO_2 + v\,CO + \left[\frac{(v+2u)}{2}\right] O_2 + 4(1+u)N_2$$

The simplest case is where $u = v = 0$:

$$C + O_2 + 4\,N_2 \rightarrow CO_2 + 4\,N_2$$

This case corresponds to complete stoichiometric combustion in air. If a diagram is constructed where the molar percentage of final O_2 is assigned to the x-axis and final CO_2 to the y-axis (Fig. 1.4) then, under these conditions, the point B ($x = 0$ and $y = 20$) can be plotted. The limiting case where u tends to infinity, i.e. where air alone is present, is indicated by the point A ($x = 20$ and $y = 0$).

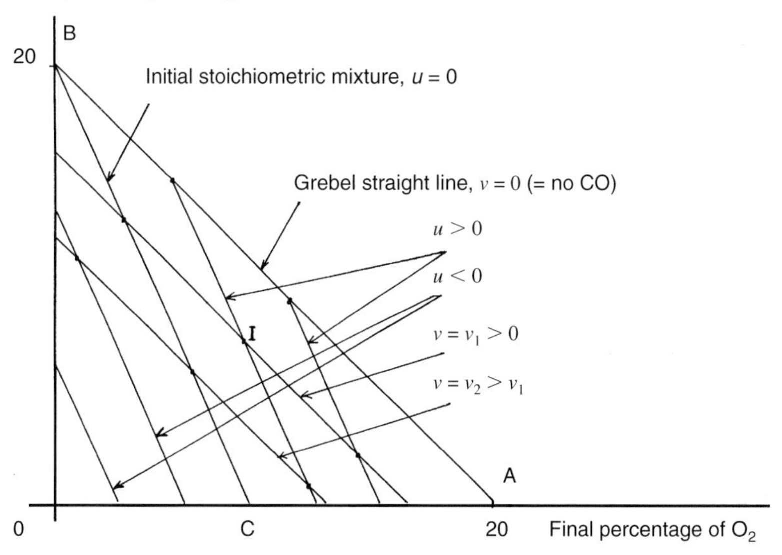

Figure 1.4 Ostwald diagram for the reaction:

$$C + (1+u)\,(O_2 + 4N_2) \rightarrow (1-v)\,CO_2 + v\,CO + \frac{1}{2}\,(v+2u)O_2 + 4(1+u)N_2$$

(in this case, OAB is an isosceles triangle).

The case $v = 0$ for the initial, non-stoichiometric system, with complete combustion:

$$C + (1 + u)(O_2 + 4\,N_2) \rightarrow CO_2 + u\,O_2 + 4(1 + u)N_2$$

results in:

$$x = \frac{100u}{5(1 + u)} \quad \text{and} \quad y = \frac{100}{5(1 + u)}$$

which are the parametric equations of the straight line, known as the Grebel line, which passes through A and B, and whose equation is $y = 20 - x$. Each point on the line has a corresponding value of u, thus of excess air.

The case $u = 0$ for a system which is initially stoichiometric, and with incomplete combustion, gives:

$$C + O_2 + 4\,N_2 \rightarrow (1 - v)\,CO_2 + v\,CO + \left(\frac{v}{2}\right)O_2 + 4\,N_2$$

and results in:

$$x = 100\frac{\frac{v}{2}}{5 + \frac{v}{2}} \quad \text{and} \quad y = 100\frac{1 - v}{5 + \frac{v}{2}}$$

which are straight-line parametric equations, passing through B and C, with the equation:

$$y = 20 - \left(\frac{11}{5}\right)x.$$

In a similar way, the case of a system which is initially non-stoichiometric, and with incomplete combustion gives:

$$x = 100\frac{u + \frac{v}{2}}{5 + 5u + \frac{v}{2}} \quad \text{and} \quad y = 100\frac{1 - v}{5 + 5u + \frac{v}{2}}$$

which, for a given v, and with u variable, gives the straight-line parametric equation:

$$y = -5x\frac{1 - v}{5 - 2v} + 100\frac{1 - v}{5 - 2v}$$

which gives the Grebel line for $v = 0$ (no CO) and straight lines which are virtually parallel to this line for small values of v.

Equally, for a given u, and with v variable, the following family of straight lines is produced:

$$y = -x\frac{11+10u}{5+4u} + \frac{100(1+2u)}{5+4u}$$

which gives the straight line BC for $u = 0$ (stoichiometric mixture), lines above BC for u greater than 0, and below for u values less than 0, but > -1, and a straight line passing through the origin when $u = -0.5$. All these straight lines are more or less parallel. When u is positive they may be extrapolated above Grebel's line, for complete combustion, to cut the y-axis at a point above point B, although evidently this cannot be achieved in practice since the reaction cannot go further than completion. Each intersection point, such as I, where a line from the first family crosses one from the second, corresponds to a pair of values of u and v. From the triangle OAB in the diagram, we can read off the CO concentration in the burnt gases and the corresponding deficit or excess of air for each CO_2 concentration produced.

The case of incomplete combustion due to insufficient air, which moreover, would not yield O_2 in the burnt gas mixture:

$$C + (1+u)(O_2 + 4N_2) \rightarrow (1+2u)CO_2 - 2u\,CO + 4(1+u)N_2$$

and where u is negative and $u + v/2$ is zero, is represented by the points on the y-axis: $x = 0$ and $y = 100(1 + 2u)/(5 + 4u)$.

WORKED EXAMPLES

■ **1)** Let the general gas-phase reaction be:

$$A + B \rightarrow 2C + D$$

occurring at a given total temperature and at a pressure, p, (although the volume is variable). Find the relationships which give, for an advancement z, the total G for the system (with upper case G) and the Δg of the reaction (with lower case g). Then the difference in G will be determined for the system when $z = 1$, (i.e. G_1), less G for the system when $z = 0$ (G_0, which is not the same as $G°$). The initial system will be stoichiometric, i.e. with n_0 moles of A and n_0 moles of B.

The numbers of moles are:

- For A and B: $n_0(1 - z)$
- For C: $2n_0z$
- For D: n_0z.

Since the total number is $n_0(2 + z)$, the partial pressures of A, B, C and D can be easily deduced, i.e. p_A, p_B, p_C and p_D:

$$p_A = p_B = p\,\frac{(1-z)}{(2+z)} \qquad p_C = \frac{2pz}{(2+z)} = 2p_D$$

which gives:

$$\Delta g = 2\,\mu_C^o + \mu_D^o - \mu_A^o - \mu_B^o + RT\ln\left(\frac{p_C^2 p_D}{p_A p_B}\right)$$

which is a function of T and, for a given p, of z, where:

$$2\,\mu_C^o + \mu_D^o - \mu_A^o - \mu_B^o = \Delta g^o$$

Moreover:

$$\frac{G}{n_0} = (1-z)(\mu_A^o + \mu_B^o + RT\ln p_A p_B) + z(2\mu_C^o + \mu_D^o + RT\ln p_C^2 p_D)$$

$$\frac{G}{n_0} = \mu_A^o + \mu_B^o + RT\ln p_A\,p_B + z\left[\Delta G^o + RT\ln\left(\frac{p_C^2 p_D}{p_A p_B}\right)\right]$$

Finally:

$$G_1 - G_0 = \Delta G^o + n_0 RT\ln\left(\frac{p_C^2 p_D}{p_A p_B}\right)$$

However, this time with the initial state being A + B for a total pressure p and the final state 2C + D for a total pressure p, it may be stated that $p_C = 2\,p_D = 2\,p/3$ and that $p_A = p_B = p/2$, the partial pressures no longer being functions of z.

Hence:

$$G_1 - G_0 = \Delta G^o + n_0 RT\ln\left(\frac{16p}{27}\right) \text{ function of } p \text{ and } T.$$

■ **2)** Write (in this case in Pascal programming language) a numerical calculation programme to determine the final temperature, T_M, produced by the adiabatic combustion of a stoichiometric mixture of hydrogen-air containing n_0 moles of H_2, $n_0/2$ of O_2 and approximately $2n_0$ of N_2 (i.e. 4 times more moles of N_2 than of O_2) in a chamber of fixed volume V. The reaction may be written:

$$H_2 + \frac{1}{2}O_2 + 2N_2 \rightarrow H_2O + 2N_2.$$

T_0 is the initial temperature on ignition. By varying n_0 or V, i.e. the pressure prior to the reaction occurring, it will be easy to check that the effect produced is indeed that predicted by the laws of thermodynamics. It may be assumed that ignition is by electrical means, and that it does not change the thermal balance.

A method must first be selected, which assumes the development of an algorithm. Now equation (1.8) gives T_M so long as z is known. However, z in turn is dependent on K_p, and thus on T, and hence on T_M, as defined by the series of equations (1.9) and (1.10). Moreover, equation (1.10) contains the total pressure p, which depends on T_M which is not known.

Let us write that the final pressure p is the product of RT_M/V and the final number of moles, with the combustion occurring in air:

$$p = n_0 \frac{7-z}{2} \frac{RT_M}{V} = f(T_M \text{ and } z)$$

$$K_p = z(1-z)^{-3/2} \left(\frac{2V}{n_0 RT} \right)^{1/2}$$

Since K_p depends only on T, this equation shows, after differentiation, that for a given T, z and n_0/V, and thus also p, vary in the same direction:

1. Knowing K_p the advancement of the reaction, z, for various values of T can be determined. It is of course desirable that T_M should fall within the range under consideration. The "procedure" used for this is called ADVANCEMT (written in Pascal, subroutine in Fortran).
2. Knowing one reaction enthalpy for one of the temperatures in this range, it is possible to calculate the others for the other temperatures. This is the HEAT "procedure".
3. We can therefore see, for each of these temperatures, to what extent the thermal balance of the relationship (1.8) is confirmed. This is the BALANCE "procedure".
4. In general this is not the case. At best it is possible to discover in which sub-field the temperature T_M lies. However, this sub-field can be further reduced by interpolating. This is the INTERPOL "procedure". The ADVANCEMT "procedure" is inserted into INTERPOL in order to determine the advancement z at each interpolation temperature.

APPROXIMATED NUMERICAL VALUES

(deduced by P. Pascal's extrapolation, *Traité de chimie minérale*. Masson I, p. 634)

T(K)	1500	2000	2250	2500	2750	3000	3250
K_p*	5.2574 10^5	3344	609	155	49.5	19.1	8.41

* All the partial pressures are in atmospheres.

$\Delta h°$ for the reaction is taken to be – 60 700 cal at 2000 K.

```
PROGRAM Tflame;
(* Giving the temperature produced by the adiabatic combustion *)
(* of n0(H2 + 0.5 O2 + 2 N2) to give H2O vapour in a volume V*)
{$N+}
Const
R1=0.082053;        (*R in litre atm*)
R2=1.9871;          (*R in cal*)
Var
Stop:Char;
M,I,J,Nbz,I1,I2,touche:Integer;
z,zinter,f,D,Diff,Temp,Tinter,Cste,enthal,T0,n0,V,
cvwater,cvH2,cvO2,cvN2,cpwater,cpH2,cpO2,cpN2,deltacp,P,Q,Qinter:Double;
T,Kp,ksi,deltaH,deltaU:Array[1..10] of Double;
ADVANCEMT PROCEDURE;
Begin
Writeln('Select  an  initial  value  of  z  to  give  Kp  preferably  false  by
default');
Writeln('Nbz=No. of the z value, e.g. 6, selected to include');
Writeln('the correct Kp value;never z=1');
Write('Nbz INTEGER = ');Readln(Nbz);
Repeat
For J:=1 to Nbz Do
Begin
Write('z=');Readln(z);
f:=(z/(1-z))*Sqrt((2*V)/((1-z)*n0*R1*T[I]));
Diff:=Kp[I]-f;Writeln('Diff=',Diff:10:3,' If>0,z is false by default');
End;Writeln;
Writeln('  If you want to select another series of values for z');
Writeln('  type 1,otherwise type 0');Write('  TYPE:');Readln(key);
Until (key=0);
Write('  A T = ',T[I]:4:1,' K,the best z=');Readln(ksi[I]);Writeln;
End;
HEAT PROCEDURE;
Begin
Writeln('      We require the REACTION HEAT at the selected temps.');
Writeln('Are the temp.for which you have for Kp those');
Writeln('for which you have the best deltaH?');
Write('The latter in K=');Readln(Temp);
Write('deltaH in cal,which is NEGATIVE=');Readln(enthal);
Write('et Kp=');Readln(Cste);
Writeln('DeltaH is calculated for the lowest of the M selected temp.');
deltacp:=cpwater-cpH2-(0.5*cpO2);
deltaH[1]:=enthal+(deltacp*(T[1]-Temp));
deltaU[1]:=deltaH[1]+(0.5*R2*T[1]);
For I:=1 to M Do
Begin
deltaH[I]:=deltaH[1]+deltacp*(T[I]-T[1]);
deltaU[I]:=deltaH[I]+(0.5*R2*T[I]);Writeln;
Writeln(' At T=',T[I]:4:1,' Kp=',Kp[I]:6:1);
Writeln('deltaH[I]=',deltaH[I]:6:0,' and deltaU[I]=',deltaU[I]:6:0,' cal');
End;Writeln;
Write('Restart by typing any letter:');Readln(Stop);Writeln;
End;
BALANCE PROCEDURE;
Begin
Writeln('HEATING UP - HEAT DUE TO THE REACTION at the selected T');
For I:=1 to M do
Begin
P:=(cvwater*ksi[I])+(cvH2*(1-ksi[I]))+(cvO2*0.5*(1-ksi[I]))+(2*cvN2);
P:=P*(T[I]-T0);Q:=-ksi[I]*deltaU[I];D:=P-Q;
```

```
Writeln('A T=',T[I]:4:0,' K heat up - heat due to react.=',D:6:1,' cal');
End;Writeln;
Write('Restart by typing any letter:');Readln(Stop);Writeln;
End;
INTERPOL PROCEDURE;
Begin
Writeln('      INTERPOLATION to give advancemt z and heat released');
Writeln('at the temp.assumed for the flame, between 2 of the M selected
temp.,');
Writeln('if sufficiently close, i.e. for 2 values of I');
Repeat
Write('I for the 1st=');Readln(I1);Write('for the 2nd=');Readln(I2);
I:=M+1;Write('T for which the program interpolates for K=');Readln(T[I]);
Kp[I]:=0.5*(deltaU[I1]+deltaU[I2])-(0.5*R2*T[I]);
Kp[I]:=Kp[I]*(T[I]-T[I1])/(R2*T[I]*T[I1]);Kp[I]:=Kp[I]+ln(Kp[I1]);
Kp[I]:=Exp(Kp[I]);
Writeln('  A T en K=T[',I,']=',T[I]:4:1,' Kp=',Kp[I]:6:3);
Writeln;ADVANCEMT;
Tinter:=T[I];zinter:=ksi[I];
Qinter:=-deltaU[I1]-(deltaU[I2]-deltaU[I1])*(Tinter-T[I1])/(T[I2]-T[I1]);
P:=(cvwater*zinter)+(cvH2*(1-zinter))+(cvO2*0.5*(1-zinter))+(2*cvN2);
P:=P*(Tinter-T0);Q:=zinter*Qinter;D:=P-Q;
Writeln('AND HEAT UP - HEAT DUE TO THE REACTION=',D:6:1,' cal');
Writeln('negative values become positive when T increases');
Writeln;Writeln('If the difference appears to be small enough for you, type
0');
Write('If not, type 1;type in?');Readln(key);Writeln;
Until(key=0);
Writeln('THE FLAME TEMPERATURE IS APPROXIMATELY:',Tinter:4:1);
End;
Begin (*MAIN PROGRAM *)
Write('The initial temp. of K is T0=');Readln(T0);
Write('the initial number of moles of H2 is n0=');Readln(n0);
Write('The volume in litres is V=');Readln(V);
Writeln('The molar cp are in cal');
Write('  for H2O vapour:');Readln(cpwater);cvwater:=cpwater-R2;
Write('  for H2:');Readln(cpH2);cvH2:=cpH2-R2;
Write('  for O2:');Readln(cpO2);cvO2:=cpO2-R2;
Write('  for N2:');Readln(cpN2);cvN2:=cpN2-R2;Writeln;
Write('Kp    is    known    at    M    temperatures,not    more    than
10;M=');Readln(M);Writeln;
Write('ADVANCEMENT,z or ksi, at the selected temp.');Writeln;Writeln;
I:=1;For I:=1 to M Do
Begin
Write('At T in K=');Readln(T[I]);
Write('Kp with p in atm=');Readln(Kp[I]);
ADVANCEMT;
End;
HEAT;
BALANCE;
INTERPOL;
End.
```

This programme can be made more general by analysing the case of a combustion reaction with a hydrocarbon C_xH_y:

$$C_xH_y + b\,O_2 + 4b\,N_2 \rightarrow c\,CO_2 + d\,CO + \left(\frac{y}{2}\right) H_2O + 4b\,N_2$$

with $x = c + d \quad 2b = 2c + d + \left(\frac{y}{2}\right)$.

However, numerical values for K_p must be known.

■ **3)** The enthalpy of formation is the enthalpy associated with the synthesis reaction for the molecule in question, where the reactant molecules are in elemental form. For example, for methanol, CH_3OH, the hypothetical reaction, whether it can actually be achieved or not, is:

$$C + 0.5\ O_2 + 2\ H_2 \rightarrow CH_3OH$$

The enthalpy of formation clearly varies depending on whether the methanol produced is a liquid or a gas. Based on the First Law, it can easily be proved that the enthalpy of reaction is equal to the enthalpies of formation of the product molecules less those of the reactant molecules, both multiplied by their stoichiometric coefficients. These enthalpies of formation are given in the tables. In accordance with the definition of the enthalpies of formation, those of elements are zero.

At 298 K and in the state known as standard (see chemical thermodynamics textbooks for details) these values are given in the following table:

	CH_3OH vapor	H_2O liquid	CO_2 gas
in $kJ{\cdot}mol^{-1}$	– 201.2	– 285.83	– 393.51

The enthalpy of vaporisation for liquid methanol, under the same conditions, with the same units, is 39.23, and its density is 790 $kg{\cdot}m^{-3}$. The maximum energy produced by the total combustion of one litre of liquid methanol at 298 K (the initial temperature of methanol and the final temperature of the products of combustion) can be calculated:

a) at constant pressure, i.e. when the final pressure is the same as the initial pressure

b) at constant volume, i.e. when the final volume is the same as the initial volume.

a) The combustion reaction in oxygen is:

$$CH_3OH \text{ liquid} + 1.5\ O_2 \rightarrow CO_2 + 2\ H_2O \text{ liquid}$$

Since the methanol must vaporise before it can burn, the total change in the enthalpy of the process is the algebraic sum of the enthalpy of vaporisation of methanol and its enthalpy of combustion, i.e.:

$$-393.51 - 2 \times 285.83 + 201.2 + 39.23 = -724.74\ kJ{\cdot}mol^{-1}$$

for one mole, i.e. 0.032 kg of methanol, or alternatively:

$$-\frac{724.74 \times 0.790}{32 \times 10^{-3}} = -1.789 \times 10^4 \text{ kJ·dm}^{-3}$$

b) It may be shown that, in approximate terms:

$$\Delta h = \Delta u + RT \times \Delta \text{v}$$

where Δv is the number of moles of gas in the products of the stoichiometric reaction minus this number in reactants, which in this case gives:

$$\Delta u^\circ = \Delta h^\circ + 0.5 \times RT$$

$$= (-724.74 + 0.5 \times 8.314 \times 10^{-3} \times 298)\frac{0.790}{32 \times 10^{-3}} = -1.786 \times 10^4$$

with the same units as above. It is clear that Δu° differs only slightly from Δh°.

■ **4)** What volume of air is required to burn stoichiometrically 1 mole of propane gas, with the initial air at 298 K and with a pressure of 0.98 bar (0.98×10^5 Pa). The combustion reaction may be written:

$$C_3H_8 + 5\ O_2 + 20\ N_2 \rightarrow 3\ CO_2 + 4\ H_2O + 20\ N_2$$

25 moles are necessary, 5 of O_2 plus 20 of N_2, making:

$$V = 25\frac{RT}{p} = \frac{25 \times 8.314 \times 298}{0.98 \times 10^5} = 0.632 \text{ m}^3$$

■ **5)** The gross calorific value of kerosene is 46 890 kJ·kg^{-1}. Kerosene contains 86% by mass of carbon and 14% of hydrogen. Knowing that the condensation of one mole of water vapour releases 41.1 kJ of heat energy, the net calorific value can easily be deduced. In fact, if the general combustion reaction for 1000 g of kerosene is written as:

$$C_{860/12}H_{140/1} + n\ O_2 \rightarrow \left(\frac{860}{12}\right)CO_2 + \left(\frac{140}{2}\right)H_2O$$

then we have:

$$\frac{\text{NCV}}{\text{GCV}} = \frac{46\,890 - \left(\frac{140}{2}\right)41.1}{46\,890} = 0.939$$

$$\text{NCV} = 44\,013 \text{ kJ·kg}^{-1}$$

■ **6)** In a boiler, fuel is burnt in air. Calculate the maximum temperature of the combustion products assuming that the combustion is complete with 30% excess air, initially at 293 K and with P = 1 atm. This excess is defined by the ratio:

$$\frac{\text{air in the initial mixture}}{\text{air in the stoichiometric mixture}}$$

The enthalpy of combustion of the fuel is:

$$\Delta h = 10^7 \text{ cal·kg}^{-1}$$

Its composition by mass is:

86% of C 12.5% of H 1.5% of unburnt reactants

The molar heat capacities vary with temperature and are given in cal·mol^{-1}·K^{-1} by:

$$c_p(CO_2) = 10.34 + 2.7 \times 10^{-3}\, T$$
$$c_p(H_2O) = 8.22 + 0.15 \times 10^{-3}\, T$$
$$c_p(N_2) = 6.5 + 10^{-3}\, T$$
$$c_p(O_2) = 8.27 + 0.26 \times 10^{-3}\, T$$

From the mass percentages, it may be deduced that for 100 g of fuel:

$$\text{number of C} = \frac{86}{12} = 7.17 \quad \text{and the number of H} = 12.5$$

which gives the number of moles produced:

$$\text{of } CO_2 = 7.17 \quad \text{and of } H_2O = \frac{12.5}{2} = 6.25$$

The number of moles of O_2 required to produce these moles of CO_2 and H_2O is therefore:

$$7.17 + \left(\frac{6.25}{2}\right) = 10.295$$

and of stoichiometric air:

$$5 \times 10.295$$

The air used thus contains:

$$1.3 \times 5 \times 10.295 = 66.92 \text{ moles}$$

including 53.54 of N_2 and 13.384 of O_2.

There are therefore (13.384 – 10.295) = 3.089 moles of uncombined, residual O_2.

Ultimately, the heat released heats 7.17 moles of CO_2, 6.25 of H_2O, 53.54 of N_2 and 3.089 of residual O_2. The sum of the c_p given above multiplied by this number of moles is:

$$499.07 + 74.64 \times 10^{-3}\, T$$

The thermal balance for 100 g of fuel is therefore:

$$10^6 = \int_{T_0}^{T_M} (499.07 + 74.64 \times 10^{-3} T)\mathrm{d}T$$

which, by setting $T_0 = 293$ K, gives the quadratic equation in T_M:

$$37.32 \times 10^{-3}\, T_M^2 + 499.07\, T_M - 1\,146\,260 = 0$$

whose positive root is $T_M = 1995$ K. Note that if 1 kg of fuel is used instead of 100 g, the two members of the thermal balance equation would have been multiplied by 10, which would not, of course, have changed T_M.

■ **7)** The compound NO is a pollutant produced by combustion reactions and, when the oxidant is air, partly by the attainment of a chemical equilibrium:

$$N_2 + O_2 \rightleftharpoons 2\,NO$$

For this equilibrium:

$$\ln K_p = -\left(\frac{10\,802}{T}\right) + 0.14 + 7.2 \times 10^{-5}\, T$$

Clearly, K_p increases as the temperature increases, as does the pollution caused by this phenomenon. This increase can be quantified by using the relationship giving ln K_p in order to deduce K_p and the number ($2x$) of moles of NO produced at various temperatures, with air as the oxidant, and initially for one mole of O_2. Once equilibrium is reached, there are $(4 - x)$ moles of N_2 for $(1 - x)$ moles of O_2, giving:

$$K_p = \frac{4x^2}{(1-x)(4-x)} \quad \text{approximating to } 4x^2 \text{ if } x \ll 1$$

which gives the table:

T (K)	300	1000	2000	5000
K_p	2.71×10^{-16}	2.52×10^{-5}	6.0×10^{-3}	0.19
$2x$	16.4×10^{-9}	5.0×10^{-3}	0.08	0.44

The number of moles of NO is obviously lower if equilibrium is not fully achieved.

■ **8)** The Ostwald diagram can be constructed for a liquid fuel whose molar percentages are:

85 for C   12 for H   2.5 for S   0.5 for N

and for which the burnt gases are, in the most general case, a mixture of CO_2, H_2O, SO_2 and N_2, (taking 20.8% as being the molar percentage of nitrogen in air).

In a similar way with the findings of section 1.5.2 in this chapter, the molar percentage of (CO_2 + SO_2) is plotted on the x-axis. Then, proceeding as indicated in section 1.5.2, a Grebel line can be drawn through the points ($x = 20.8$ and $y = 0$) and ($x = 0$ and $y = 15.64$). The straight line for $u = 0$ and v not equal to 0 passes through this same point and the point ($x = 7.2$ and $y = 0$).

CHAPTER 2

CHEMICAL KINETICS APPLIED TO COMBUSTION

2.1 GENERAL POINTS AND DEFINITIONS

Thermodynamics is based primarily on parameters, such as temperature or concentrations, and for this reason can only be used to analyse systems containing a sufficiently high number of individual elements to ensure that the statistical assumptions on which thermodynamics are based are valid. The chemical entities, H, H_2, H_2O, CH_4, etc., thus correspond unambiguously to moles and not to atoms or molecules (which is required for atomic notation). For certain applications, however, for example in chemical kinetics, the individual species themselves must be considered: atoms, ions, radicals or molecules, with the analysis then being achieved using the standard atomic notation.

Consider a chemical reaction, for example:

$$CH_4(\text{gas}) + 2\,Cl_2(\text{gas}) \rightarrow C\,(\text{solid}) + 4\,HCl(\text{gas}) \tag{2.1}$$

which can take place under certain conditions. By definition, this equation states that 16 g of CH_4 reacting with 142 g of Cl_2 would ultimately produce 12 g of C and 146 g of HCl. The left-hand side of the equation is a snapshot of the system prior to the reaction, and the right-hand side is a snapshot of the final system. These two snapshots provide no information about what actually happens in between, i.e. regarding changes occurring within the system, or about what happens during the transition from the initial system ($CH_4 + 2\,Cl_2$) to the final system (C + 4 HCl). Equation (2.1) is a **balance reaction**.

At a microphysical level, the equation assumes that the methane's 4 C—H bonds and the chlorine's 2 Cl—Cl bonds have been broken and that 4 H—Cl bonds have been formed in the 4 HCl molecules produced. There has also been formation of bonds linking each C atom with its closest neighbours to produce the solid C lattice. In atomic notation, the symbol C stands for one mole of carbon, (i.e. 12 g), thus in the case of a solid, a single, giant molecule containing a large number of interlinked C atoms. Each C atom will therefore have *n* close neighbours, linked to them by *n* bonds, with each C atom contributing half of each bond). It is easy to imagine that the breaking of a bond occurs in one step, synchronised with the production of another. However, as soon as the mechanisms involved in the combustion process become numerous, as is the case here, it is fair to consider that they could not occur during a single step. Several steps are required, each called an **elementary reaction**, and together forming **the kinetic mechanism**.

It should be noted that whilst eliciting the concepts of elementary reactions and kinetic mechanisms is fairly straightforward, their inherent characteristics are less well understood. The notion of kinetic mechanisms has been formulated based on hypotheses which have yet to be validated with experimental data, particularly the values of the reaction rates. Even though this comparison may be satisfactory, the mechanism still has to be considered as being theoretical, and there is no reason to suppose that another theory, yet to be discovered, would not also be compatible with the same experimental data. However, it is true that as the amount of validating data increases, so the number of reasonable theories reduces, and a single solution tends to emerge to explain the observed phenomena. The problems associated with kinetic mechanisms lie at the heart of chemical kinetics.

In order to study the purely thermodynamic aspects of combustion, and to calculate the adiabatic temperatures reached, there is no real need to understand the actual chemical mechanisms involved in the combustion process. However, these mechanisms must be analysed in order to determine whether or not combustion can actually occur under the given conditions, and whether it can actually reach complete chemical equilibrium in the time available. These two important questions essentially require knowledge of the elementary reactions involved in the combustion process. These reactions will be considered in this chapter. Part one will discuss the definition and study of elementary reactions, the "speed" at which they occur, and the laws which enable them to be calculated. Part two introduces the concept of **chain reactions**, which is the basic, fundamental mechanism in all combustion processes. The detailed mechanisms involved in the combustion of hydrogen with oxygen or with air, or also of certain non-complex hydrocarbons, are now reasonably well understood, with the calculations agreeing reasonably well with experimental results. Only the fundamentals of these mechanisms will be described here; a more detailed discussion and the use of the numerical calculation tools are beyond the scope of this book.

By definition, **molecularity** is the number of chemical species on the reactants side of an elementary reaction. It is therefore a theoretical concept insofar as the concept of an elementary reaction is itself a theoretical concept. The molecularity has a value of one for what are known as unimolecular reactions of the type:

$$A \rightarrow ...$$

and has a value of 2 for bimolecular reactions of the type:

$$A + B \rightarrow ... \quad \text{or} \quad A + A \rightarrow ...$$

and 3 for termolecular reactions of the type:

$$A + B + C \rightarrow ... \quad \text{or} \quad 2\,A + B \rightarrow ... \quad \text{or even} \quad 3\,A \rightarrow ...$$

In the first case, for unimolecular reactions, A is assumed to be unstable under the considered conditions; which may be the case for a transient species which is modified (for example by isomerisation) or broken down once produced. The second case assumes that two chemical species are brought together, and the third that three species are involved, which is much rarer (since it assumes that during the generally brief interval starting when the two species are brought together and ending at their separation, that a third arrives). The probability of a quadruple encounter, with a molecularity equal to four, is so unlikely that it has never been considered. Thus the elementary reactions that we shall consider will concern mainly bimolecular reactions.

2.2 RATES OF ELEMENTARY REACTIONS: DEFINITION AND MEASUREMENT

2.2.1 Definition

2.2.1.1 In a closed system

Let the general **elementary reaction** be written as:

$$a\,A + b\,B + ... \rightarrow a'A' + ... \tag{2.2}$$

and occur in a **closed system** (i.e. with no transfer of matter with the surroundings). Let the advancement of the reaction be termed z (see chapter 1) and thus the differential dz has the value given by the following ratios, which are numerically all the same as a result of the way in which the reaction is written:

$$-\frac{dn_A}{a}, \quad -\frac{dn_B}{b}, \quad +\frac{dn_{A'}}{a'}, \text{ etc.}$$

If the reaction is elementary, and only in this case, then during the same elementary duration dt, n_A will vary by dn_A, n_B by dn_B, $n_{A'}$ by $dn_{A'}$, etc. Thus:

$$-\frac{dn_A}{a dt} = -\frac{dn_B}{b dt} = +\frac{dn_{A'}}{a' dt} = \frac{dz}{dt}$$

and **the rate, or extent of the reaction** can be defined as:

$$w = \frac{dz}{dt} \tag{2.3}$$

which is not dependent on the species under consideration (A, B, A′, etc.). So long as the volume V does not vary during the reaction, we can divide by this volume, V, to give the differentials of the concentrations, d[A], d[B], d[A′]... instead of the differentials of the numbers of moles dn_A, dn_B, $dn_{A'}$ ... In this case, the rate is in moles per volume per unit time. The specific mass may also be considered instead of concentration, and thus the masses of A, B, A′, etc. instead of their numbers of moles; in this case the rate is in mass per volume per unit time.

Insofar as the magnitude defined in this way is not truly a rate, in m/s, it would be more correct to refer to it as the "extent" of the reaction, which is thus molar or specific (i.e. per unit mass). However, the term rate is commonly used.

2.2.1.2 Open systems: relationship between flow of matter and rate, or extent of reaction

By idealising the situation, simply in order to show that there must be a relationship of uncertain complexity between the flow of matter and the rate of reaction, and between concentration profile and rate of reaction profile along the tube, let us consider a reactive flow, such as, for example, a one-dimensional flame inside a cylindrical tube. The assumptions required for a one-dimensional model are that across a straight section (Fig. 2.1a and 2.1b) the flow speed is the same at all points, that there is no lateral diffusion and that the heat transfer is independent of distance from the wall, which in practice is unlikely to be the case. In order to ensure that the flame is stable in the tube, it must have a **spatial velocity** towards the unburnt gas which is equal and in the opposite direction to the velocity of the unburnt gases in the tube. Under these conditions, for a species i in the gas above the plane P where combustion begins, and thus in partly-burnt gas, the flow j_i ($mol \cdot cm^{-2} \cdot s^{-1}$) is related to the concentration c_i ($mol \cdot cm^{-3}$) (where c_i, like [I], is the concentration of component I) and to the velocity V ($cm \cdot s^{-1}$) of the flow at a height x, referred to a coordinate system fixed on the tube:

$$j_i = c_i V$$

Moreover, due to the chemical reactions occurring in the medium, c_i at a height x becomes $c_i + dc_i$ at a height $x + dx$; n_i become $n_i + dn_i$ and j_i become $j_i + dj_i$ (Fig. 2.1a).

For a given x, in steady state $\partial n_i/\partial t = 0$ irrespective of x.

On the contrary, since dt is the time required for dn_i moles of component i to rise from a height x to a height $x + dx$ (Fig. 2.1b), dn_i/dt (which is related to

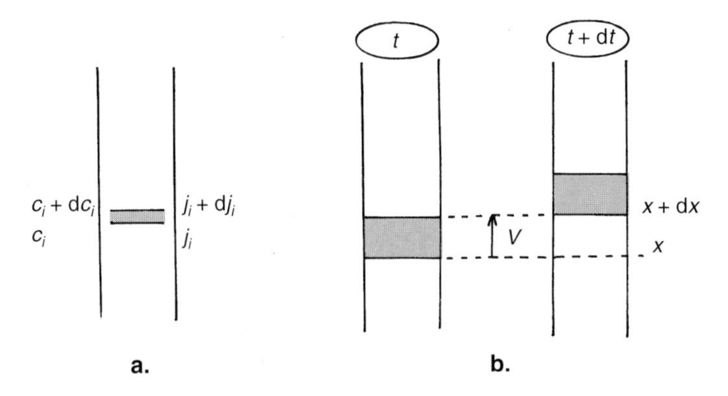

Figure 2.1 Idealised representation of a burner with a planar flame front, stabilised inside a cylindrical tube (one-dimensional, steady-state, reactive flow).

a. The flow j_i corresponds to the concentration values c_i, and the concentrations $c_i + dc_i$ to the flow $j_i + dj_i$. At a given height x, c_i is not dependent on t ($\partial c_i/\partial t = 0$ for any value of x).

b. The differential dt is the time taken by a particle to rise from a height x to a height $x + dx$; dx/dt is thus its velocity between the heights x and $x + dx$.

the rate of reaction), is non-zero and the following equation can be written, where A is the cross-sectional area:

$$j_i = \frac{1}{A}\frac{dn_i}{dt} \quad \text{and} \quad dj_i = \frac{1}{A}\frac{d(dn_i)}{dt}$$

And:

$$dn_i = c_i A\, dx \quad \text{and} \quad d(dn_i) = A\, dc_i\, dx$$

by balancing the two values found for $d(dn_i)$:

$$\left(\frac{dj_i}{dx}\right)_{\text{molar}} = \left(\frac{dc_i}{dt}\right)_{\text{molar}} \quad \text{(for example mol·cm}^{-3}\text{·s}^{-1}\text{)}$$

If instead of molar flow, the mass flow is considered, then in a similar way:

$$\left(\frac{dj_i}{dx}\right)_{\text{unit mass}} = \left(\frac{dc_i}{dt}\right)_{\text{unit mass}} \quad \text{(for example g·cm}^{-3}\text{·s}^{-1}\text{)}$$

Thus dj_i/dx is a measurement of the rate or extent of the reaction of the species i, w_i, for an open system. Since dj_i/dx is negative for a reactant and positive for a product, then:

$$w_i = \pm \frac{dj_i}{dx} \tag{2.4}$$

with the + used if i is a product (i.e. a final species) and the – used if i is a reactant (i.e. an initial species). In this way, equation (2.4) always gives a positive value for w_i. If the entire kinetic mechanism can be described by a single reaction written in the same form as reaction (2.2), a single rate of reaction can be defined, w, which is analogous to that defined by equation (2.3), i.e.:

$$-\frac{w_A}{a} = -\frac{w_B}{b} = +\frac{w_{A'}}{a'} = \ldots = w \tag{2.3'}$$

Although, as we shall see later, the assumptions made above in reaching this relationship are broad approximations, it is clear even at this stage that dj_i/dx and dc_i/dt are linked. It is also apparent that the rate of reaction of the species i, dc_i/dt, is not directly linked to the velocity V of the flow, but rather to the derivative dj_i/dx, and thus to $d(c_iV)/dx$, where c_i tends to be more dependent on x than V. In conclusion, the existence of reactions results in flow variations in the various sections of the flame.

2.2.2 Measuring

2.2.2.1 Measuring in a closed system

The methods implemented are the standard methods used in chemical kinetics. Although straightforward in principle, the analysis of the measured values is often complex. For gaseous systems, variations in the total pressures may be monitored, although the individual sources of the pressure variations need to be identified. When monitoring the radiation emitted by the reacting medium, it should in theory be possible to identify the emitters. However, it is not an easy task when there are multiple emitters, having more than four or five atoms, etc. Unlike the process which occurs in traditional, isothermal kinetics, the temperature in pre-explosive reactions also varies, and to an even greater extent in explosive reactions. Whilst the temperature measurements may be of practical interest, it is difficult to draw conclusions from them regarding the kinetic mechanism involved.

2.2.2.2 Measuring in an open system

When analysing combustion processes, the systems studied tend to be open systems: classical chemical kinetic systems or specific systems such as burners.

- The easiest apparatus to set up and analyse is undoubtedly the laminar flame burner, with the oxidant and fuel premixed prior to arriving at the burner orifice. If the flame is stable then the system is in steady-state, and unless it is affected by an external influence will continue in this way indefinitely and can be studied at leisure. The flame is nonetheless dependent on both the chemical kinetics and on the transfer of matter and heat.

When operating at low pressures, between 10 and 100 torr, the reaction zone is enlarged, thus extending the spatial distribution of the temperature and concentrations, and simplifying their study. Furthermore, the flow can be channelled to create uniform velocity at all points over the same straight section, creating "flat" flames, which are virtually one-dimensional, and for which the reaction zone thickness may exceed 10 mm. An example of a flat-flame burner is shown in Figure 2.2. If necessary, corrections can be included in the analysis if the effects of thermal diffusion are taken into consideration.

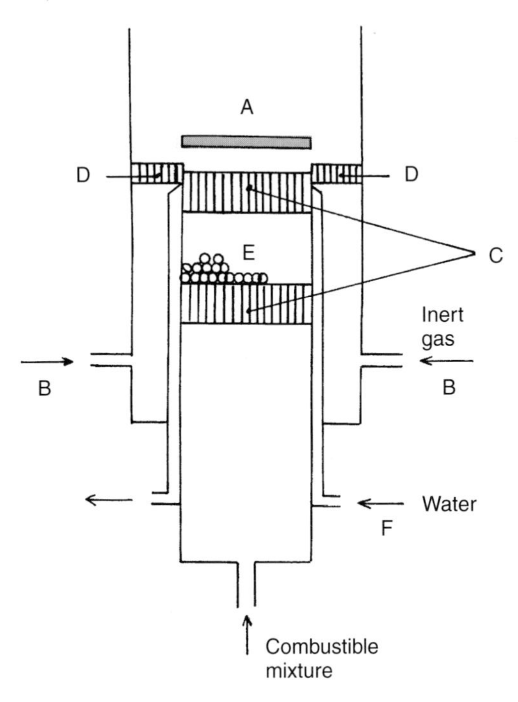

Figure 2.2 Flat-flame burner. The flat flame is located at A; and is often anchored to a metal grill. The combustible mixture is supplied via the inlet pipe at the bottom of the diagram. Above the burner, the burnt gases are extracted into the chimney by a pump which regulates the flow. An inert gas is pumped in via B. The combustible mixture is channelled through thin, parallel guides (C), and the inert gas via guides (D). A stack of glass beads are located at E. Water is circulated (F) to regulate the temperature.

This type of burner provides a good simulation of the ideal case discussed in section 2.2.1.2, and thus of the conditions required for the application of equation (2.4). The only difference is that diffusional flow must be added in order to correct for the convectional flow $c_i V$. The rates w_i can thus be measured. Figure 2.3 shows the flow profile for methane, and for several radicals, for a mixture of 9.5% by volume CH_4 with 90.5% O_2, at 40 torr with a flow speed of 67 cm·s^{-1}. It is apparent that the maximum value of the methane disappearance rate, $dj_{methane}/dx$, is reached 0.34 cm above the mouth of the burner, and has a value at this position of 8.4×10^{-5} mol·cm^{-3}·s^{-1}. It is also possible to measure the flows of the radicals O$^•$, OH$^•$, etc. and, consequently, the rates of reaction:

$$CH_4 + OH^• \rightarrow CH_3^• + H_2O$$

$$CH_3^• + O^• \rightarrow CH_2O + H^•$$, which is the principal of $CH_3^•$

$$CH_2O + OH^• \rightarrow CHO^• + H_2O$$, etc.

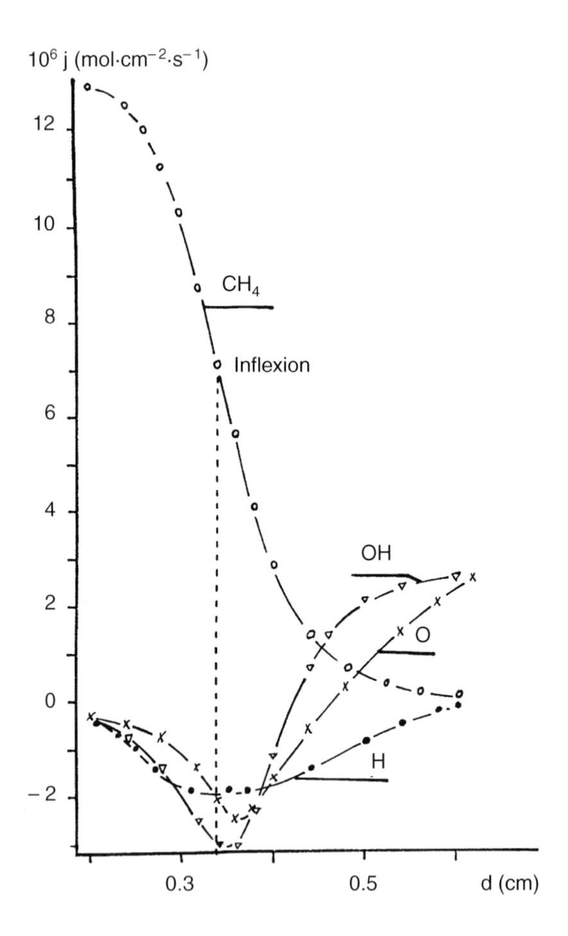

Figure 2.3 Combustion of a CH_4/O_2 mixture whose percentages by volume are 9.5% CH_4 and 90.5% O_2, at 40 torr and where the unburnt gas speed is 67 cm·s^{-1}. Variations in the molar flow of the various species (CH_4, O, H and OH) as a function of distance above the burner. On the curve relating to CH_4 the point of inflexion corresponds to the maximum rate of disappearance of methane, $w_{methane} = (dj_{methane}/dx)$ i.e. 8.4×10^{-5} mol·cm^{-3}·s^{-1} at the point under consideration (After Peeters and Manhem (1972), *14th Symposium (International) on Combustion,* p. 133, The Combustion Institute).

This apparatus is thus of interest in studying the kinetic mechanism. However, due consideration must be given to the fact that its operation differs greatly from that of industrial burners. In practice, for an industrial burner, large quantities of combustible mixtures must be burnt, requiring high flow rates and speeds, which leads to turbulence phenomena. A later chapter in this book considers turbulent flames.

• In a piston or plug-flow reactor the apparatus is designed such that the mixture advances in parallel slices between which there is no transfer of material, and thus negligible diffusion. For this reason, instead of the various stages in a complex combustion reaction occurring over a relatively long time period, there is a spatial distribution of the parameters of the system following on from the rapid mixing of the reactants at a height $x = 0$ (since the subsequent mixing of reactant and products is itself slow). In order for this to be achieved, the reactant mixture must have the same axial velocity at every radial and axial coordinate value, which must be equal to the mean axial velocity of the carrier gas. If the velocity V of the flow is constant, the residence time is the ratio x/V, where x is the distance between the mixing plane and the analysis plane. More generally, temperature and flow velocity only depend on the height x. For turbulent flow, the turbulence eliminates the fluid mechanics problems encountered with laminar flows, such as the dependence of the flow velocities on radial distance. Dilution with a carrier gas limits the interactions between turbulence and the chemical kinetics. Since the carrier gas is the principal component, it is this gas which sets, at least in principle, the physical properties of the gaseous flow: pressure, temperature, velocity, heat capacity, thermal conductivity, etc.

• Figure 2.4 shows a mixing system which in practice simulates fairly well the concept of a plug-flow reactor, at least at low pressures. It has been used to investigate chemiluminescent phenomena, resulting from the reactions of atomic C with OCS or other molecules, with He as the carrier gas. The first reactant, atomic C, enters via the central tube A whereas the second arrives via the mixing head B, made of Kelf, and fitted with side burner jets I and axial jets J. The mixer head can slide up and down a Teflon tube to vary the distance between the plane in which the reactants are mixed and the window through which the luminous emission is studied. A pump circulates the gases. The use of Kelf and Teflon restricts the phenomenon of radical recombination on the walls. If necessary, when the conditions produced do not exactly create those of the plug-flow reactor, the various errors due to diffusion, wall losses, etc. should be calculated.

• On the contrary, in a well-stirred *(or continuously-stirred)* tank reactor, (WSTR), the rate at which the reactants and products are mixed is very rapid compared with the rate of reactions. The time taken by one molecule to travel a certain distance within the volume of the reactor is negligible compared with its residence time. For this reason the temperature and concentrations in the reactor have the same value at all points and the emerging mixture has the

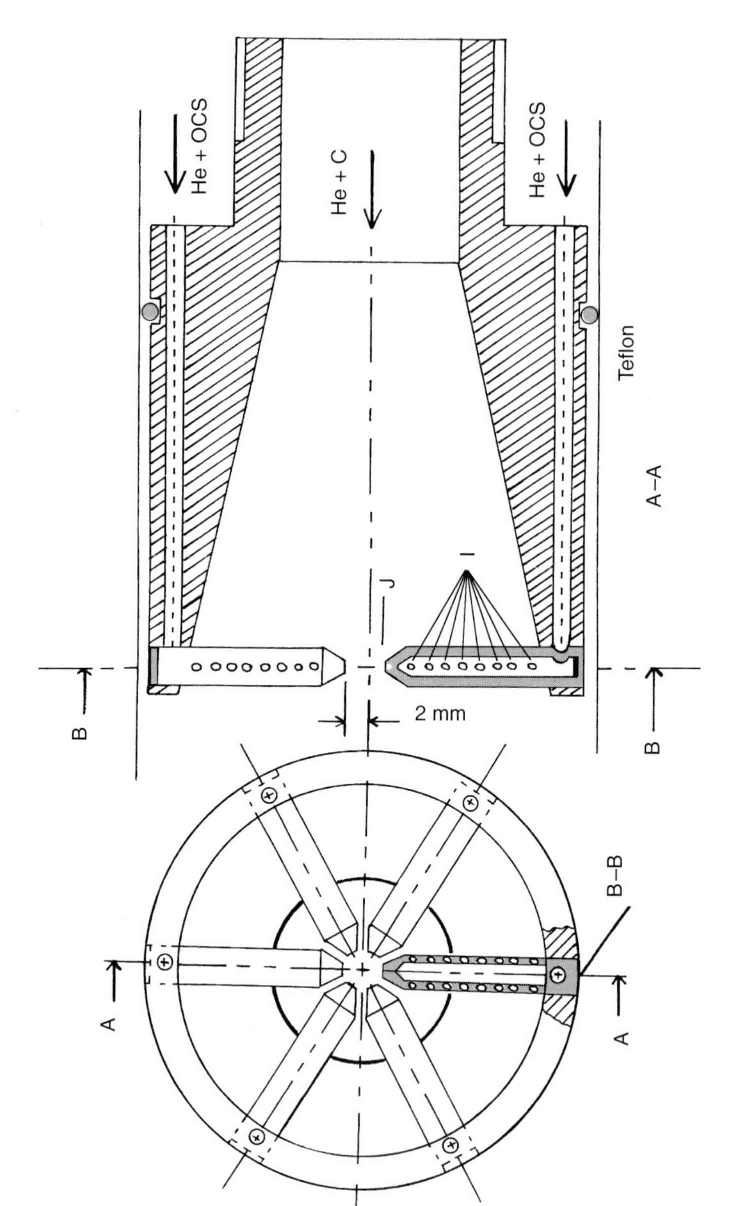

Figure 2.4 Mixing device, for mounting on a reactor designed to operate under conditions similar to those of a plug-flow reactor, and for reactants which only react when mixed. Section B shows the plane of the mixing head, on the left in the figure; the reaction thus occurs after this plane (after Dorthe *et al.* (1991) *J. Phys. Chem.*, 95, 5109).

same temperature and the same instantaneous composition as the reaction mixture in the reactor. By varying the inlet mass flow rate, the average residence time can be varied. In the model used by the MIT (*Massachussetts Institute of Technology*) (Fig. 2.5a) the reaction chamber is toric with the reactive mixture introduced via a series of uniformly-arranged side injectors, which create a rapid circular movement enabling the reactants to mix quickly with the recirculated burnt gases. The conditions inside the reactor are thus virtually the same at all points. Radial channels direct the gas towards a central exhaust. A measurement sensor is fitted in the wall of the torus. Thermal insulation ensures that adiabatic conditions are maintained. Variations have been designed, such as that shown in Figure 2.5b. It has thus been possible to study the NO, soots, etc. generated by combustion.

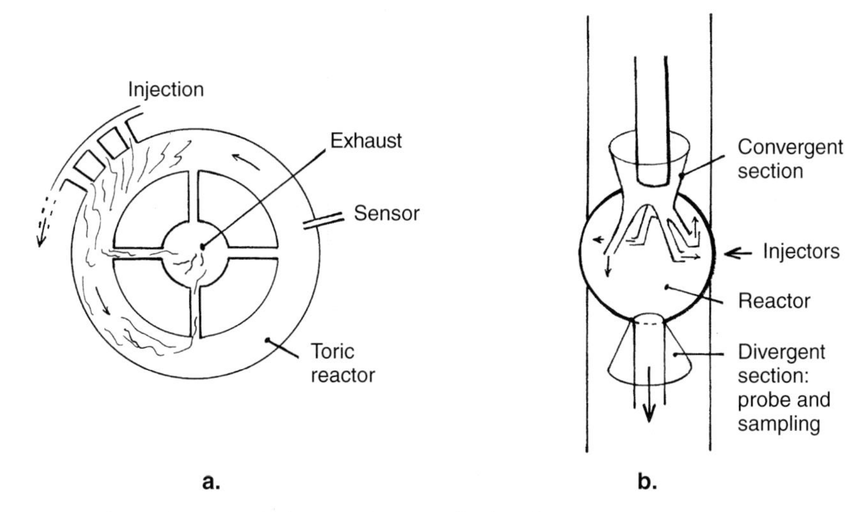

Figure 2.5 Practical examples of well-stirred reactors.
a. Example of a reactor used by Longwell, in which the mixture is injected at the periphery and exhausted via the centre. The inverse can also be imagined (mentioned by Miller and Fisk (1987) *Chem. Eng. News,* 22).
b. Example of the reactor used by Cathonnet *et al.* at the CNRS (National Centre for Scientific Research) specialising in combustion at Orléans, France. (Analysing the operation of flow reactors and operating conditions in plug-flow reactors, see Lede and Villermaux (1979) *J.Chim. Phys.,* 74, 459).

• In a shock tube, a sheet of metal foil initially separates a volume of a gas at a known pressure (called the "driver" gas) from a "driven section" containing the reactant mixture under study at low pressure. The sudden tearing of the metal foil triggers a shock wave into the gas being studied, which raises the temperature and pressure. By varying, for example, the initial value of the driver gas pressure, it is possible to control the pressure produced in the gas being studied. The phenomena which arise following the passage of the incident wave or after the wave reflects off the end wall of the tube can be investigated. In the first case, the conditions are very similar to those created in the plug-flow reactor, whereas in the second the gas is virtually immobile, although its temperature is very high. It is therefore an ideal method to use when studying the dissociation of stable molecules.

The shock wave can only be used to raise the temperature of the mixture being studied, the reactive species being produced by flash photolysis. The method is well adapted to the study of radical-molecule and radical-radical elementary reactions. A. Fontijn used this method to study the reactions of atomic oxygen, produced by the photolysis of CO_2, with acetylene and ethylene.

- A combustion chamber in an I.C. engine or in a turbine is also an open system. A combustible mixture enters and burnt gases are exhausted. In this case, it is always possible to define the global rate at which the combustible mixture is converted into burnt gases per second or based on the converted mass or volume of the combustible mixture. However, it is difficult to relate this conversion rate to the chemical kinetics of the reactions which occur in the combustion chamber, since the mixture does not have the same characteristics at all points.

2.3 EXPERIMENTAL RATE LAWS FOR ELEMENTARY REACTIONS

Experimentally it has been observed that the rate of elementary reactions varies as a function of the product of the concentrations of the species on the reactants side of the equation, each raised to a power equal to their stoichiometric coefficient in the reaction. For example, in the reaction:

$$A + B \rightarrow ...$$

the rate, at a given temperature, is $k[A][B]$, where k is a constant, known as the **rate constant** or, more rarely, as the "specific rate".

2.4 RATES OF REACTION FOR BALANCE REACTIONS: THE STEADY STATE APPROXIMATION

In our first approach, we took the precaution of defining the rate of reaction (in the singular) for an elementary reaction only. This approach is not valid for a reaction, such as a balance reaction, which in general is not elementary. Consider now the balance reaction:

$$A \rightarrow B$$

which occurs in two stages, both elementary:

$$A \rightarrow X$$

followed by:

$$X \rightarrow B$$

leading to:

$$\frac{dn_X}{dt} = -\frac{dn_A}{dt} - \frac{dn_B}{dt} \tag{2.5}$$

dn_A is negative, dn_B is positive and dn_X/dt is thus positive, negative or possibly zero. Initially, dn_X/dt can only be positive and X is formed. Consequently, n_X must rise to a maximum and then decrease. $-dn_A/dt$ is only equal to $+dn_B/dt$ at the instant when n_X passes through its maximum. At any other time, **the rate defined at A is not the same as that defined at B,** thus it should be clearly stated whether the rate is defined at A or at B. There is no question of it being **the** rate of reaction of the overall process A → B.

Having made this point, at the instant when n_X reaches its maximum, the rate of the first step, w_1, is equal to that of the second, w_2. The two steps are said to be **synchronised**. If the maximum is relatively flat, then the two steps are virtually synchronised both slightly before and after the maximum, but not towards the end of the reaction nor, in particular, at the start where w_2 is zero, although w_1 is not. There is, however, a range, whose extent may vary, within which dn_X/dt is approximately zero. The generalisation of this approximation to all the transient species such as X is **the pseudo-steady-state approximation** (abbreviated to SSA). The approximation is more accurate if the concentration of X remains low from beginning to end; which is the case for a highly reactive species. Thus the accuracy of the approximation varies with the species under consideration. Once all the factors have been taken into consideration, it may be observed (when it is possible to measure the concentrations of the transient species) that its range of validity is often fairly large, even for mixtures reacting rapidly (which is the case in combustion reactions).

The advantages offered by this approximation are not simply its accuracy, but also that it considerably simplifies the calculations. Returning to the case of the mechanism with two consecutive reactions, where k_1 is the rate constant for the first stage, k_2 the constant for the second and $[A]_0$ the initial concentration at A:

$$A \xrightarrow{1} X \xrightarrow{2} B \tag{2.6}$$

If the three equations are written for the three unknowns [A], [X] and [B]:

$[A] + [X] + [B] = [A]_0$ which takes into consideration the mass balance,

$-d[A]/dt = k_1 [A]$ the rate of the first step,

$+d[B]/dt = k_2 [X]$ the rate of the second.

The system can be solved fairly simply, giving:

$$[X] = k_1[A]_0 \frac{\exp(-k_1 t) - \exp(-k_2 t)}{k_2 - k_1} \tag{2.7}$$

and similarly for [A] and [B]. However, it is easy to see how much more difficult this calculation would be for 3 steps, or worse still for a great many more: a situation often encountered. In this case the steady-state approximation simplifies the calculations significantly. Nevertheless, by returning to the system:

$$A \rightarrow X \rightarrow B$$

it is clear that by assuming steady state for X, (i.e. $d[X]/dt = 0$), we are adding a fourth equation to a three-equation system with three unknowns [A], [X] and [B]. There must therefore be an inherent incompatibility. Moreover, simply deriving the equation given above for [X] clearly highlights the problem since the solution $d[X]/dt = 0$ is not obtained. In more general terms, when there are n stages with m transient species, the steady state approximation would produce $(n + m)$ equations for "only" n unknowns. The SSA introduces errors which require evaluation, which is not easy. Here, writing equation (2.5) as:

$$\frac{d[X]}{dt} = k_1[A] - k_2[X]$$

shows that $d[X]/dt$ may be considered as being close to zero if $k_1[A]$ and $k_2[X]$ are large in relation to their difference, in which case $[X] = k_1[A]/k_2$. It is straightforward enough to check whether this is the value given by equation (2.7) if $k_2 \gg k_1$ and $t \gg 1/k_2$.

Having said this, the analytical solution can be passed over in favour of a numerical one. Algorithms have been created and used as the basis for programmes which can be easily obtained. One of the first to be developed, the Runge-Kutta algorithm, turned out not to give valid results when used in the common case of mechanisms containing steps with very different rates. However, programmes are now available which do not suffer from this failing. The power and flexibility of these programmes will ultimately make the use of SSA redundant, except in cases where not only the numerical rate values, but also their analytical expression is of interest.

When, in a kinetic mechanism, the numerical values of several rate constants are known, and where others, preferably fewer, are not known, it is possible to carry out **simulations**. Reasonable numerical values are attributed to the unknown rate constants. The programme displays on the screen both the concentration variation curves, measured with the apparatus used, and those deduced from the values assigned to the unknown rate constants. With a little skill and experience, the user can soon determine the "correct" values, i.e. those which best fit one curve on to the other. This method offers an extremely useful tool for analysing elementary reactions which cannot in practice be studied in isolation, separated from the complex mechanism in which they occur.

2.5 EXPERIMENTAL RATE LAWS FOR BALANCE REACTIONS

Consider those reactions which may or may not be elementary. This uncertainty is always present when the reaction is studied for the first time. There is no reason why one should not study the disappearance of a species from the reactant side or the appearance of species on the product side. Strictly speaking, as discussed above, these rates are only rates of reaction for elementary reactions, and it is an incorrect use of terminology to continue to call them rates of reaction. It is often observed, at least to a first approximation, that they may be written in terms of the various parameters, total pressure p, temperature T, concentrations c_i etc. as a product of the functions of p, T, c_i, etc. In general, the parametrisation of the recorded measurements leads to very varied relationships from which it is difficult to deduce general laws. Moreover, these relationships are only valid within the range of recorded measurements. We will consider separately the dependence of the rates on concentration, at a given temperature, and on temperature, at given concentrations.

2.5.1 Concentration dependence

Experimental results have produced widely-varying laws from which an initial analysis only allows general trends to be deduced.

Example 1:

Consider the gas-phase reaction (which may lead to combustion):

$$2\,NO + 2\,H_2 \rightarrow N_2 + 2\,H_2O \tag{2.8}$$

Hinshelwood and Green found that between 700 and 1000°C, for an equimolar mixture of NO and H_2, the total pressure varied in accordance with the relationship:

$$-\frac{dp}{dt} = k_{\text{apparent}}\,[NO]^2\,[H_2] \tag{2.9}$$

where dp is negative, and k_{apparent} is a function of temperature. This is not truly a rate constant, since p is not a concentration. The reaction is said to have a **partial order** equal to 2 with respect to NO, and 1 with respect to H_2; and an **overall order** equal to the sum (2 + 1). This terminology has been agreed on through common usage and provides no further information other than the relationship found, in fact it says rather less. The only certainty provided by this relationship is that the reaction is not elementary. At first sight, the reaction as it is written, would appear to be tetramolecular, which is impossible. Dividing all the numbers of moles by 2 to produce a seemingly bimolecular reaction would give:

$$-\frac{d[NO]}{dt} = k\,[NO][H_2]$$

which is incompatible with the experimental relationship.

Example 2:

The synthesis reaction of HI in the gas phase:

$$H_2 + I_2 \rightarrow 2\,HI \tag{2.10}$$

approximates to:

$$\frac{d[HI]}{2dt} = k_s\,[H_2][I_2] \tag{2.11}$$

and, in the opposite direction, for its decomposition:

$$-\frac{d[HI]}{2dt} = k_d\,[HI]^2 \tag{2.12}$$

with, in both directions, the overall order equal to 2, as if the two reactions were elementary. However, this does not prove that this is the case. In the direction of the synthesis, HI produced at a rate w_s, is subsequently decomposed at a rate w_d, with the overall rate being:

$$w = w_s - w_d \tag{2.13}$$

and similarly, in the direction of the decomposition, H_2 and I_2 produced at a rate w_d and recombining at a rate w_s, the global rate would be:

$$w' = w_d - w_s \tag{2.14}$$

Example 3:

It would be convenient to believe that in terms of the synthesis of HBr the rate law is similar to that determined above for HI. However, this is not the case. Consider:

$$\frac{d[HBr]}{2dt} = \frac{k[H_2][Br_2]^{1/2}}{1 + k'\dfrac{[HBr]}{[Br_2]}} \tag{2.15}$$

which can no longer be presented as a simple concentration-dependent monomial. It is not possible to assign it an order, rather we say "the reaction has no order". This formulation with "defamatory" connotations, implies that it is a slightly unusual reaction, and thus "deviant". In fact, the greater the accuracy of the measurements, the more it becomes apparent that, other than for the elementary reactions, the rate laws are rarely monomial. Many laws considered as being monomials are in fact only so to a first approximation. In all cases, attempts should be made to find the most likely mechanism behind the determined laws.

The rate law found for the synthesis of HBr also shows that the rate may depend on the concentration of the molecules produced by the reaction, in this case HBr. The concentrations may also be given raised to negative powers or be non-integers.

On the contrary, as shown in section 2.3, elementary reactions are always governed by simple laws: monomial functions of the concentrations of the species of the reactants raised to powers which are the stoichiometric coefficients of the reaction; which leads to the assertion that "for elementary reactions, the order is equal to the molecularity".

2.5.2 Temperature dependence

Like concentration dependence, the laws deduced for temperature vary greatly. The most common involve, at least over a limited range, T^n, with n a positive or negative integer or non-integer; more often $\exp(-E/RT)$, where E designates a quantity known as the **activation energy;** or $T^n \exp(-E/RT)$. The $\exp(-E/RT)$ law is related to the rate constant k by:

$$k = A \exp\left(\frac{-E}{RT}\right) \tag{2.16}$$

(often written in the form $A \exp(-T/T_a)$, where T_a is the activation temperature). This is Arrhenius' law, where A is a constant, often called the **pre-exponential factor** or **frequency factor.** This law is shown schematically in Figures 2.6a and 2.6b: 2.6a for variations of k with T, 2.6b for variations of $\ln k$ with $1/T$. The point of inflexion which occurs at a temperature T_i in Figure 2.6a is not of major importance in practice. In fact, with the standard values of E, it occurs at high temperatures, much greater than those encountered in standard chemistry: at approximately 12 500 K for $E = 50\ \text{kcal·mol}^{-1}$, as may be readily verified. If, therefore, only the lowest temperatures are taken into consideration, Arrhenius' law predicts a very rapid increase in the rate constant as the temperature increases.

2.5.3 Total-pressure dependence

In the gas phase, if the molecules A and B react within a bath gas G, which is unchanged at the end of the reaction, it is not surprising that their rate of reaction is a function of their partial pressures in the mixture, since these vary in turn with their concentration.

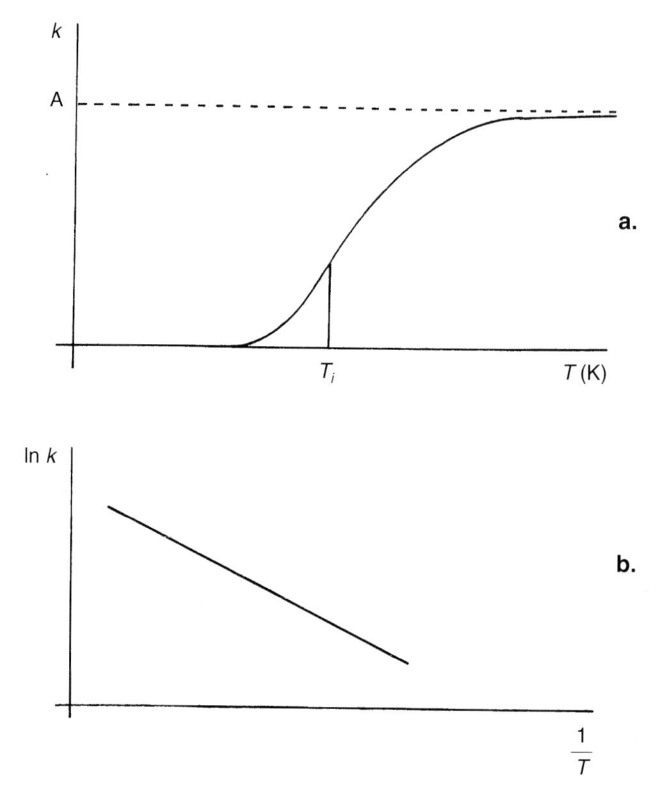

Figure 2.6 Graph of the variation with temperature T of a rate constant which follows Arrhenius' law.
a. $k = f(T)$. T_i is the temperature at the point of inflexion. k tends towards the frequency factor A when the temperatures becomes very high. In practice, reaction kinetics are studied well below this limit.
b. $\ln k = g(1/T)$.

However, they may, moreover, depend on the total pressure, i.e. that of the bath gas G. In this case, it must be assumed that in some way this gas is involved in the kinetic mechanism. If the pressure of the bath gas is sufficient, i.e. in practice between 1 and 100 bar or more, the rate constants for the recombination reactions (radical-radical or radical-molecule type) depend to only a small extent on temperature and their values vary only slightly over the range extending from 10^8 to 10^{11} $dm^3 \cdot mol^{-1} \cdot s^{-1}$, the upper limit being more accurate than the lower limit.

2.6 THEORETICAL LAWS FOR ELEMENTARY REACTION RATES

2.6.1 Molecular collision theory

Clearly the minimum condition required to enable an elementary reaction of the type:

$$A + B \rightarrow ...$$

to occur is that A and B come into contact. Its rate cannot therefore be greater than the **number of collisions** between A and B, Z_{AB}, which, for gases, is given by the kinetic theory of gases:

$$Z_{AB} = n_A n_B (r_A + r_B)^2 \left[\frac{8\pi k_B T (m_A + m_B)}{m_A m_B} \right]^{1/2} \tag{2.17}$$

where

- n_A is the number of molecules of A per unit volume, i.e. its molecular (and not molar) concentration,
- n_B the number of molecules of B,
- r_A the radius of A, r_B that of B (where A and B are assumed to be spherical and rigid),
- m_A the molecular (and not molar) mass of A, m_B that of B, and
- k_B Boltzmann's constant.

This equation predicts (correctly) that the rate must vary as a function of the product of the concentrations of A and B. On the contrary, it does not predict for the temperature a relationship of a type similar to Arrhenius' law, even though the bimolecular reactions often follow this law. Moreover, the actual rate is much less; 10^{10} times (or more) less than the collision frequency Z_{AB}. The collisions can be thought of as being effective only if their energy exceeds a threshold value, E, which conveniently introduces the exp($-E/RT$) term found in Arrhenius' Law. One may criticize the concept of collisions and consider reacting molecules as rigid spheres; which should be thought of more as an improvement to the theory of molecular collisions than a truly different approach. This perspective will be developed below.

2.6.2 Activated complex theory

It is widely accepted that a collision occurs over an extremely brief period of time. It is more appropriate to consider it as an event during which the molecules approach each other, interact and are mutually modified. A transient state is passed through which differs from the initial state and which influences the final state.

Consider the exchange reaction:

$$X + YZ \rightarrow XY + Z \tag{2.18}$$

If, from the start to the end of the process, the three atoms X, Y and Z are considered as being aligned, which from a quantum point of view is in fact the most likely situation, three parameters are required to describe the system and how it will change during the reaction. The simplest solution is to select the distance between Y and Z, d_{YZ}, and assign it to the x-axis (see Figure 2.7) and the distance between X and Y, d_{XY}, to the y-axis. The potential energy of the system may then be represented vertically, (i.e. out of the page), or more simply, as in Figure 2.7, by "contour" lines formed by lines of equal potential energy (in the Born-Oppenheimer approximation, this potential energy is the total electronic energy for the various system configurations). Stability is represented by a "well" in the potential energy contours for a molecule, and by extending the analogy, by a "valley" for the initial system and a second valley for the final system.

The initial system corresponds to a point such as a in the first valley, where d_{YZ} is small and d_{XY} large. The final system is shown as a point such as b, where d_{YZ} is large and d_{XY} small. The transition from the initial to the final system, according to this logic, requires "travelling" from the first valley to the second, with the most "energy-efficient" pathway passing via the pass or "saddle" at point c. This is not, however, the only possible route. Crossing the pass, or "potential barrier", can only occur if the system receives an amount of energy; heat energy insofar as this book is particularly concerned, which must be equal to or greater than the difference (E_c – E_a), i.e. the energy at point c minus the energy at point a (E_c – E_b in the opposite direction). The rate of reaction calculation shows that this energy is represented by the presence of an exp $(-(E_c - E_a)/RT)$ term. This links back to Arrhenius' law, whilst clarifying the concept of an effective collision. A transition state occurs at the pass, represented symbolically by X ... Y ... Z, called the **activated complex.** However, even if the two approaches, molecular collisions and transition state, share a common basis, the rate of reactions predicted by both are not exactly the same, particularly in terms of temperature-dependence. In both cases, it is not only present in the exponential term, even if this term in general predominates. Theoretically, based on a quantum calculation, so long as it is possible to predict the potential energy values for Figure 2.7 for the various system configurations, then the pass transition energy is automatically known, which may still be called activation energy. The pass may also be crossed, as shown in Figure 2.8, with non-uniform, cyclic variations in d_{XY} and d_{YZ}, i.e. with molecular vibrations, which are not envisaged in the theory of collisions.

The curve of least energy shown on Figure 2.7, can be transferred to Figure 2.9, which shows the variation in potential energy along this pathway, as a function of a curvilinear coordinate called the "reaction coordinate". Figure 2.9 shows, reading from left to right, that the energy level of the final state is lower than that of the initial state, and the reaction is thus exoenergetic. Clearly the inverse is true in the opposite case.

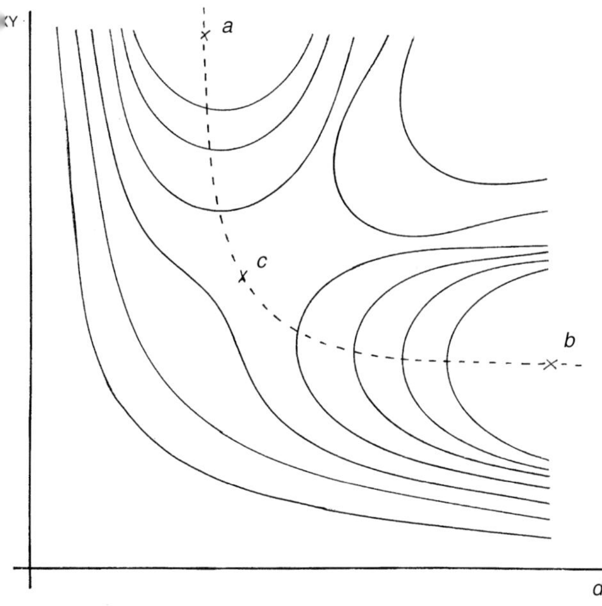

Figure 2.7 Lines of equal potential energy for the linear system XYZ as a function of the internuclear distances d_{XY} and d_{YZ}. *a* is a point in the valley X + YZ, *b* a point in valley XY + Z, *c* is the summit of the pass between the two valleys. The dashes show the path of least energy across the pass. Reaction X + YZ → XY + Z occurs when passing from point *a* to point *b*, and the reverse reaction in the opposite direction.

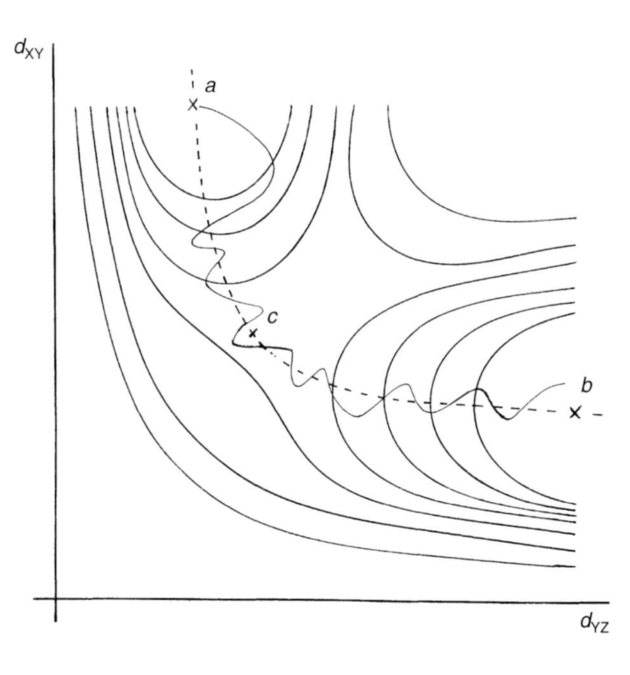

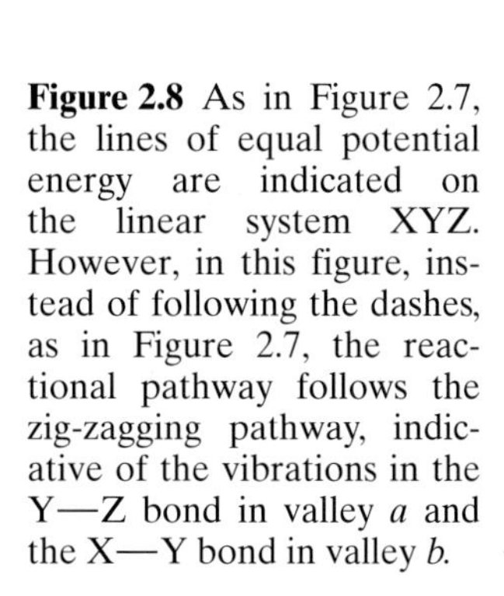
Figure 2.8 As in Figure 2.7, the lines of equal potential energy are indicated on the linear system XYZ. However, in this figure, instead of following the dashes, as in Figure 2.7, the reactional pathway follows the zig-zagging pathway, indicative of the vibrations in the Y—Z bond in valley *a* and the X—Y bond in valley *b*.

In a previous work, dating back to 1927, Tolman showed that the activation energy of a bimolecular elementary reaction was also the total average energy, whether translational or vibrational, of all the pairs of effectively reacting reactants reduced by the total average energy of all the pairs of reactants. This virtually unknown piece of work has been the subject of more research recently undertaken by Truhlar[1].

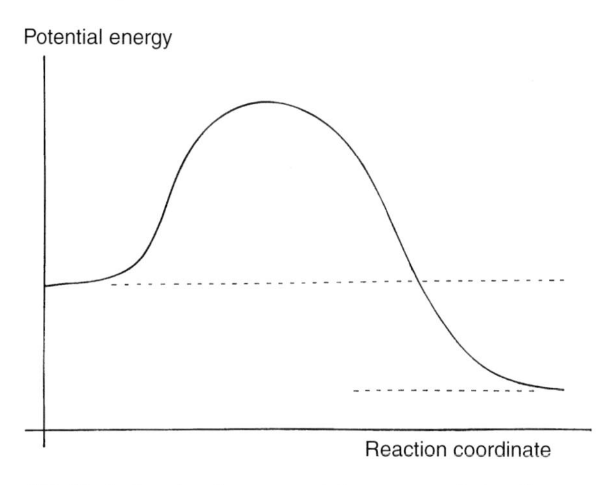

Figure 2.9 Variation in the potential energy of the linear system XYZ along the path shown by the dashes in Figures 2.7 and 2.8. The curvilinear coordinate which follows this path is the **reaction coordinate,** which should not be confused with the advancement of the reaction.

The case of the dissociation of diatomic molecules, shown in Figure 2.10, is clearly very different. Two parameters are sufficient to describe the system: potential energy and the distance between the molecule's two atoms. By comparing Figure 2.10 with Figure 2.9, it is apparent that for the molecule to dissociate it must leave the potential energy well (as opposed to the potential energy valleys mentioned above). The increase in interatomic distance from its value within the molecule to an infinitely large value assumes that the potential barrier has been crossed. On the contrary, at least from an initial analysis, the recombination of the separated atoms occurs with a zero activation energy: "the atoms fall into the wells". The dissociation of a molecule into two radicals can be treated in the same way, for example from C_2H_6 to $2CH_3$, as well as their association. This case is said to be "quasi-diatomic". Consequently, the rates of recombination reactions must be only slightly affected by temperature, unlike those of dissociation reactions.

Let us consider the dissociation reaction of the H_2 molecule into two H atoms, which corresponds with Figure 2.10 reading from left to right. The two

1. Truhlar D.G. (1978) *Journal of Chemical Education*, 55, p. 309.

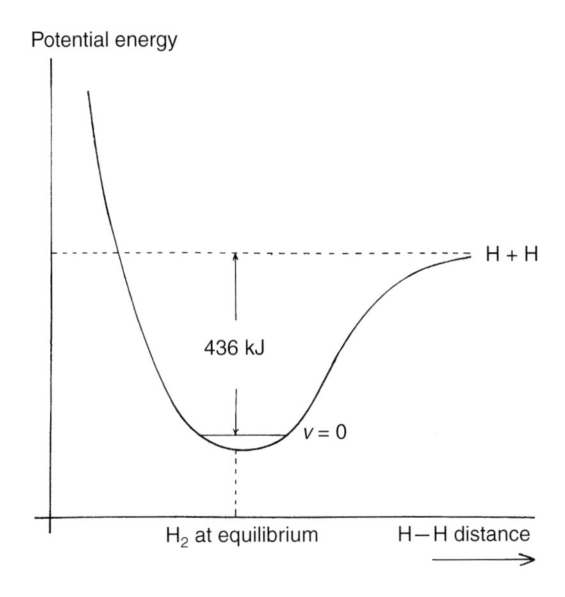

Figure 2.10 A straight cross section through the potential energy surface from Figure 2.7, passing through point *a* and at a given d_{XY} would represent the potential energy curve of the molecule YZ as a function of distance d_{YZ}. Here, the figure relates to the case Y = Z = H, thus to the stability of the H_2 molecule. The distance H—H at equilibrium = 0.75 Angstrom. In order to dissociate H_2 when it is found in its lowest vibrational state (vibration level $v = 0$) a minimum energy must be transferred equal to 436 kJ·mol^{-1} (104.2 kcal·mol^{-1}).

H atoms released have an energy not less than the potential barrier. This minimum energy required to cross the barrier may be produced by the collision with another molecule in the medium, (with a H_2 or other molecule) or the wall itself of the reaction vessel or, more generally, with what is known as a third body, M. Thus, for the dissociation:

$$H_2 + M \rightarrow H^{\bullet} + H^{\bullet} + M \tag{2.19}$$

(the dot in $H^{\bullet}$ indicates a non-paired electron, meaning that the hydrogen atom is a "free radical"). In the opposite direction, for termolecular recombination:

$$H^{\bullet} + H^{\bullet} + M \rightarrow H_2 + M \tag{2.20}$$

the H_2 molecule, produced with an energy equal to or greater than that of the H + H system, and thus of the dissociation energy, may however not redissociate, and thus be stabilised, by reducing its energy by colliding with M. If the third body is a gas, the dissociation and recombination rates increase with the concentration of the gas, and thus with its pressure, but not indefinitely. Ultimately, a pressure is reached at which the collisions become much more frequent than the purely chemical transition from the initial to the final system achieved through the creation or breaking of chemical bonds: the process which is principally collisional (limited by the collision rate), and thus intermolecular at low pressure, becomes principally chemical and thus intramolecular at high pressures.

The recombination of fairly large radicals into molecules may occur with or without a third body, the many internal degrees of freedom, vibrations and internal rotations of the large molecule produced may "soak up" their energy to such an extent that they fall to less than that of dissociation.

2.7 CHAIN REACTIONS

The balance reaction:

$$H_2 + Cl_2 \rightarrow 2\,HCl \tag{2.21}$$

forming HCl is known for its ability to be initiated by photons of sufficiently high frequency. Experiments have shown that more than a million molecules of HCl can be produced following the absorption of a single photon by the mixture; the quantum yield is said to be elevated. This observation has led to the following mechanism being suggested:

$$Cl_2 + \text{photon} \rightarrow 2\,Cl^\bullet \quad (\text{photon absorption by } Cl_2) \tag{2.22}$$

(as before, the dot associated with $Cl^\bullet$ indicates an unpaired electron and hence that the chlorine atom is a free radical. The same is true for the other radicals mentioned below). Following this reaction:

$$\left.\begin{array}{l} Cl^\bullet + H_2 \rightarrow HCl + H^\bullet \quad (2.23) \\ H^\bullet + Cl_2 \rightarrow HCl + Cl^\bullet \quad (2.24) \end{array}\right\} \text{many times}$$

...

$$2\,Cl^\bullet + M \rightarrow Cl_2 + M \tag{2.25}$$

$$2\,H^\bullet + M \rightarrow H_2 + M \tag{2.26}$$

The first reaction in this sequence is the **initiation** stage of the chain reaction, and is photochemical in this example. The second and third reactions, which can be repeated many times and hence produce the large number of HCl molecules observed for one initiation photon, are chain **propagation** steps (they constitute a "link" in the chain). The fourth and fifth reactions are chain **termination** reactions. The free radicals, such as H and Cl, which promote propagation of the chain reaction are known as **active centres** or **propagating centres**.

Chain reactions play a critical role in combustion. Starting with hydrogen, the mechanism of the process approximately follows the outline given below:

$H_2 + O_2 \rightarrow 2\ OH^{\bullet}$	initiation	(2.27)
$OH^{\bullet} + H_2 \rightarrow H_2O + H^{\bullet}$	straight-chain propagation (same number of radicals on reactant side as on product side)	(2.28)
$H^{\bullet} + O_2 \rightarrow OH^{\bullet} + O^{\bullet}$	chain branching (more radicals on product side than reactant side)	(2.29)
$O^{\bullet} + H_2 \rightarrow OH^{\bullet} + H^{\bullet}$	chain branching as for (2.29)	(2.30)

followed by radical recombination reactions, i.e. chain termination processes such as:

$$H^{\bullet} + H^{\bullet} + M \rightarrow H_2 + M$$

$$H^{\bullet} + OH^{\bullet} + M \rightarrow H_2O + M, \text{ etc.}$$

For further details about chain reactions and their chemical kinetics, the reader may wish to consult the works of Semenov[2], Sochet[3], or Destriau, Dorthe and Ben-Aim[4].

2.8 CHEMICAL KINETICS OF COMBUSTION

2.8.1 General points

The chemical mechanisms associated with combustion reactions normally involve many steps, the rates of which are not known accurately, and hence it is often difficult to know which are the principal steps. As the temperature increases, those steps for which the activation energy is low become increasingly less important compared with higher activation energy steps. Unimolecular steps are dissociation reactions either of molecules, in which case the activation energy is generally high, or of radicals of varying stability, in which case the activation energy will be lower. Bimolecular steps are the most common and those which proceed without any major rearrangement of the atoms

2. Semenov N.N. (1959) *Chemical Kinetics and Reactivity*, Pergamon, London.
3. Sochet L.R. (1971) *La cinétique des réactions en chaînes*, Dunod, Paris.
4. Destriau M., Dorthe G., Ben-Aim R. (1981) *Cinétique et dynamique chimiques*, Éditions Technip, Paris.

involved tend to have large pre-exponential factors and low activation energies. This is the case for "abstraction" reactions such as:

$$H^{\bullet} + CH_4 \rightarrow H_2 + CH_3^{\bullet} \quad (2.31)$$

or:

$$OH^{\bullet} + CH_4 \rightarrow H_2O + CH_3^{\bullet} \quad (2.32)$$

Termolecular reaction steps are much less common.

2.8.2 The H_2—O_2 reaction system

This reaction occurs mainly in accordance with the chain reaction mechanism (2.28), (2.29) and (2.30) described above. The reaction can become violent and lead to an **explosion** (see chapter 4 and Figure 4.6), within a given temperature range if the pressure exceeds a lower limit p_1 whilst remaining below an upper limit p_2. Alternatively, the reaction may occur if the pressure exceeds an even higher value p_3. The pressures p_1, p_2 and p_3 are called the first, second and third explosion limits respectively. As combustion proceeds, the reverse reactions of steps (2.28), (2.29) and (2.30) also occur. Another limiting effect is the recombination reaction:

$$H^{\bullet} + O_2 + M \rightarrow HO_2^{\bullet} + M \quad (2.33)$$

The $HO_2^{\bullet}$ radical is in fact quite stable, and its lifetime is sufficiently long to enable it to diffuse to the walls of the container holding the reactants, where it is destroyed. This process can have a negative influence on the conditions under which the H_2—O_2 reaction system may become explosive (see chapter 4 and Figure 4.6).

2.8.3 The CO—O_2 reaction system

CO—O_2 mixtures can also explode and have three associated explosion limits. Additionally, in between the conditions which produce a slow reaction and explosion, there is a pressure/temperature range in which a luminous emission (or *glow*) is produced, which lasts longer than the glow associated with the explosive reaction. In all probability, in this range, although not necessarily only in this range, electronically-excited CO_2 is produced, undoubtedly in the $^3B_2$ state, according to the reaction sequence:

$$CO + O_2 \rightarrow CO_2 + O^{\bullet} \quad (2.34)$$

$$O^{\bullet} + CO \rightarrow CO_2 \text{ (excited)} \quad (2.35)$$

and which leads to the branching reaction:

$$CO_2 \text{ (excited)} + O_2 \rightarrow CO_2 + 2O^{\bullet} \quad (2.36)$$

At low pressures, just above the first explosion limit, the diffusion of oxygen atoms towards the container walls, followed by their deactivation at this location occurs, sufficiently quickly to allow a transition to take place from within the explosion limits to that of slow reaction. On the contrary, the second limit could be due to the following gas-phase recombination reaction:

$$CO + O^\bullet + M \rightarrow CO_2 + M \tag{2.37}$$

Furthermore, the rate of the CO reaction with O_2 is greatly accelerated by the presence of traces of water or hydrogen. At high temperatures, the presence of water leads to the formation of the $OH^\bullet$ radical due to the branching reaction:

$$CO + O^\bullet + H_2O \rightarrow CO_2 + H^\bullet + OH^\bullet \tag{2.38}$$

which is followed by the very fast reaction:

$$CO + OH^\bullet \rightarrow CO_2 + H^\bullet \tag{2.39}$$

2.8.4 Combustion of hydrocarbons

The explosion limits associated with the combustion of hydrocarbons (see chapter 4) are extremely complicated and, unlike H_2—O_2 and CO—O_2 mixtures, the oxidation of hydrocarbons often produces no detectable effect over quite a long period (several minutes or even longer). This duration is known as the **induction period** or **induction delay**. An explosion may eventually result without any apparent prior indicator. This phenomenon is often explained by the **degenerate-branching** theory, involving the formation of a stable intermediate, Z, (e.g. the relatively stable $HO_2^\bullet$ radical, RCHO, ROOH, etc.), which can either lead to branching (2.41a below) or to the formation of stable compounds (2.41b below):

fuel + O_2 giving, through straight-chain reaction, Z (2.40)

then Z giving, through chain branching, propagating centres (2.41a)

or, without chain formation, giving stable products (2.41b)

It is also possible to have:

fuel + O_2 giving Z through straight-chain reaction,
which gives radicals and hence branching (2.42a)

or fuel + O_2
which gives stable products through straight chain reaction (2.42b)

In general, at high temperatures, initiation is undoubtedly caused by decomposition of fuel to give alkoxy radicals $R^\bullet$. However, at lower temperatures this process is slow and other initiation mechanisms must then be considered. For example, for the alkane RH:

$$RH + O_2 \rightarrow R^\bullet + HO_2^\bullet \tag{2.43}$$

The alkoxy radicals can then either decompose in turn, for example:

$$RCH_2CHR' \rightarrow R^{\bullet} + CH_2{=}CHR' \tag{2.44}$$

or react with O_2, following the most likely pathway:

$$R^{\bullet} + O_2 \rightarrow \text{alkene} + HO_2^{\bullet} \tag{2.45}$$

or:

$$R^{\bullet} + O_2 \rightarrow RO_2^{\bullet} \tag{2.46}$$

or isomerise.

$RO_2^{\bullet}$ radicals can themselves either decompose by the reverse reaction to (2.46), or isomerise

$$RO_2^{\bullet} \rightarrow QO_2H \rightarrow \text{stable product (possibly cyclic)}, + OH^{\bullet} \text{ or } HO_2^{\bullet} \tag{2.47}$$

where the radical Q has one hydrogen atom fewer than the radical $R^{\bullet}$; or react with a hydrocarbon molecule:

$$RO_2^{\bullet} + R'H \rightarrow ROOH + R'^{\bullet} \tag{2.48}$$

The molecular hydroperoxides, ROOH, and their peroxide analogues, ROOR′, can decompose:

$$ROOH \rightarrow RO^{\bullet} + OH^{\bullet} \tag{2.49}$$

$$ROOR' \rightarrow RO^{\bullet} + R'O^{\bullet} \tag{2.50}$$

These, and many other reactions, can occur during degenerate branching.

The mechanism associated with hydrocarbons reacting with air has been studied in depth over the past few years, and the reader can find additional details in the book written by Gardiner[5].

WORKED EXAMPLES

■ **1)** Consider a reaction which, at the same concentrations, is 5 times faster at 60°C than at 40°C. If it obeys Arrhenius' law, then its activation energy can be given by:

$$E = \left(\frac{RT_2T_1}{T_2 - T_1}\right) \ln 5 = 16.7 \text{ kcal·mol}^{-1} \ (69.73 \text{ kJ·mol}^{-1})$$

5. Gardiner W. C. (1984) *Combustion Chemistry*, Springer.

■ **2)** The pyrolysis (or thermal decomposition) of ethane plays an important role in its combustion, mainly occurring according to the mechanism outlined below and in which it is assumed that the rate constant k_i of each elementary reaction obeys Arrhenius' law, $k_i = A_i \exp(-E_i/RT)$:

	A_i	E_i
(1) $C_2H_6 \rightarrow 2\,CH_3^{\bullet}$	$10^{16.9}$ s^{-1}	374.0 kJ·mol^{-1}
(2) $CH_3^{\bullet} + C_2H_6 \rightarrow CH_4 + C_2H_5^{\bullet}$	$10^{9.6}$ dm^3·mol^{-1}·s^{-1}	73.2 kJ·mol^{-1}
(3) $C_2H_5^{\bullet} \rightarrow H^{\bullet} + C_2H_4$	$10^{14.4}$ s^{-1}	171.0 kJ·mol^{-1}
(4) $H^{\bullet} + C_2H_6 \rightarrow C_2H_5^{\bullet} + H_2$	$10^{11.1}$ dm^3·mol^{-1}·s^{-1}	39.2 kJ·mol^{-1}
(5) $2\,C_2H_5^{\bullet} \rightarrow C_4H_{10}$	$10^{8.5}$ dm^3·mol^{-1}·s^{-1}	0 kJ·mol^{-1}

Note that the dimensions of the k_i constant change depending on the molecularity of the reaction (i) being considered, which explains the different units used for A_i. In fact:

$A \rightarrow ...$ corresponds to: $-\dfrac{d[A]}{dt} = k[A]$

$A + B \rightarrow ...$ to: $-\dfrac{d[A]}{dt} = k'[A][B]$

$A + B + C \rightarrow ...$ to: $-\dfrac{d[A]}{dt} = k''[A][B][C]$

Experimentally it can be shown that the overall rate of ethane disappearance, occurring at around 800 K and with a partial pressure of around 1.5×10^4 Pa, varies roughly as a function of $[C_2H_6]^{1/2}$; why should this be so ?

The rate of ethane disappearance can be written as:

$$-\frac{d[C_2H_6]}{dt} = k_1[C_2H_6] + k_2[CH_3][C_2H_6] + k_4[H][C_2H_6] \qquad \text{(a)}$$

but, since neither [H] nor $[CH_3]$ is known then, rather than writing all the differential equations for the rates of change of concentration of all the species present, the steady-state approximation is written for each of the transitory species H, CH_3 and C_2H_5, with the rate of disappearance equal to that of appearance.

For:

$$H \;:\; k_3[C_2H_5] = k_4[H][C_2H_6] \qquad \text{(b)}$$

$$CH_3 \;:\; 2\,k_1[C_2H_6] = k_2[CH_3][C_2H_6] \qquad \text{(c)}$$

$$C_2H_5 : k_2[CH_3][C_2H_6] + k_4[H][C_2H_6] = k_3[C_2H_5] + 2\,k_5[C_2H_5]^2 \qquad \text{(d)}$$

combining (b) and (d) gives:

$$k_2[CH_3][C_2H_6] = 2k_5[C_2H_5]^2 \quad \text{(e)}$$

hence:

$$-\frac{d[C_2H_6]}{dt} = k_1[C_2H_6] + 2k_1[C_2H_6] + k_3[C_2H_5]$$

$$= 3k_1[C_2H_6] + k_3\left(\frac{k_1[C_2H_6]}{k_5}\right)^{1/2} \quad \text{(f)}$$

Considering C_2H_6 to be an ideal gas at 800 K with a partial pressure of 1.5×10^4 Pa, $[C_2H_6]$ can be calculated to have a value of 2.25 mol·m^{-3}. Under these conditions, the first term on the right-hand side of equation (f) is some 4000 times smaller than the second term, thus explaining the approximate rate relationship involving $[C_2H_6]^{1/2}$.

It is then easy to determine the steady-state concentrations of the radicals at the various C_2H_6 partial pressures, which are found to be low. For example:

Pressure of C_2H_6	100 torr (1.33×10^4 Pa)	113 torr (1.5×10^4 Pa)
	mol·dm^{-3}	mol·dm^{-3}
$[CH_3]$	9.4×10^{-13}	9.4×10^{-13}
$[C_2H_5]$	4.5×10^{-10}	4.8×10^{-10}
$[H]$	1.1×10^{-12}	0.95×10^{-12}

A more complete calculation would show that these concentrations are reached relatively quickly after pyrolysis begins and, insofar as the fastest rates are those of reactions (3) and (4), the balance reaction can be approximated to:

$$C_2H_6 \rightarrow C_2H_4 + H_2$$

although this does not necessarily mean that the rate of disappearance of C_2H_6 is equal to that of the appearance of C_2H_4 or H_2.

The following synthesis reaction for HBr can be treated in a similar way:

$$H_2 + Br_2 \rightarrow 2\,HBr$$

based on this mechanism:

$$Br_2 \rightarrow 2\,Br^\bullet \text{ (by thermal initiation)}$$
$$Br^\bullet + H_2 \rightarrow HBr + H^\bullet$$
$$H^\bullet + Br_2 \rightarrow HBr + Br^\bullet$$
$$Br^\bullet + Br^\bullet + M \rightarrow Br_2 + M$$
$$H^\bullet + H^\bullet + M \rightarrow H_2 + M$$

This mechanism is similar to that discussed for HCl formation, although the cycle involving the second and third steps are not repeated as often. Using the steady-state approximation for H and Br, an equation similar to (2.15) above is obtained and it can be shown that:

$$k = 2k_2\left(\frac{k_1}{k_5}\right)^{1/2} \quad \text{and} \quad k' = \frac{k_4}{k_3}$$

■ **3)** The frequency factors, A_i, and activation energies, E_i, of the principal reactions involved in the combustion of H_2 in O_2, are known:

	A_i	E
(1) $H_2 + O_2 \rightarrow 2\,OH^\bullet$	2.5×10^9 dm³·mol⁻¹·s⁻¹	163.0 kJ·mol⁻¹
(2) $H^\bullet + O_2 \rightarrow OH^\bullet + O^\bullet$	2.2×10^{11} dm³·mol⁻¹·s⁻¹	70.3 kJ·mol⁻¹
(3) $O^\bullet + H_2 \rightarrow OH^\bullet + H^\bullet$	1.7×10^{10} dm³·mol⁻¹·s⁻¹	39.5 kJ·mol⁻¹
(4) $OH^\bullet + H_2 \rightarrow H_2O + H^\bullet$	2.2×10^{10} dm³·mol⁻¹·s⁻¹	21.5 kJ·mol⁻¹
(5) $H^\bullet + O_2 + M \rightarrow HO_2^\bullet + M$	1.8×10^9 dm⁶·mol⁻²·s⁻¹	0

For a stoichiometric mixture at 3000 K and at a total pressure of 1 atm, and if (as has been indicated by several studies) the concentration of each transitory species X is 100 times less than the total concentration of species, then the rate of reaction (1), w_1, is given by:

$$w_1 = [H_2][O_2]\,A_1 \exp\left(-\frac{E_1}{RT}\right)$$

$$= \left(\frac{p}{3\,RT}\right)\left(\frac{2p}{3\,RT}\right)A_1 \exp\left(-\frac{E_1}{RT}\right) = 13.6 \text{ mol·dm}^{-3}\text{·s}^{-1}$$

$$w_2 = [H][O_2]\,A_2 \exp\left(-\frac{E_2}{RT}\right)$$

$$= \left(\frac{p}{3\,RT}\right)\left(\frac{p}{100\,RT}\right)A_2 \exp\left(-\frac{E_2}{RT}\right) = 724 \text{ mol·dm}^{-3}\text{·s}^{-1}$$

Similarly, with the same units:

$$w_3 = 384 \qquad w_4 = 1023 \qquad w_5 = 0.4$$

which shows that the rate varies greatly depending on the reactions considered.

■ **4)** The reaction of the methyl peroxyl radical, CH_3O_2, with itself has been studied and has been found to proceed by two elementary channels:

$$CH_3O_2^\bullet + CH_3O_2^\bullet \rightarrow 2\ CH_3O^\bullet \text{ (unstable radical)} + O_2 \qquad (1a)$$

and:

$$CH_3O_2^\bullet + CH_3O_2^\bullet \rightarrow CH_3OH + HCHO + O_2 \text{ (stable molecules)} \qquad (1b)$$

ß is defined as the ratio of the rate constants of the two channels, k_{1a}/k_{1b}. In addition there is the subsequent reaction:

$$CH_3O^\bullet \text{ (produced in (1a))} + O_2 \rightarrow HCHO + HO_2^\bullet \qquad (2)$$

where HO_2 can then in turn react with CH_3O_2:

$$CH_3O_2^\bullet + HO_2^\bullet \rightarrow \text{stable molecules which can react no further} \qquad (3)$$

Reactions (2) and (3) are fast enough to ensure that each CH_3O radical produced in (1a) immediately gives an HO_2 radical through reaction (2), and that the HO_2 itself reacts instantaneously with CH_3O_2 through reaction (3). The disappearance of the CH_3O_2 radical as a function of time can be analysed by UV spectrometry, but because of the chemical mechanism and hence the disappearance of CH_3O_2 by not only reactions (1a) and (1b) but also by reaction (3), the rate of disappearance of CH_3O_2 leads to the measurement of an apparent rate constant, $k_{apparent}$, defined as:

$$-\frac{d[CH_3O_2]}{dt} = k_{apparent}[CH_3O_2]^2$$

which is not the true rate constant for reaction (1) equal to $(k_{1a} + k_{1b})$. In fact:

1 CH_3O gives 1 HO_2 which leads to reaction (3) once

2 CH_3O give 2 HO_2 which lead to reaction (3) twice,

which means in reality that twice as much CH_3O_2 is consumed than by reaction (1) alone, hence:

$$k_{apparent} = 2k_{1a} + k_{1b}$$

By varying the temperature it is found that the branching ratio, k_{1a}/k_{1b}, varies, especially through k_{1a}, according to the equation:

$$\beta = 45 \exp\left(-\frac{1463}{T}\right)$$

giving 0.21 at 273 K and 3.5 at 573 K. At 573 K, the apparent rate constant, $k_{apparent}$, is approximately 4.2×10^{-13} $cm^3 \cdot molecule^{-1} \cdot s^{-1}$ (units used by specialists in the field), and hence:

$$k_{true} = k_{apparent}\left(\frac{k_{1a} + k_{1b}}{2k_{1a} + k_{1b}}\right) = 2.4 \times 10^{-13} = 2.4 \times 10^{-13}$$

in the same units.

CHAPTER 3

MASS AND ENERGY TRANSPORT BY CONVECTION AND DIFFUSION

3.1 RELEVANCE OF TRANSPORT PHENOMENA TO FLAMES

As stated in the introductory chapter, flames involve not only chemical phenomena but also the physical and mechanical gas-transport phenomena of diffusion and convection. A good example is provided by the flame produced by a gas cigarette lighter.

Cigarette lighters produce a jet of gas (mainly CH_4) which escapes at a speed of around 1 to 2 ms^{-1} due to the pressure in the reservoir. Through friction, the gas entrains the surrounding air, and the two gases then mix together in the jet plume. Figure 3.1a is a schematic representation of the mixing zone produced. A section through this zone shows that the concentration of the air-gas mixture varies from pure gas at the centre of the jet to pure air at the edge. When this jet is ignited (via a spark from the flint) a reaction zone is established in the middle of the mixing zone. This reaction zone will be sustained so long as the air/gas mixture continues to be supplied. Moreover, as indicated in Figure 3.1b, the reaction products themselves and the heat "given off" also tend to mix in the gaseous region, diffusing away from the reaction zone in which they are produced. Flames produced by this type of lighter, in which fuel and oxidant are supplied separately, are called diffusion flames.

This chapter will discuss the characteristics of the transport phenomena associated with diffusion and convection, and will introduce the basic laws and equations which govern them.

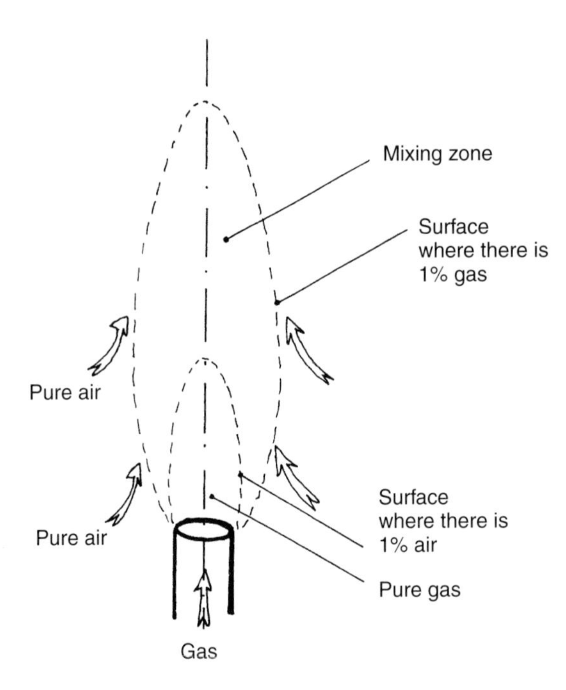

Figure 3.1a A jet of gas mixing with entrained air.

The proportions shown in the diagram are not to scale since, in reality, the mixing zone is over 10 times longer than it is wide.

The ratio of gas to air varies across the mixing zone.

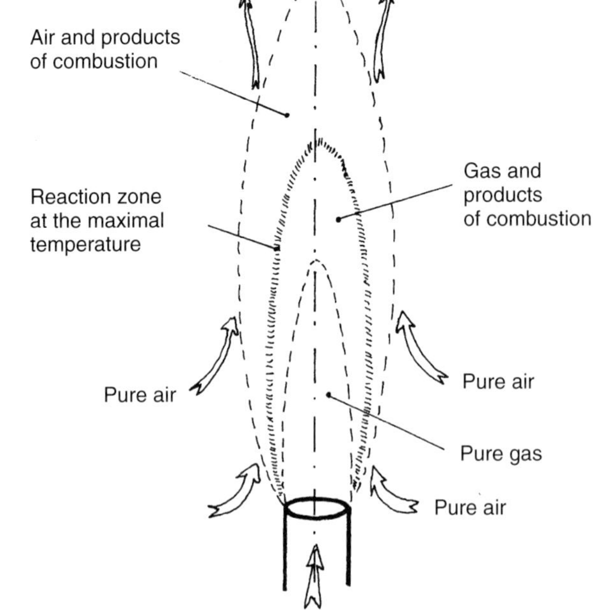

Figure 3.1b The mixing and flame produced in an ignited gas jet.

The luminous zone containing soot is not shown, but is situated in the hottest part of the "air and products" zone.

The extent of the mixing zone (containing the products, air and gas) is greater than for the un-ignited mixture shown in Figure 3.1a.

Firstly, we shall present the transport phenomena involving mass and energy diffusion resulting from molecular motion; known as "molecular diffusion". Convection, i.e. the motion of gases on a macroscopic scale, is a well known phenomenon in fluid mechanics. We will not dwell on its physical effects, but will concentrate rather on the basic equations which together govern the various diffusion and convection phenomena concerning mass and energy. These equations are known as the balance aerothermodynamic equations which, when added to considerations of chemical processes, become the aerothermochemistry equations. We will then see that through applying a simplifying assumption (that chemical reactions occur infinitely rapidly compared to all other processes) the balance equations enable a virtually complete theoretical description to be given (neglecting radiation) of the example selected above, i.e. a diffusion flame produced by a gas lighter. Finally, at the end of the chapter, the effect of turbulence on mass transport by diffusion and heat transfer will be studied since turbulence is the principal motor for diffusion in the majority of industrial flame applications.

3.2 MOLECULAR DIFFUSION OF MASS, ENERGY AND MOMENTUM

3.2.1 A simplistic description of molecular diffusion

The molecular diffusion of mass from one species into another is most clearly demonstrated by the following example. Consider a reservoir initially containing oxygen in its left half and nitrogen in its right. With time, the two components will gradually "diffuse" one into the other to ultimately produce, after an infinitely long period, a homogeneous mixture in the reservoir.

This process will occur even in the absence of any disturbance or macroscopic motion of the gases. Even for a gas at rest, i.e. for which the average velocity of the molecules is zero, diffusion will occur since each individual molecule is not at rest (except at absolute zero temperature) and its "natural tendency", known from experiment, is to occupy all of the space available to it. The same process occurs even if the **same** gas is initially contained on both sides of the reservoir, at the same pressure, although the phenomenon is then impossible to observe since the individual molecules obviously cannot be identified on a macroscopic scale.

Theoretical and experimental studies into this phenomenon carried out by physicists for over a century have produced a series of linear laws, which provide a first approximation. To illustrate these laws, let us again consider the reservoir with two chambers, of equal volume, and initially with gas A on the right and gas B on the left. Suppose that the left and right halves are separated by a porous wall which allows both A and B to pass, but slowly enough to allow the compartments A and B to be considered as uniform (Fig. 3.2a).

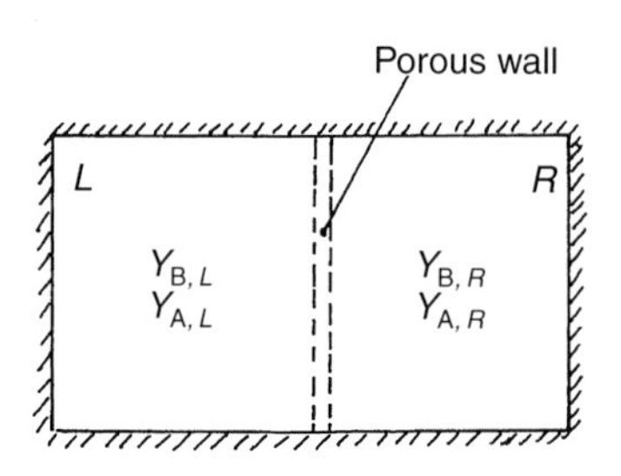

Figure 3.2a The diffusion of species A and B between two homogeneous compartments.

Under these conditions, if at time $t = 0$ the mass fraction of gas A in the right-hand compartment is 100%, and that of B on the left 100%, then at a later time these fractions will have reduced and will tend towards 50% after an infinite period of time. This fraction will be designated Y, and hence Y_B on the left at $t = 0$, $Y_{B,L}(t = 0) = 1$, Y_A on the right at $t = 0$, $Y_{A,R}(t = 0) = 1$. Although it is clear that $Y_{A,L} + Y_{B,L} = 1$ and $Y_{A,R} + Y_{B,R} = 1$, for any value of t, how can we calculate $Y_{A,R}(t)$ and $Y_{B,L}(t)$?

The linear law assumes that the flux of A, designated as j_A, crossing the porous wall in unit time is proportional to the difference $Y_{A,R}(t) - Y_{A,L}(t)$, and that this flux occurs such that the compartment with the lowest fraction of A is supplied with A. Thus:

$$\frac{dY_{A,R}}{dt} = j_A = +K(Y_{A,L} - Y_{A,R}), \quad \text{where} \quad K > 0 \text{ and is constant} \tag{3.1}$$

and, of course, $d(Y_{A,L})/dt = -j_A$, such that the flow from left to right is effectively opposite to that from right to left; thus $Y_{A,L} + Y_{A,R}$ is constant and equal to 1.

Of course, since $Y_{A,R} + Y_{B,R} = 1$ we have $d(Y_{A,R})/dt = -d(Y_{B,R})/dt$, and the same on the left, so that $d(Y_{B,R})/dt = j_B = -j_A$. The relationship described by equation (3.1) for j_A can also be used to calculate j_B.

Integrating the differential equation obtained from this linear relationship (for $Y_{A,R}$ for example) enables $Y_{A,R}(t)$ to be determined, and hence $Y_{A,L}(t)$, $Y_{B,L}(t)$ and $Y_{B,R}(t)$. Exponential profiles are obtained with the concentrations tending towards the same value of 1/2 as t tends towards infinity.

If the porous wall is now removed, and the reservoir is filled initially with gas A only on the right and gas B on the left, a similar linear law enables the evolution of the fraction of gas A in the mixture to be calculated as a function of time, this fraction being locally defined at each point along an axis x normal to the plane initially separating gases A from B (Fig. 3.2b).

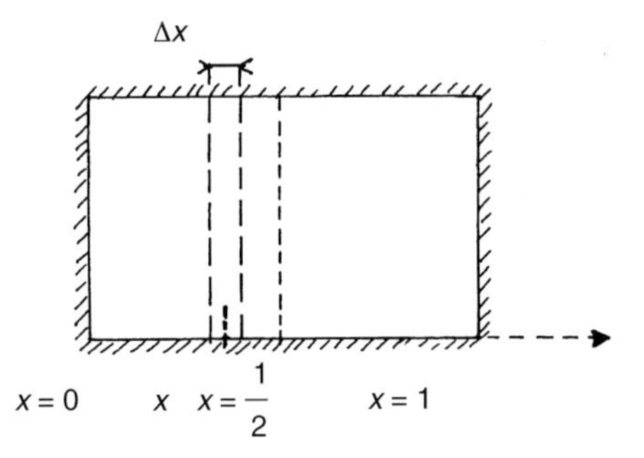

Figure 3.2b Diffusion of the species A and B in a continuous medium.

If the assumption is now made that in this case the two gases have the same molar mass, and that $Y_A(x, t)$ represents the mass fraction of gas A in a very small volume (which nonetheless contains billions of molecules) located at some value x on the x axis, then at time t, the linear approach can be applied more generally by predicting that the mass flux crossing a plane at x is proportional to the gradient along the x-direction of Y_A:

$$j_{A,x} = -d_A \frac{\partial Y_A(x,t)}{\partial x} \qquad d_A > 0 \tag{3.2}$$

where d_A is the coefficient of diffusion of gas A in B. This equation is called Fick's Law of Diffusion.

In fact, this law can be derived from (3.1) if the reservoir is divided by an imaginary plane at x acting as the porous wall and if one considers in (3.1) the mass fractions $Y_{A,L}$ and $Y_{A,R}$ immediately to the left and to the right of this plane at points separated by a distance Δx.

Equation (3.1) then gives:

$$j_{A,x} = K\left[Y_A\left(x - \frac{\Delta x}{2}\right) - Y_A\left(x + \frac{\Delta x}{2}\right)\right]$$

or, in other words, if Δx is small:

$$j_{A,x} = -K\,\Delta x \frac{\partial Y_A}{\partial x}$$

then d_A becomes simply $K\Delta x$.

Following the same procedure as that outlined above, this law can be used to determine the evolution of $Y_A(x, t)$, starting from $Y_A(x, 0)$, although the actual calculation is more complicated. The method involves balancing the mass entering and leaving each small volume represented by a slice with a cross-sectional area S in the reservoir (Fig. 3.2b). For this slice located at x, the balance is calculated as:

$$\frac{\partial}{\partial t}(\Delta x S Y_A) = S(j_A(x) - j_A(x + \Delta x))$$

or, using (3.2):

$$\frac{\partial Y_A}{\partial t} = -\frac{\partial}{\partial x}(j_{A,x}) = \frac{\partial}{\partial x}\left(d_A \frac{\partial Y_A}{\partial x}\right) \tag{3.3}$$

Integrating this partial differential equation for Y_A (with d_A = a constant), with the boundary conditions at the reservoir's side walls:

$$\frac{\partial Y_A}{\partial x} = 0$$

(since there is no flux out of or into the reservoir) and with the initial condition $Y_A = 1$ on the right side and $Y_A = 0$ on the left, once again yields curves for $Y_A(x, t)$ which tend to $Y_A = 1/2$ after an infinite period of time and whose profile is given in Figure 3.3.

The above derivations cannot be considered as exact representations of the laws governing the diffusion of mass. The phenomenological thermodynamics of irreversible processes (Onsager, Prigogine, De Groot) or statistical thermodynamics (Hirschfelder, Curtiss) allow a more rigorous treatment, through specifying more completely the assumptions required to obtain the linear laws such as equation (3.2).

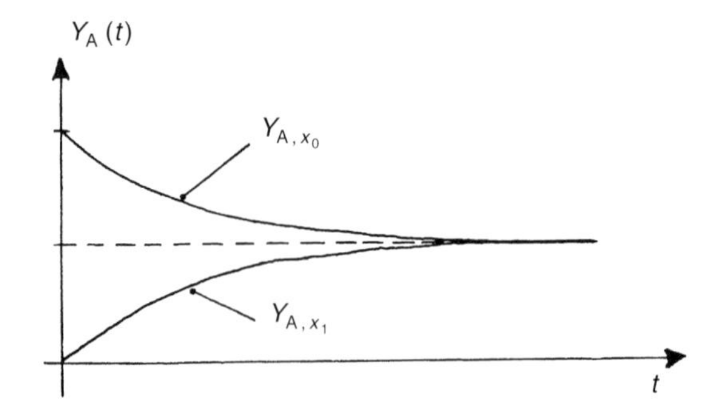

Figure 3.3 Change in mass fraction of Y_A for a value on the right side ($x = x_0 > 1/2$) and on the left side ($x = x_1 < 1/2$).

These more accurate treatments show that temperature and pressure gradients can result in mass fluxes, in addition to the mass fraction gradient of (3.2). A further result is that, in reality, it is the **mole fraction** gradient rather than the mass fraction gradient which is important. However, the effects due to pressure gradients (barodiffusion) and temperature gradients (thermodiffusion or the Soret effect) are very often negligible. Equally, a further approximation is to consider the diffusion of one species in a complex mixture as being the diffusion of a small amount of one species in the "remainder" of the medium where the "remainder" is of known composition and of constant molar mass. An example where this approach is particularly appropriate is when considering combustion in air, since air always contains approximately 75% nitrogen.

Thus Fick's Law of Diffusion, in the form shown in (3.2), is still used but in its general form, where the flux of i in a given direction is proportional to the gradient of Y_i in this direction, and where the density is taken into consideration in the coefficient of diffusion:

$$\vec{j_i} = -\rho D_i \, \overrightarrow{\text{grad}} \, Y_i \tag{3.4}$$

D_i is the binary coefficient of diffusion of species i in the remainder of the mixture, and is dependent on the nature of i, on the "remainder" of the mixture and on the thermodynamic conditions. D_i can be calculated by statistical thermodynamics and measured by experiment.

Clearly, since we have seen that $\vec{j_A} = -\vec{j_B}$ in a binary diffusion process, and that $Y_A + Y_B = 1$ in this case, then $\vec{j_A} = -\vec{j_B}$ suggests that $D_A = D_B$, i.e. the coefficient of diffusion of A in B is the same as that of B in A. Furthermore, like K in (3.1) and d_A in (3.2), D_i must be positive to allow mass to diffuse towards those zones where its concentration is lower, in agreement with the Second Law of Thermodynamics.

Since the diffusion process is related to molecular motion, clearly it must be dependent on the size of the molecules in question, and thus D_i must depend to a large extent on the nature of the species involved. The following table gives coefficient of diffusion values for some common compounds at $p = 1$ bar ($\approx 10^3$ hPa now taken as the reference value) and $T = 298$ K:

CO_2—N_2	1.7×10^{-5}	(SI units)
H_2—N_2	7.8×10^{-5}	
CH_4—N_2	1.6×10^{-5}	
H_2O—N_2	9.9×10^{-5}	

A basic theory of statistical thermodynamics, which considers the various molecules as being hard spheres and which assumes that there are no molecular interactions other than collisions between molecules (the case for a gas whose pressure is not excessively high), can be used to calculate the dependence of coefficient of diffusion on temperature and pressure. Since the dimensions of the coefficient of diffusion are metres squared per second (L^2T^{-1}) its value must be proportional to the product of a velocity and a characteristic length of molecular motion. Considering that the average velocity of motion of a molecule is the speed of sound (a) in the medium and the characteristic length of collision is the mean free path (l), it can be concluded from a simple dimensional analysis that D must be proportional to the product $l \times a$:

$$D = \alpha l \times a$$

where a varies as $\sqrt{T}$ and l is inversely proportional to the density of the medium, which itself varies with p/T. Thus:

$$D = \alpha p^{-1} T^{3/2}$$

is the approximate variation with p and T of the coefficient of diffusion, for each species, α being a constant which depends on the nature of the species.

3.2.2 Heat conduction

Heat conduction is another molecular diffusion process, resulting from the random motion of molecules in a fluid (a gas or even a liquid). The internal energy of a gas is related to the kinetic energy of motion of the molecules, both in terms of their random motion, which itself results in the diffusion of mass, and in terms of the internal energy of intramolecular motion: rotational kinetic energy, vibrational energy (for a species comprising more than one atom), and electronic energy arising from the deformation of the electronic clouds surrounding the nuclei. This explains why when molecules move they "carry" internal energy with them which can be transferred to neighbouring molecules through collisions, and which may be distributed in this way throughout the medium.

For this reason, the form of the law of molecular heat diffusion looks very like the law of mass diffusion. Fourier's Law, which can be written as:

$$\overrightarrow{j_q} = -\lambda \ \overrightarrow{\text{grad}} \ T \qquad \lambda > 0 \tag{3.5}$$

is in fact the analogue of Fick's Law of Diffusion. In this case, j_q, the heat flux, is the energy flux carried by molecular diffusion across a given surface area (in joules per metre squared per second), and T is the local temperature of the medium. The coefficient λ is called the thermal conductivity and is positive for the same reason as D_i is positive.

This heat flux is the only energy flux present when the gas (or liquid) medium consists of only one type of molecule. However, it is not the only flux if the gas is a non-homogeneous mixture of several components and where the mass of each component is able to diffuse through space. In this case, internal energy is transmitted not only through collisions between molecules but also through the diffusion of the molecules themselves.

The diffusion flux of internal energy, j_u, is the sum of the heat flux defined above and the energy flux associated with the mass fluxes. The latter is not simply the sum of the mass fluxes of each species (multiplied by the internal energy by unit mass of these species) but also includes the work done by pressure forces. Each diffusing molecule carries not only internal energy but also leaves room for the remaining molecules to expand and thus further reduces the internal energy of all the remaining molecules. In thermodynamics, this effect is taken into consideration by using the enthalpy per unit mass, h, rather than the internal energy per unit mass and thus by using h_i in the equation instead of u_i.

We should therefore write:

$$\vec{j_u} = \vec{j_q} + \sum_1^n h_i \vec{j_i} \tag{3.6}$$

Fourier's Law can be written in a slightly different form to take into account the fact that for a mixture of ideal gases at thermal equilibrium, $h = \sum_1^n Y_i h_i(T)$

Then:

$$\overrightarrow{\text{grad}}\ h = \sum_1^n h_i\ \overrightarrow{\text{grad}}\ Y_i + \sum_1^n c_{p,i} Y_i\ \overrightarrow{\text{grad}}\ T,$$

where:

$$\frac{\mathrm{d}h_i(T)}{\mathrm{d}T} = c_{p,i} \tag{3.7}$$

It follows, therefore, that:

$$\vec{j_q} = -\frac{\lambda}{\sum_1^n Y_i c_{p,i}}\ \overrightarrow{\text{grad}}\ h + \frac{\lambda}{\sum_1^n Y_i c_{p,i}} \sum_1^n h_i\ \overrightarrow{\text{grad}}\ Y_i \tag{3.8}$$

and the quantity $a_T = \lambda / \left(\rho \sum_1^n Y_i c_{p,i} \right)$ is called the thermal diffusivity, whose dimensions are the same as those of diffusivity D_i, namely metres squared per second ($L^2 T^{-1}$).

In gases, the order of magnitude of a_T is very often the same as the coefficient of diffusion D_i, except for certain species which deviate greatly from the average, such as hydrogen, a small molecule which diffuses easily, or alternatively species which diffuse poorly because of their high molar mass. Moreover, the ratio $Le_i = a_T / D_i$, known as the Lewis number, is quite close to one for gases, when expressing a_T and D_i in the same system of units; Le is a dimensionless number (we shall consider other dimensionless number in later chapters). For liquids, the effect of the diffusion of mass is generally much less than the effect of heat conduction (with collisions becoming increasingly important) and the Lewis numbers are thus much greater than one (by a factor of 10 or more). This is shown even more clearly for solids; the molecules hardly move at all through the medium, but rather vibrate around an equilibrium configuration. In many cases only electrons or ions are mobile (if the solid is a conductor).

If (3.8) and Fick's Law of Diffusion (3.4) are used to calculate $\overrightarrow{j_u}$ then we can introduce a_T and Le_i, and (3.6) becomes:

$$\overrightarrow{j_u} = \rho a_T \overrightarrow{\text{grad}}\ h + \rho a_T \left(\sum_1^n h_i \overrightarrow{\text{grad}}\ Y_i \right) \left(1 - \frac{1}{Le_i} \right) \tag{3.9}$$

When the approximation $Le_i \approx 1$ is valid, the flux of energy is much simpler to treat, since only the gradient of h is involved.

3.2.3 The diffusion of momentum

The diffusion of momentum by molecular motion in a fluid is governed by the same principles as those which apply to mass or energy diffusion and cannot be neglected if the latter two are included in an analysis. The study of momentum diffusion is more complicated than that for mass or energy since momentum is a vector.

The diffusion flux of momentum in a gaseous mixture at rest is, on average, simply the pressure, which can be proved using statistical thermodynamics. When the medium is in motion, this diffusion flux also includes a factor known as "friction".

For an idealised, two-dimensional flow, with only one component of velocity v in a direction x, which varies only in relation to the perpendicular direction y, and where the density is constant, the linear law analogous to Fick's or Fourier's is called Newton's Law and is given by:

$$j_{v,x} = p - \mu \frac{\partial v}{\partial y}$$

Under these conditions, friction has only one component which is proportional to the single component of the gradient of v multiplied by the coefficient μ, called the dynamic viscosity.

However, a three-dimensional flow must automatically have three velocity components and thus three fluxes, each flux in turn having three components. The momentum flux is thus a matrix, denoted as $j_{\alpha\beta}$, $\alpha, \beta = 1, 2, 3$.

Pressure is involved on the diagonal of this matrix, expressed in a similar way for each term, thus

$$j_{\alpha\beta} = p\delta_{\alpha\beta} - \tau_{\alpha\beta} \tag{3.10}$$

$\delta_{\alpha\beta}$ is Kronecker's symbol, which is equal to zero if $\alpha \neq \beta$ and equal to one if $\alpha = \beta$, and $\tau_{\alpha\beta}$ is called the "friction tensor". Like $j_{\alpha\beta}$, this friction tensor is also a matrix, and each of its elements must therefore be expressed as a linear combination of the element of matrix $\partial v_\alpha / \partial x_\beta$, consisting of gradients in each direction of the velocity components $\vec{v} = v_\alpha$, $\alpha = 1, 2, 3$. The corresponding law is:

$$\tau_{\alpha\beta} = \mu\left(\frac{\partial v_\alpha}{\partial x_\beta} + \frac{\partial v_\beta}{\partial x_\alpha}\right) - \frac{2}{3}\mu\left(\sum_{\gamma=1}^{3}\frac{\partial v_\gamma}{\partial x_\gamma}\right)\delta_{\alpha\beta} \tag{3.11}$$

where μ must be positive, like D_i and λ, in order to satisfy the Second Law of Thermodynamics. Equation (3.11) shows that the matrix $\tau_{\alpha\beta}$ is symmetrical, and that $\tau_{11} + \tau_{22} + \tau_{33} = 0$, a result demonstrated by statistical thermodynamics. In general terms, these results can also be proved experimentally, with the exception of certain special fluids (known as non-Newtonian fluids). The last term in equation (3.11) is related to the effect called the "viscosity in volume" and, as we will see later, using equation (3.16), is equal to zero if the density is constant in the flow.

By defining $\nu = \mu/\rho$, where ν is the coefficient of kinematic viscosity (with dimensions the same as a_T and D_i,), then two other dimensionless numbers can be defined:

Prandtl's number: $$Pr = \frac{\nu}{a_T} = \frac{\mu c_p}{\lambda}$$

Schmidt's number: $$Sc_i = \frac{\nu}{D_i}$$

For gases, these numbers are quite close to unity, again due to the fact that it is molecular motion which is responsible for the diffusion of heat, mass, and momentum.

3.3 THE BALANCE EQUATIONS FOR AEROTHERMOCHEMISTRY

The diffusion phenomena discussed above should now be included in the general equations which describe the changes occurring in a non-isothermal, non-isobaric medium containing several chemical species and in motion. These equations simply involve considering the overall variations of each of the variables of interest in a small, fixed volume of the medium.

Equation (3.3) is one such balance equation, which considers the mass fraction Y of the species A in the medium. In order to further our analysis, this equation must be generalised and similar equations derived for the other quantities of interest.

Particularly simple systems will be considered below to explain the principles involved in deriving the equations. A more detailed treatment would be outside the scope of this book but is covered in depth by Barrère and Prud'homme [1].

3.3.1 Balance equations in a one-dimensional medium

The first step in deriving the equations is to consider a one-dimensional medium, i.e. a medium in which the quantities of interest may vary with time and with one spatial variable. This is the case considered in equation (3.3) which can be applied to the medium shown in Figure 3.2b (the characteristics of the medium are taken as being the same in each plane perpendicular to the x-axis).

The same reasoning which gave equation (3.3) is applied, except that the following additional parameters are considered for the medium (gaseous or liquid): the possibility of a velocity v, and an extent (or rate) of reaction w, both of which may be functions of x and time. The density of the medium is ρ, and is also a function of x and time.

For the slice of thickness Δx shown in Figure 3.2b, the process of analysing the overall changes in the quantity Y_i (the mass fraction of a species i) involves stating that the variation with time of the quantity contained in the volume ($\Delta x S$) is due to convection at a velocity v, and to diffusion and to the reaction. In this case, the use of "and" implies that the effects of these three phenomena are additive and thus:

$$\frac{\partial(\Delta x S \rho(x) Y_i(x))}{\partial t} = S(\rho(x)\,v(x)\,Y_i(x) - \rho(x+\Delta x)\,v(x+\Delta x)\,Y_i(x+\Delta x)) + S(j_i(x) - j_i(x+\Delta x)) + S\Delta x \rho(x)\,w_i(x)$$

The first two phenomena are fluxes: they are the differences between the values entering and leaving the volume $S\Delta x$, whereas the rate of the chemical reactions applies for the entire volume, and thus the specific rate w_i (mass of i per unit mass of the mixture and per unit time) must be multiplied by the density and the volume $S\Delta x$. The values of w_i and ρ are taken at x, instead of at $x + \Delta x/2$, the centre of gravity of the volume, but this second order error does not affect the accuracy since Δx is very small.

1. Barrère and Prud'homme (1973) *Équations générales de l'aérothermochimie*, Masson et Cie.

By dividing by Δx, which tends to zero, the balance equation in Y_i becomes a partial derivatives equation:

$$\frac{\partial}{\partial t}(\rho Y_i) + \frac{\partial}{\partial x}(\rho v Y_i) = \frac{\partial}{\partial x}(-j_i) + \rho w_i \tag{3.12}$$

This equation is true for all species i. All the equations can be summed for each i:

$$\frac{\partial}{\partial t}\left(\rho\left(\sum_{i=1}^{n} Y_i\right)\right) + \frac{\partial}{\partial x}\left(\rho v\left(\sum_{1}^{n} Y_i\right)\right) = \frac{\partial}{\partial x}\left(-\sum_{1}^{n} j_i\right) + \rho\sum_{1}^{n} w_i$$

However $\sum_{1}^{n} Y_i = 1$, since Y_i is a mass fraction; moreover, $\sum_{1}^{n} j_i = 0$, since all the diffusion fluxes for all the species must cancel overall (as discussed in section 3.2 where $j_A = -j_B$). Finally, $\sum_{1}^{n} w_i = 0$ (as determined in section 2.2 of chapter 2) since, as originally stated by Lavoisier, "nothing can be created or destroyed" during one or a series of chemical reactions. In this way the so-called "continuity" equation is obtained:

$$\frac{\partial}{\partial t}\rho + \frac{\partial}{\partial x}\rho v = 0 \tag{3.13}$$

The same reasoning is used to obtain the balance equation for the total energy (i.e. the sum of the internal and kinetic energies) which, for $u_t = u + v^2/2$, and $h_t = u_t + p/\rho$ (called the total or stagnation enthalpy), is given by:

$$\frac{\partial}{\partial t}\rho u_t + \frac{\partial}{\partial x}(\rho v h_t) = \frac{\partial}{\partial x}\left(-j_q - \sum_{i=1}^{n} h_i j_i + v\tau\right) + vF \tag{3.14}$$

where F is an external force per unit volume which can act externally on the medium in the x direction; gravity for example.

No chemical reaction term is required here. Indeed the internal energy u is a function of temperature as well as of medium composition. For an ideal gas mixture, as discussed in chapter 1:

$$u = \sum_{i=1}^{n} Y_i u_i = \sum_{i=1}^{n} Y_i \int_0^T c_{vi}(T)\,\mathrm{d}T + \sum_{i=1}^{n} Y_i u_{i,0}$$

and:

$$h = \sum_{i=1}^{n} Y_i h_i = \sum_{i=1}^{n} Y_i \int_0^T c_{pi}(T)\,\mathrm{d}T + \sum_{i=1}^{n} Y_i h_{i,0}$$

where, for gases assumed to behave in a manner similar to that of an ideal gas, $u_{i,0} = h_{i,0}$. This relationship is the same, when adapted to correspond to a unit mass of mixture, as equation (1.3) in chapter 1.

By defining $\sum_{i=1}^{n} Y_i \int_0^T c_{vi}(T)\,\mathrm{d}T = \overline{c_v}T$, and similarly for h with $\overline{c_p}$, $\overline{c_v}$ and $\overline{c_p}$ as average values in the medium, (3.14) can be re-written in the following way:

$$\frac{\partial}{\partial t}\rho\left(\overline{c_v}T+\frac{v^2}{2}\right)+\frac{\partial}{\partial t}\left(\rho v\left(\overline{c_p}T+\frac{v^2}{2}\right)\right)=\frac{\partial}{\partial x}\left(-j_q-\sum_{i=1}^{n}h_i j_i+v\tau\right)+vF$$
$$-\sum_{i=1}^{n}\left[\frac{\partial}{\partial t}(\rho Y_i)+\frac{\partial}{\partial x}(\rho v Y_i)\right]h_{i,0}$$

and, by taking into consideration (3.12), another form is produced which introduces a reaction term (involving w_i):

$$\frac{\partial}{\partial t}\rho\left(\overline{c_v}T+\frac{v^2}{2}\right)+\frac{\partial}{\partial x}\left(\rho v\left(\overline{c_p}T+\frac{v^2}{2}\right)\right)=\frac{\partial}{\partial x}\left(-j_q-\sum_{i=1}^{n}(h_i-h_{i,0})j_i+v\tau\right)$$
$$-\rho\sum_{i=1}^{n}h_{i,0}w_i+vF \qquad (3.14')$$

The balance equation for momentum is obtained in the same way and, without entering into detail, finally yields the following result:

$$\frac{\partial}{\partial t}(\rho v)+\frac{\partial}{\partial x}(\rho v^2)=-\frac{\partial p}{\partial x}+\frac{\partial \tau}{\partial x}+F \qquad (3.15)$$

In fact, this equation is the general form of the fundamental law of dynamics, with the right-hand side corresponding to the forces applied and the left to the acceleration of the volume of fluid in question.

For a steady-state system, neglecting friction τ and all long-range forces, F, the following simple equation is obtained:

$$\frac{\partial(p+\rho v^2)}{\partial x}=0 \qquad (3.15')$$

which also indicates that the quantity $(p + \rho v^2)$ is constant in the direction of flow. This equation, not to be confused with Bernoulli's equation (which is similar except for a factor of 1/2, but has no relevance here), will be used frequently in the following analyses.

3.3.2 Equations for a three-dimensional medium

When analysing a flow in three-dimensional space, the three components of each flux must be considered: $j_{i,\alpha}$, $j_{q,\alpha}$, for $\alpha = 1, 2, 3$ in addition to the 9 components of $\tau_{\alpha\beta}$ which affect the 3 components of the momentum ρv_α, where $\alpha = 1, 2, 3$ (or ρv_β, where $\beta = 1, 2, 3$).

The intermediate equations obtained by applying the same principles as above will not be shown. By considering a small volume ΔV rather than $S\Delta x$, the resulting equations are:

$$\frac{\partial}{\partial t}(\rho) + \sum_{\alpha=1}^{3} \frac{\partial \rho v_\alpha}{\partial x_\alpha} = 0 \tag{3.16}$$

$$\frac{\partial}{\partial t}(\rho Y_i) + \sum_{\alpha=1}^{3} \frac{\partial}{\partial x_\alpha}(\rho v_\alpha Y_i) = \sum_{\alpha=1}^{3} \frac{\partial}{\partial x_\alpha}(-j_{i,\alpha}) + \rho w_i \tag{3.17}$$

$$\frac{\partial}{\partial t}(\rho u_t) + \sum_{\alpha=1}^{3} \frac{\partial}{\partial x_\alpha}(\rho v_\alpha h_t) = \sum_{\alpha=1}^{3} \frac{\partial}{\partial x_\alpha}\left(-j_{q,\alpha} - \sum_{i=1}^{n} h_i j_{i,\alpha} + \sum_{\beta=1}^{3} v_\beta \cdot \tau_{\alpha\beta}\right) + \sum_{\beta=1}^{3} v_\beta \cdot F_\beta \tag{3.18}$$

$$\frac{\partial}{\partial t}(\rho v_\beta) + \sum_{\alpha=1}^{3} \frac{\partial}{\partial x_\alpha}(\rho v_\alpha v_\beta) = -\frac{\partial p}{\partial x_\beta} + \sum_{\alpha=1}^{3} \frac{\partial}{\partial x_\alpha}(\tau_{\alpha\beta}) + F_\beta \quad \beta = 1, 2, 3 \tag{3.19}$$

In these equations, the fluxes $j_{i,\alpha}$, $j_{q,\alpha}$ and $\tau_{\alpha\beta}$ must be expressed using the equations given in section 3.2.

The solution to these equations is used to calculate, for each point in the reacting fluid medium and at any time, the characteristic parameters of this medium, namely ρ, v_β, Y_i, u_t and h_t. The solutions depend upon the conditions at the medium's boundaries, since they are specific to each flame under consideration.

The energy equation (3.18) may also be written by introducing explicitly the reaction term. If the relationships $u = \overline{c}_v T + \sum_1^n u_{i,0} Y_i$ and $h = \overline{c}_p T + \sum_1^n h_{i,0} Y_i$, are also used, the general form of (3.14′) is obtained:

$$\frac{\partial}{\partial t}\left(\rho\left(\overline{c}_v T + \sum_1^3 \frac{v_\beta v_\beta}{2}\right)\right) + \sum_1^3 \frac{\partial}{\partial x_\alpha}\left(\rho v_\alpha\left(\overline{c}_p T + \sum_1^3 \frac{v_\beta v_\beta}{2}\right)\right)$$

$$= \sum_1^3 \frac{\partial}{\partial x_\alpha}\left(-j_{q,\alpha} - \sum_{i=1}^{n}(h_i - h_{i,0}) j_{i,\alpha} + \sum_1^3 v_\beta \tau_{\alpha\beta}\right) - \rho \sum_1^n h_{i,0} w_i + \sum_1^3 v_\beta F_\beta \tag{3.18′}$$

3.3.3 Specific simple forms of the balance equation for energy

For certain specific cases, the balance equation for energy can be conveniently simplified to facilitate subsequent calculations.

The simplest situation of this type is that of steady-state, non-viscous flow ($\mu = 0$), with no long-range forces and where the transfer of mass and heat can be neglected. Equation (3.18) then becomes:

$$\sum_1^3 \frac{\partial(\rho v_\alpha h_t)}{\partial x_\alpha} = 0$$

Since (3.16) indicates that the product ρv_α also obeys the same equation, it follows that the rest enthalpy h_t is constant throughout the flow. This simple result will be used in chapters 5 and 7.

In cases when the transfer of heat and mass cannot be neglected, (due to their importance in the process), the assumption that the Lewis number is equal to one is extremely useful. By using the simple form (3.9) of the energy flux equation, with $Le_i = 1$, the following equation can be obtained for the enthalpy in (3.14) and (3.18):

$$\frac{\partial(\rho u_t)}{\partial t} + \sum_1^3 \frac{\partial(\rho v_\alpha h_t)}{\partial x_\alpha} = \sum_1^3 \frac{\partial\left(\rho a_T \frac{\partial h}{\partial x_\alpha} + \sum_1^3 v_\beta \tau_{\alpha\beta}\right)}{\partial x_\alpha} + \sum_1^3 v_\beta F_\beta \qquad (3.18'')$$

For slow-speed flows (well below sonic), kinetic energy and the effects of friction can be neglected and h_t can be simplified to h. The case of a steady flow with no long-range forces enables the previous equation to be reduced further to:

$$\sum_1^3 \frac{\partial}{\partial x_\alpha}\left(\rho v_\alpha h - \rho a_T \frac{\partial h}{\partial x_\alpha}\right) = 0 \qquad (3.18''')$$

In an adiabatic medium, i.e. with no heat exchange across any walls (note that radiation has been neglected throughout) and such that the gas flux entering is uniform in terms of enthalpy, it can be shown that the solution to this equation is simply that the enthalpy h is constant throughout, even when heat and mass are transferred and when chemical reactions occur in the bulk flow.

Equation (3.18‴) may also be written by introducing a reaction term, and by using mean $\overline{c_p}$ values, as in (3.18′). The process used to derive (3.18′) from (3.18) is applied to give:

$$\sum_1^3 \frac{\partial}{\partial x_\alpha}(\rho v_\alpha \overline{c_p} T) = \sum_1^3 \frac{\partial}{\partial x_\alpha}\left(\rho a_T \frac{\partial(\overline{c_p} T)}{\partial x_\alpha}\right) - \rho \sum_1^n h_{i,0} w_i \qquad (3.18'''')$$

For the case which considers that only one global reaction occurs, the reaction term can again be simplified since all the w_i are then proportional to a single w, as seen in chapter 2. Using the equation $w_i = \nu_i w$, where ν_i are the stoichiometric coefficients for this reaction but expressed by mass rather than by number of moles (those corresponding to the left-hand side being negative), this term can be written directly as $\rho q w$.

By definition, $q = -\sum_{1}^{n} \nu_i h_{i,0}$ is then the "heat released" by the reaction.

However, in practice, q does not have a set, known value since many reactions occur and, as discussed in chapter 1, it can only be calculated if the temperature attained is known, which affects product composition. However, this temperature can only be calculated using equation (3.18'''') which is in turn dependent on q... The usual adopted approach gives an approximate result by attributing an empirical value to q corresponding to an experiment performed under conditions similar to those of interest. This equation will be employed repeatedly in the following chapters.

3.4 SIMPLIFIED CALCULATION OF A LAMINAR DIFFUSION FLAME

Enough information is now known about diffusion and convection phenomena, and how to construct the corresponding equations, to be able to calculate, at least approximately, the features of the diffusion flame produced by the cigarette lighter mentioned at the beginning of this chapter. This will be achieved by simplifying as much as possible the chemical phenomena actually involved, but without eliminating them completely. Chemical reactions will be assumed to occur so rapidly that the oxidant (the oxygen in the air) and the fuel (methane) cannot exist simultaneously in the same fluid particle without immediately forming the combustion products (CO_2 and H_2O).

To develop this theory in such a way as to allow the equations to be solved analytically, further approximations need to be made. These will be made with no greater justification than that we are looking to produce an exemplary, albeit unreal flame. "Unreal" in this context means that although it does not correspond exactly to a real flame, it does have most of its characteristics.

This theoretical calculation was one of the first of its kind to attempt to describe a diffusion flame, and was proposed by Burke and Schumann in 1928. Rather than considering a round jet (of gas) the jet is considered to be planar, emerging from an infinitely long slit. The geometry of the system is thus totally two-dimensional with one plane of symmetry instead of an axisymmetrical geometry. Moreover, assume that the jet of gas and the surrounding air have the same unit flow rate ρv (which implies that they will have the same velocities if their densities are equal), thus greatly simplifying the calculations.

3.4.1 Principal equations

The general equations of aerothermochemistry will provide the principal equations, but under certain specific additional conditions.

Firstly, consider that only three reactive chemical species are present, the fuel (CH_4) denoted as K, the oxidant (O_2) denoted as Ox, and the products (a mixture of CO_2 and H_2O) denoted as P. Assume that these species react in accordance with a single reaction :

$$\mathrm{K} + \nu\, \mathrm{Ox} \rightarrow \mathrm{P}$$

where ν is the molar stoichiometric coefficient "relating to Ox" in the reaction. A mass stoichiometric coefficient can be defined as $\nu_s = \nu M_{\mathrm{Ox}}/M_{\mathrm{K}}$, which incorporates the molar masses of Ox and K respectively, and which will appear in the following calculations.

Three equations can then be written for Y_i, where i = K, Ox, P. They are written with the assumption that the system is in steady state, i.e. $\partial\phi/\partial t = 0$ for any variable ϕ.

Molecular diffusion is presumed to follow Fick's Law of Diffusion, and the assumption is made that all the coefficients of diffusion (for all the species) are equal, *i.e.* $D_{\mathrm{Ox}} = D_{\mathrm{K}} = D_{\mathrm{P}} = d$. The rate of reaction for each species is ρw_i, which will be examined in more detail later.

Combining (3.17), and Fick's Law, gives:

$$\sum_1^3 \frac{\partial}{\partial x_\alpha}(\rho v_\alpha Y_i) = \sum_1^3 \frac{\partial}{\partial x_\alpha}\left(\rho d \frac{\partial Y_i}{\partial x_\alpha}\right) + \rho w_i \quad i = \mathrm{Ox, K, P} \qquad (3.17')$$

The continuity equation should also be brought in since the flame can be assumed to exist in the presence of an inert species (nitrogen in this case) such that $Y_{\mathrm{Ox}} + Y_{\mathrm{K}} + Y_{\mathrm{P}} \neq 1$. Thus, according to (3.16):

$$\sum_{\alpha=1}^3 \frac{\partial}{\partial x_\alpha}(\rho v_\alpha) = 0$$

In the energy balance, the kinetic energy is assumed to be negligible, as are any long-range forces. Furthermore, the hypothesis that $Le_i = 1$ can be used, which assumes $a_T = d$. Equation (3.18‴) thus gives:

$$\sum_1^3 \frac{\partial}{\partial x_\alpha}(\rho v_\alpha h) = \sum_1^3 \frac{\partial}{\partial x_\alpha}\left(\rho d \frac{\partial h}{\partial x_\alpha}\right)$$

Finally, the momentum equation must be added, written in a form similar to (3.19) but neglecting only long-range forces. Again, the assumption is made that the coefficient of viscosity $\nu = d$, i.e. that the Schmidt numbers are equal to one.

The approach now requires the introduction of a hypothesis which assumes that the flow velocity at all points is always in a direction parallel to the plane of symmetry (see Fig. 3.4). The equations will therefore neglect all flows per-

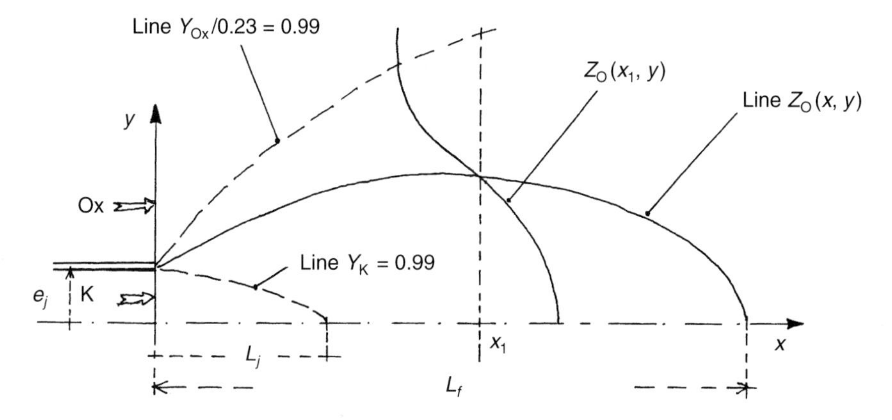

Figure 3.4 Schematic representation of a Burke-Schumann flame with a transversal profile of $Z_O = Y_K - Y_{Ox}/\nu_s$. e_j = thickness.

pendicular to the plane of symmetry. Neglected also will be the diffusion fluxes (of mass, energy and momentum) in the flow direction (x), presuming that the effect of convection predominates in this direction. This cannot be the case for fluxes in a direction normal to the flow (y) since the transversal velocity is zero.

The equations resulting from (3.16), (3.17′), (3.18‴), and (3.19) are, respectively:

$$\left.\begin{aligned} &\frac{\partial}{\partial x}(\rho v) = 0 \\ &\frac{\partial}{\partial x}(\rho v Y_i) = \frac{\partial}{\partial y}\left(\rho d \frac{\partial Y_i}{\partial y}\right) + \rho w_i, \quad i = \text{Ox, K, P} \\ &\frac{\partial}{\partial x}(\rho v h) = \frac{\partial}{\partial y}\left(\rho d \frac{\partial h}{\partial y}\right) \\ &\frac{\partial}{\partial x}(\rho v^2) = -\frac{\partial p}{\partial x} + \frac{\partial}{\partial y}\left(\rho d \frac{\partial v}{\partial y}\right) \end{aligned}\right\} \tag{3.20}$$

These equations are not completely accurate, as was later discovered. The inaccuracy arises from ignoring the transversal velocity which should not be neglected when considering its gradient in the y direction (that gives the so-called "boundary layer equations"). However, since the results obtained from the Burke and Schumann approach are not significantly erroneous, equations (3.20) will be retained.

Integrating the first equation immediately produces the result that ρv is a function of y alone. Due to the entry conditions selected for the flame, (i.e. constant ρv), ρv does not depend on y either and so is truly a constant.

Consequently, it is possible to break down the equations for v, for h, and those for Y_i:

$$\rho v \frac{\partial Y_i}{\partial x} = \frac{\partial}{\partial y}\left(\rho d \frac{\partial Y_i}{\partial y}\right) + \rho w_i, \quad i = \text{Ox, K, P}$$

$$\rho v \frac{\partial h}{\partial x} = \frac{\partial}{\partial y}\left(\rho d \frac{\partial h}{\partial y}\right)$$

$$\rho v \frac{\partial v}{\partial x} = \frac{\partial}{\partial y}\left(\rho d \frac{\partial v}{\partial y}\right) - \frac{\partial p}{\partial x}$$

Furthermore, experiments have shown that pressure varies very little in the flame (confirmed by full theory) which means that the equation for v becomes obsolete. If we can solve the equations for Y_i and h, the equations of state could be used to calculate ρ and hence v (note that this does not imply that v is a constant, which would be unrealistic).

The remaining equations for Y_{Ox}, Y_{K}, Y_{P} and h, are partial differential equations.

They require boundary conditions to be set at $x = 0$, $y = 0$ and at $y = +\infty$ (see Fig. 3.4), as given below:

at $x = 0$ if $y < e_j$ $\quad Y_{\text{Ox}} = 0, \; Y_{\text{P}} = 0, \; Y_{\text{K}} = 1, \; h = h_{\text{jet}}, \; v = v_j, \; d = d_j, \; \rho = \rho_j$

if $y > e_j$ $\quad Y_{\text{Ox}} = 0.23$ (O_2 in pure air), $\; Y_{\text{P}} = 0, \; Y_{\text{K}} = 0, \; h = h_{\text{ext}}$

for $y = 0$, due to the symmetry of the system, all the $\partial/\partial y$ derivatives are zero,

for $y = +\infty$, $\quad Y_{\text{Ox}} = 0.23, \; Y_{\text{P}} = 0, \; Y_{\text{K}} = 0, \; h = h_{\text{ext}}$

3.4.2 Specific relationships

Some specific relationships exist between the various functions Y_i and h which will allow the problem to be simplified.

First of all, note that since the reaction which consumes K and Ox, is the same as that which produces P, then the rates of reaction by mass for these species are related by their specific stoichiometric coefficients:

$$w_{\text{Ox}} = \nu_s w_{\text{K}}, \quad \text{and} \quad w_{\text{P}} = -w_{\text{K}}$$

Consequently, if two functions are defined, $Z_{\text{O}} = Y_{\text{K}} - Y_{\text{Ox}}/\nu_s$ and $Z_{\text{P}} = Y_{\text{K}} + Y_{\text{P}}$, then these two functions will satisfy the following differential equation, obtained by combining the equations for Y_{K}, Y_{Ox}, and Y_{P} as indicated in the definitions of Z_{O} and Z_{P}:

$$\rho v \frac{\partial Z_{\text{O}}}{\partial x} = \frac{\partial}{\partial y}\left(\rho d \frac{\partial Z}{\partial y}\right) \qquad Z = Z_{\text{O}} \text{ or } Z_{\text{P}}$$

which no longer includes a reaction term.

Such functions are called Schvab-Zeldovitch functions, and only exist where there is a single reaction and where the coefficients of diffusion are identical.

Note also that Z_O, Z_P and h satisfy the same differential equation, and that consequently, if we let:

$$Z_O = \alpha_o h + \beta_o, \qquad Z_P = \alpha_P h + \beta_P$$

where α and β are constants, then the equations for Z_O and Z_P reproduce the equation for h. Therefore, if such relationships satisfy the limiting conditions, they can be considered as being the first integrals of the equations for Z_O and Z_P (which, like h, have a unique solution for a given ρv, p and d).

It is easy to check that these linear equations are compatible with the boundary conditions and that they allow α_o, α_P and β_o, β_P to be determined in a unique way. By way of an example, the following equations may be obtained:

$$\left.\begin{aligned} 1 &= \alpha_o h_{jet} + \beta_o \\ -\frac{0.23}{\nu_s} &= \alpha_o h_{ext} + \beta_o \end{aligned}\right\} \rightarrow \left\{\begin{aligned} \alpha_o &= (1 + 0.23/\nu_s)/(h_{jet} - h_{ext}) \\ \beta_o &= 1 - (1 + 0.23/\nu_s)\, h_{jet}/(h_{jet} - h_{ext}) \end{aligned}\right.$$

3.4.3 Very fast chemical reactions

The hypothesis of infinitely rapid chemical reactions makes it very easy to determine Y_K without actually solving the equation for Y_K. Indeed, since K and Ox cannot coexist, Z_O represents either Y_K (when positive) or $-Y_{Ox}/\nu_s$ (when negative). Thus Y_K can be determined simply by knowing Z_O!

The function Z_O (x, y) is equal to $-0.23/\nu_s$ at the extremities of the flame, far from the central axis, and is equal to 1 at $x = 0$ in the jet. Its profile is shown in Figure 3.4. Since it is positive on one side and negative on the other, the function must pass through the zero on a line, somewhere within the flame. This line in fact represents the positions where the chemical combustion reaction can take place. Outside this line there will be no more fuel and thus the reaction cannot occur, and inside the line there will be no more oxygen, again making reaction impossible.

This line of reaction (or rather reaction surface since the flame has planar symmetry and $Z_O(x, y)$ is simply a section through a surface $Z_O(x, y, z)$) will cut the plane of symmetry at a specific distance, where $Z_O(x, 0) = 0$, as indicated in Figure 3.4. For this point, the flame length can be defined as being the distance L_f such that $Z_O(L_f, 0) = 0$. Beyond this point there will not be any methane present in the flame, since it will have been totally burnt.

All that now remains to fully solve the problem is to calculate $Z_O(x, y)$.

3.4.4 The final step

Solving $Z_O(x, y)$ is now simply a question of mathematics. The following equation must be solved:

$$\rho v \frac{\partial Z_O}{\partial x} = \frac{\partial}{\partial y}\left(\rho d \frac{\partial Z}{\partial y}\right)$$

with:

for $x = 0$, if $y < e_j$ $Z_O = 1$ and if $y > e_j$ $Z_O = -0.23/\nu_s$

for $y = 0$, $\partial Z_O/\partial y = 0$ and for $y = +\infty$ $Z_O = -0.23/\nu_s$

Assuming (again using a hypothesis specific to this purpose, but one which does not greatly alter the result) that ρd is approximately constant and equal to $\rho_j d_j$, then the equation for Z_O is simply the "heat equation":

$$\frac{\partial Z_O}{\partial x} = \alpha \frac{\partial^2 Z_O}{\partial y^2}, \text{ where } \alpha = \frac{\rho_j d_j}{\rho v} = \frac{\rho_j d_j}{\rho_j v_j} = \frac{d_j}{v_j}$$

Its solution is given in all the relevant specialised texts, but only, however, for the case of a function $f(x, y)$ which, when $x = 0$, is equal to one in the gas jet and zero outside. If it is noted that such a function can be constructed based on Z_O using:

$$f(x, y) = \frac{Z_O(x, y) + 0.23/\nu_s}{1 + 0.23/\nu_s}$$

then it is possible to deduce that:

$$\frac{Z_O(x, y) + 0.23/\nu_s}{1 + 0.23/\nu_s} = \frac{1}{2\sqrt{\pi \alpha x}} \int_{-e_j}^{+e_j} \exp\left(\frac{-(y - y')^2}{4\alpha x}\right) dy'$$

In cases where the values of x are much greater than e_j, an approximation of the following form is valid:

$$\frac{Z_O(x, y) + 0.23/\nu_s}{1 + 0.23/\nu_s} = \frac{2e_j}{2\sqrt{\pi \alpha x}} \exp\left(\frac{-y^2}{4\alpha x}\right)$$

It is then possible to calculate all the required values; the flame surface is that where $Z_O = 0$ and the profiles for Y_{Ox}, Y_K, h, T, ρ, v, can be deduced from the solution $Z_O(x, y)$.

$Z_O(x, y)$ has the form shown in Figure 3.4, whereas the profiles of Y_K, Y_{Ox}, Y_P and T on the axis are given in Figure 3.5a (where $Z_O(x, 0)$ is also traced). The temperature is deduced from the enthalpy through knowing the values of Y_i, with the enthalpy itself being linearly related to Z_O.

Similarly, the radial profiles of Y_K, Y_{Ox}, Y_P, T, at a given value of x, will be as shown in Figure 3.5b or 3.5c where the shape already shown in Figures 3.1a

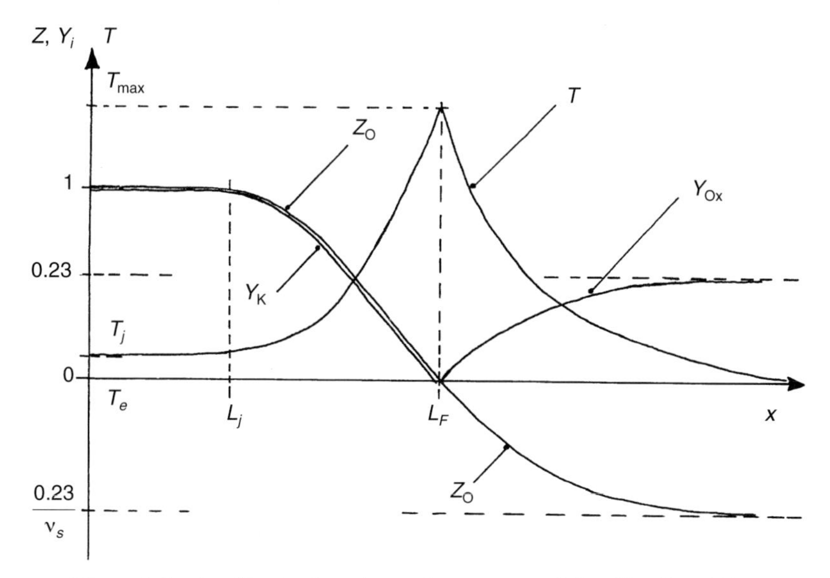

Figure 3.5a Profiles in the plane of symmetry of Z_O, Y_K, Y_{Ox}, T in a Burke and Schumann flame.

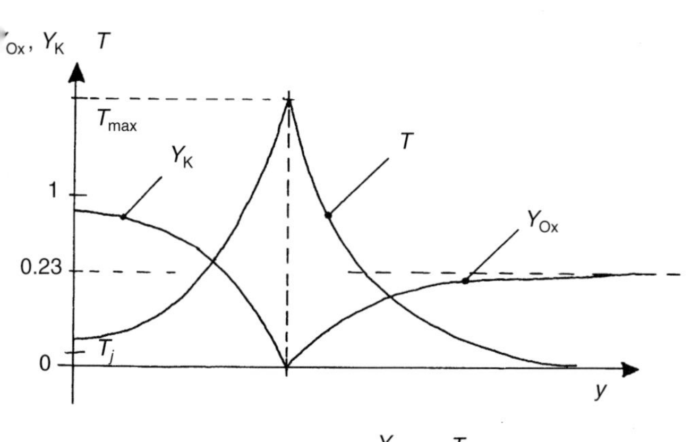

Figure 3.5b Transversal profiles of Y_K, Y_{Ox}, T, in a Burke and Schumann flame, for $L_j < x < L_F$ (if $x < L_j$, the value of Y_K at $y = 0$ is equal to one, and that at T equal to T_j).

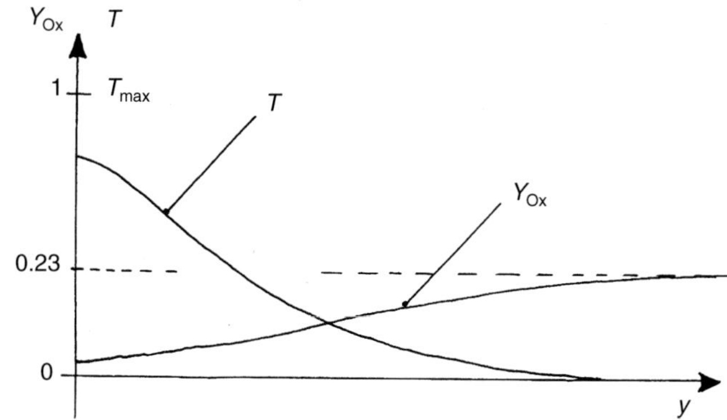

Figure 3.5c Transversal profiles of Y_{Ox} and T for a Burke and Schumann flame, for $x > L_F$ ($Y_K = 0$ everywhere).

and 3.1b is repeated, and with the additional factor that the reaction zone at this point is infinitesimally thin, since the chemical reaction is assumed to occur infinitely quickly.

The length L_F is particularly easy to find, since it is simply the value of x in the previous formula which corresponds to $y = 0$ and $Z_O = 0$.

This gives (in air):

$$L_F = \frac{e_j^2 v_j}{\pi d_j}\left(\frac{v_s + 0.23}{0.23}\right)^2$$

Note that L_F increases with the square of e_j, is proportional to v_j and is inversely proportional to d_j. In other words, the greater the coefficient of diffusion, the shorter the flame; and the greater the jet velocity, the longer the flame. The stoichiometric coefficient ν_s, in addition to the fraction of oxygen in the surrounding air, also affects L_F, but in a more complicated manner. The chemistry of the phenomena, except in relation to ν_s, is not implicated since the reaction is assumed to occur infinitely quickly.

In fact, a simple analysis of the dimensionless parameters involved in the problem would have shown, without any need for complex calculations, that L_F/e_j ought to be a function of the Peclet number for the jet $(v_j e_j/d_j)$. This calculation shows that L_F/e_j is directly proportional to Peclet number. On the other hand, L_F/e_j should also be expected to be a function of ρ_j/ρ_{ext} and h_j/h_{ext}, the other dimensionless parameters involved. However, in this case they are not present as a consequence of the approximations used.

3.5 TURBULENT TRANSPORT

3.5.1 What is turbulent flow?

In the large majority of real cases, the gas flow in flames and their surroundings is turbulent. What does this mean for the flow and what implications does it have for diffusion phenomena and the equations which describe them?

A turbulent flow is one for which the trajectory of the "fluid particles" is not "smooth", but is instead characterised by sudden changes in direction. Two fluid particles very close together at a given moment will suddenly adopt very different trajectories and will move increasingly farther apart. Even under apparently stationary limiting conditions the trajectories themselves are not stationary. In turbulent flow, the fluid particles behave in a similar way to the random motion exhibited by molecules.

A flow is not necessarily always turbulent, since "turbulence" is a phenomenon which appears more or less suddenly when the flow velocity is increased above a given value. The effect was discovered at the end of the 19th century by O. Reynolds, whilst studying fluid flow in a tube. Below this critical velocity the flow (known as "laminar" flow) is "calm", the trajectories of the fluid particles are very smooth and they are all similar. As soon as the critical velocity is reached, the nature of the flow changes, becoming turbulent and displaying the characteristics described above. Turbulence is now known to be caused by the laminar flow becoming unstable, which is itself triggered by small-scale perturbations, and that the "turbulent" state is the non-steady state situation towards which unstable flow tends.

The point at which a flow changes from laminar to turbulent is given by the Reynolds number: $v_r l_r / \nu$ (where v_r is a reference velocity, l_r a characteristic flow length, and ν the kinematic viscosity). As the Reynold's number increases the flow becomes unstable, and totally turbulent at very high values.

The balance equations discussed in paragraph 3.3 and the phenomenological laws governing diffusion phenomena are valid for both laminar and turbulent flow. Even though the variations in fluid trajectory are abrupt, they nonetheless occur on a much larger scale than the mean free path of molecules, and even the concept of fluid particles continues to be valid for turbulent flow. However, the balance equations are of virtually no practical use, partly because they cannot yield steady-state solutions and more importantly because it is impossible to define the boundary conditions required for solving the equations with sufficient accuracy to ensure the "reproducibility" of the results.

Experiments have nonetheless shown that even though the fluid particle trajectories are not reproducible, certain "average values" for these trajectories are reproducible. Quantities such as velocities, energies, average concentrations etc. can be determined based on this principle and balance equations derived which satisfy the experimentally-observed phenomena. This is known as the "Reynolds approach" and will now be discussed in detail. The results will describe a "mean" turbulent flow which in many respects is analogous to laminar flow; i.e. with coefficients of "turbulent diffusion" for mass, energy, and momentum.

3.5.2 Balance equations describing "mean" turbulent flow

Following Reynolds's principles, but using stricter, more modern definitions, we will consider the mean values of extensive and intensive parameters involved in the balance equations of paragraph 3.3. These means are statistical averages, i.e. they are derived from a large number of identical experiments (or at least identical at the macroscopic scale). If the mean value of a parameter g is noted as $\overline{g}$, then:

$$\overline{g} = \lim_{N \to \infty} \sum_{k=1}^{N} \frac{g^{(k)}}{N} \tag{3.21}$$

where $g^{(k)}$ is the value of g obtained in experiment number k. The variance is thus described as:

$$\overline{g'^2} = \lim_{N \to \infty} \sum_{k=1}^{N} \frac{(g^{(k)} - \overline{g})^2}{N} \tag{3.22}$$

and, for two different parameters f and g:

$$\overline{fg} = \lim_{N \to \infty} \sum_{k=1}^{N} (f^{(k)} g^{(k)}/N) \tag{3.23}$$

or:

$$\overline{f'g'} = \lim_{N \to \infty} \sum_{k=1}^{N} (f^{(k)} - \overline{f})\,(g^{(k)} - \overline{g})/N \tag{3.23'}$$

Hence it is possible to derive the balance equations for $\overline{v}$, $\overline{u}$, $\overline{Y_i}$, $\overline{\rho}$, based on the equations given in section 3.3. These derivations and the associated explanations are too long to be given here and the reader is recommended to consult one of the fundamental texts treating turbulence and turbulent flow (for example, Tennekes and Lumley[2] or Favre et al.[3]).

The original form of these equations relates to averages in the form defined above, and assumes that the flow has uniform density and that turbulent diffusion fluxes predominate over molecular fluxes. This is the simplest case, and was the first to be studied in detail. The equations obtained are:

– the mean continuity equation, derived from (3.16):

$$\sum_{\alpha=1}^{3} \frac{\partial}{\partial x_\alpha} \overline{v_\alpha} = 0 \tag{3.24}$$

– the mean diffusion equations, derived from (3.17):

$$\frac{\partial}{\partial t} \overline{Y_i} + \sum_{\alpha=1}^{3} \frac{\partial}{\partial x_\alpha} (\overline{v_\alpha} \overline{Y_i}) = \sum_{\alpha=1}^{3} \frac{\partial}{\partial x_\alpha} (-\overline{v'_\alpha Y'_i}) + \overline{w_i} \qquad i = 1 \ldots n-1 \tag{3.25}$$

– the mean energy equation, derived from (3.18):

$$\frac{\partial}{\partial t} \overline{u_t} + \sum_{\alpha=1}^{3} \frac{\partial}{\partial x_\alpha} (\overline{v_\alpha} \overline{h_t}) = \sum_{\alpha=1}^{3} \frac{\partial}{\partial x_\alpha} (-\overline{v'_\alpha h'_t}) + \sum_{\beta=1}^{3} \frac{\overline{F_\beta v_\beta}}{\rho} \tag{3.26}$$

2. Tennekes and Lumley (1974) *A first course in turbulence*, 3rd ed., M.I.T. Press.
3. Favre et al. (1976) *La turbulence en mécanique des fluides : Bases théoriques et expérimentales*, Gauthier-Villars.

– the mean Navier Stokes equations, also known as the Reynolds' equations, are derived from (3.19):

$$\frac{\partial}{\partial t}\overline{v_\beta} + \sum_{\alpha=1}^{3}\frac{\partial}{\partial x_\alpha}(\overline{v_\alpha}\,\overline{v_\beta}) = -\frac{\partial}{\partial x_\beta}\left(\frac{\overline{p}}{\rho}\right) + \sum_{\alpha=1}^{3}\frac{\partial}{\partial x_\alpha}(\overline{v'_\alpha v'_\beta}) + \overline{F_\beta}/\rho \quad (3.27)$$

The surprising result is that the mean equations are similar to the original equations, except for the appearance of a different diffusion flux term :

- $\overline{v'_\alpha Y'_j}$ is a turbulent mass diffusion flux of the species i, replacing $j_{i,\alpha}$ in equation (3.17).
- $\overline{v'_\alpha\, h'_t}$ is a turbulent total (or rest) enthalpy diffusion flux, replacing the energy fluxes in equation (3.18).
- $\overline{v'_\alpha v'_\beta}$ is a turbulent momentum diffusion flux (also known as the Reynolds tensor) which can be represented by a matrix in the same way as $\tau_{\alpha\beta}$ is in equation (3.19).

There is no real reason to be surprised by these new flux terms since, in reality, fluid particles in turbulent flow behave much like molecules. If we were to define a new fluid particle, a mean fluid particle with velocity $\overline{v_\alpha}$, composition $\overline{Y_i}$ and enthalpy $\overline{h}$, then the particle would be governed by equations which take into account the random motion of the fluid particles in the same way as the earlier equations considered the random effect of molecules. A simple change of scale therefore yields the same result.

The assumption that ρ is constant does not hold for flames and hence the previous equations must be modified. It is possible to recalculate them by taking into account the variation of ρ, $\overline{\rho}$, $\overline{\rho'^2}$, $\overline{\rho' v'_\alpha}$, etc. However, it is more convenient simply to change the definition of the mean values, switching to A. Favre's principles rather than O. Reynolds'. Let us define the mean parameters weighted by ρ for all extensive parameters, but keep $\overline{\rho}$ and $\overline{p}$, which are already volume-related parameters.

Thus:

$$\tilde{g} = \frac{1}{\overline{\rho}}\lim_{N\to\infty}\sum_{k=1}^{N}\frac{\rho^{(k)}g^{(k)}}{N} \quad (3.28)$$

and the new variance is:

$$\widetilde{g'^2} = \frac{1}{\overline{\rho}}\lim_{N\to\infty}\sum_{k=1}^{N}\frac{\rho^{(k)}(g^{(k)}-\tilde{g})^2}{N} \quad (3.29)$$

Similarly, for two variables f and g:

$$\widetilde{f'g'} = \frac{1}{\overline{\rho}}\lim_{N\to\infty}\sum_{k=1}^{N}\frac{\rho^{(k)}(f^{(k)}-\tilde{f})(g^{(k)}-\tilde{g})}{N} \quad (3.30)$$

(remembering that $f^{(k)}-\tilde{f} \neq f^{(k)}-\overline{f}$).

The balance equations for $\widetilde{v}_\beta$, $\widetilde{Y}_i$, $\widetilde{h}_t$ then become:

$$\frac{\partial}{\partial t}\bar{\rho} + \sum_{\alpha=1}^{3} \frac{\partial}{\partial x_\alpha}(\bar{\rho}\,\widetilde{v}_\alpha) = 0 \tag{3.31}$$

$$\frac{\partial}{\partial t}\bar{\rho}\widetilde{Y}_i + \sum_{\alpha=1}^{3} \frac{\partial}{\partial x_\alpha}(\bar{\rho}\,\widetilde{v}_\alpha\widetilde{Y}_i) = \sum_{\alpha=1}^{3} \frac{\partial}{\partial x_\alpha}(-\bar{\rho}\,\widetilde{v'_\alpha Y'_i}) + \bar{\rho}\,\widetilde{w}_i \tag{3.32}$$

$$\frac{\partial}{\partial t}\bar{\rho}\widetilde{u}_t + \sum_{\alpha=1}^{3} \frac{\partial}{\partial x_\alpha}(\bar{\rho}\,\widetilde{v}_\alpha\widetilde{h}_t) = \sum_{\alpha=1}^{3} \frac{\partial}{\partial x_\alpha}(-\bar{\rho}\,\widetilde{v'_\alpha h'_t}) + \sum_{\beta=1}^{3} \overline{F_\beta \cdot v_\beta} \tag{3.33}$$

$$\frac{\partial}{\partial t}\bar{\rho}\,\widetilde{v}_\beta + \sum_{\alpha=1}^{3} \frac{\partial}{\partial x_\alpha}(\bar{\rho}\,\widetilde{v}_\alpha\widetilde{v}_\beta) = -\frac{\partial \bar{p}}{\partial x_\beta} + \sum_{\alpha=1}^{3} \frac{\partial}{\partial x_\alpha}(-\bar{\rho}\,\widetilde{v'_\alpha v'_\beta}) + \overline{F_\beta} \tag{3.34}$$

In fact, this new notation has the same degree of physical significance as that used previously: $\bar{\rho}\,\widetilde{v}_\alpha = \overline{\rho v_\alpha}$ represents the mean momenta per unit volume, whereas $\overline{v_\alpha}$ represented the mean momenta per unit mass. Since the basis of the balance equations is a balance analysis for finite volumes (see section 3.3) this method appears to be even more physically-based than the first.

The mean total energy $\widetilde{u}_t$ is made up of the mean internal energy and the mean kinetic energy, in addition to the kinetic energy of the turbulence:

$$\widetilde{u}_t = \widetilde{u} + \frac{1}{2}\sum_\alpha \widetilde{v_\alpha v_\alpha} = \widetilde{u} + \frac{1}{2}\sum_\alpha \widetilde{v}_\alpha\widetilde{v}_\alpha + \frac{1}{2}\sum_\alpha \widetilde{v'_\alpha v'_\alpha}$$

The last term is also found in $\widetilde{h}_t$.

As can be seen in equations (3.32) to (3.34), new turbulent diffusion flux terms appear:

- $\widetilde{v'_\alpha Y'_i}$ for mass i,
- $\widetilde{v'_\alpha h'_t}$ for the total enthalpy, equal to $-\ \widetilde{v'_\alpha h'} - \sum_\beta \widetilde{v}_\beta \widetilde{v'_\alpha v'_\beta}$,

since:

$$h'_t = h_t - \widetilde{h}_t = h + \frac{1}{2}\sum_\beta(\widetilde{v}_\beta + \widetilde{v}'_\beta)(\widetilde{v}_\beta + \widetilde{v}'_\beta) - h - \frac{1}{2}\sum_\beta \widetilde{v}_\beta\widetilde{v}_\beta = h' + \sum_\beta v'_\beta\widetilde{v}_\beta$$

- $\widetilde{v'_\alpha v'_\beta}$ for the momentum,

and the balance equations are again similar to the initial equations.

Since ρ is not constant, the equations of state which link ρ, p, u, and T must not be omitted. However, they cannot be used in this form since we require equations relating the mean parameters $\widetilde{u}, \bar{p}, \bar{\rho}$. First of all:

$$\bar{p} = \overline{\rho R_m T}$$

where strictly speaking $R_m = \sum_1^n Y_i R_i$ is not a constant and therefore one must write $\overline{p}$ with the introduction of correlations for the fluctuations in Y_i and T and not simply $\widetilde{Y}_i$ and $\widetilde{T}$. However, in the case of flames where air enters as an oxidant, it is not a serious error to consider R_m as being constant (since air contains 75% nitrogen), giving the simple equation:

$$\overline{p} = \overline{\rho} R_m \widetilde{T}$$

The mean equation of thermal state is given by:

$$\widetilde{u} = \frac{1}{\overline{\rho}} \sum_1^n \overline{\rho Y_i u_i}$$

Once again, the form of this equation is complex, since it does not involve simply $\widetilde{T}$ et $\widetilde{Y}_i$. However by assuming that $u_i = c_v T + u_{i,0}$, where c_v and $u_{i,0}$ are constants (and where c_v is independent of i and T), a simplified form can be derived:

$$\widetilde{u} = \frac{1}{\overline{\rho}} c_v \overline{\left(\sum_1^n Y_i\right) \rho T} + \frac{1}{\overline{\rho}} \sum_1^n \overline{\rho Y_i u_{i,0}} = c_v \widetilde{T} + \sum_1^n u_{i,0} \widetilde{Y}_i$$

Note, however, that h is related to u, in a very simple way, and than it can be shown in all cases that:

$$\overline{\rho} \widetilde{h} = \overline{\rho} \widetilde{u} + \overline{p}$$

3.5.3 Turbulent diffusion fluxes

All that now remains is to derive relationships for these new fluxes, in the same way as they were derived for molecular diffusion. If this step were not performed, these new balance equations could not be used.

The similarity was highlighted above of the basic balance equations and the averaged balance equations, and it is tempting to believe that this similarity should allow similar phenomenological laws to be proposed, with appropriate coefficients of "turbulent diffusion" defined by phenomenological laws for turbulent flows like Fick's and Fourier's laws for laminar flows. This approach was considered by Boussinesq and Reynolds and leads to the following equations instead of (3.4), (3.9) and (3.11):

$$\widetilde{v'_\alpha Y'_i} = -D_t \frac{\partial \widetilde{Y}_i}{\partial x_\alpha}, \qquad D_t = \nu_t / Sc_t$$

$$\widetilde{v'_\alpha h'} = A_t \frac{\partial \widetilde{h}}{\partial x_\alpha}, \qquad A_t = \nu_t / Pr_t$$

$$\widetilde{v'_\alpha v'_\beta} - \left(\sum_{\alpha=1}^3 \widetilde{v'_\alpha v'_\alpha}\right)\delta_{\alpha\beta} = -\nu_t\left(\frac{\partial \widetilde{v}_\alpha}{\partial x_\beta} + \frac{\partial \widetilde{v}_\beta}{\partial x_\alpha}\right) + \frac{2}{3}\nu_t\left(\sum_{\alpha=1}^3 \frac{\partial \widetilde{v}_\alpha}{\partial x_\alpha}\right)\delta_{\alpha\beta} \qquad (3.35)$$

where ν_t is known as the "turbulent viscosity" coefficient, Sc_t and Pr_t are the Schmidt and Prandtl numbers for turbulent flow, D_t is the turbulent mass diffusivity, and A_t the turbulent thermal diffusivity.

With this approach, the differences between the two sets of balance equations for turbulent and laminar flow are even further reduced, confined to the ν_t, Sc_t, Pr_t coefficients alone. This even allows the application of the results obtained for laminar flows, with Fick diffusion (for diluted solutions), for a Lewis number of unity and at low velocities (as derived in section 3.4) to turbulent flows where it is assumed that $Sc_t = Pr_t = 1$ and ν_t is constant. Indeed, a first approximation to turbulent flow can be obtained by considering it as a more diffusive and viscous laminar flow with uniform viscosity (and diffusivity). The approximations that $Sc_t = Pr_t = 1$ are therefore, except in certain cases, not unrealistic.

However, unlike molecular coefficients of diffusion, these turbulent coefficients of diffusion are not known quantities, characteristic of fluid flow. In fact it is the flow itself which creates the turbulence and not the fluid, and hence the values of ν_t, D_t, and A_t are not independent of the flow. Simple reasoning enables the orders of magnitude to be obtained for the coefficients upon finding a characteristic velocity and flow length at which turbulence occurs. For the gas jet in the diffusion flame discussed in section 3.4 these are of course the jet velocity v_j and its thickness e_j. Hence, using a purely dimensional argument, $D_t = A_t = \nu_t = C_\nu v_j e_j$ where C_ν is a constant; (experiment has shown that taking $C_\nu \approx 0.05$ provides a rough approximation).

If this reasoning is applied to the calculation of the flame length, following the same procedure as that outlined in section 3.4, then the following result is obtained:

$$\frac{L_F}{e_j} \simeq \frac{20}{\pi}\left(\frac{\nu_s + 0.23}{0.23}\right)^2 \simeq 1000$$

with ν_s as defined in section 3.4.2.

It is an experimental fact that the length of a turbulent diffusion flame is not dependent on the velocity of the jet producing the flame. The value of 1000 calculated by this formula is only an approximate estimation of the order of magnitude. Finally, note that D_t is a characteristic of the flow and not of the fluid and must be independent of the nature of the diffusing species, as confirmed by experiment.

Current theories of turbulence and turbulent diffusion suggest more precise rules for expressing the coefficients of turbulent diffusion, and even how to calculate turbulent fluxes directly. These approaches will be discussed later in this book in the context in which they arise. For the time being, it is sufficient to note the definition of turbulent fluxes and their similarity to molecular diffusion fluxes, and the fact that they both produce very similar effects.

WORKED EXAMPLES

■ 1) Heat conduction

The balance equation for internal energy (3.14′) can also be applied to solids. In such cases, ρ is constant, $j_i = 0$, $w_i = 0$ and if we also assume that c_p and c_v are constant, then:

$$\frac{\partial T}{\partial t} = \frac{\partial}{\partial x}\left(\frac{\lambda}{\rho c_v}\frac{\partial T}{\partial x}\right)$$

This equation is used to calculate the "heat propagation" in a solid, as a function of the dimension x and of time, so long as we know $T(x, 0)$ and the boundary conditions at $x = -1$ and $x = +1$. Considering the heat flux to be zero at the extremities and that :

$$T(0, x) = T_0 \cos\left(k\pi \frac{x}{l}\right)$$

with k a positive integer, show that $T(x, t)$ is given by:

$$T(t, x) = T_0 \cos\left(k\pi \frac{x}{l}\right) \exp\left(-k^2\pi^2 \frac{\alpha t}{l^2}\right) \quad \text{where} \quad \alpha = \frac{\lambda}{\rho c_v}$$

In the case where $T(0, x)$ is some function which can be broken down into a Fourier series, show that:

$$T(t, x) = \sum_{k=1}^{\infty}\left(b_k \cos\left(k\pi \frac{x}{l}\right) + a_k \sin\left(k\pi \frac{x}{l}\right)\right) \exp\left(-k^2\pi^2 \frac{\alpha t}{l^2}\right)$$

where a_k et b_k are the coefficients for the Fourier deconvolution of $T(0, x)$.

■ 2) Mass diffusion

Consider a cylindrical jar of height $2l$, filled with water to a level l. The whole system is at a uniform temperature T_0 so that, at the surface of the liquid, the mass fraction of the water vapour Y_0 is such that the partial pressure of the vapour is equivalent to the saturated vapour pressure and is a function of T_0. Outside the jar, a current of air ensures that the air is kept dry and hence $Y = 0$ across the top of the jar. Calculate the mass fraction of water vapour Y as a function of the height x above the liquid in the jar. Consider the evaporation to be steady state in relation to a coordinate system based on the liquid surface, and that vaporisation is sufficiently slow to allow l to be considered as being constant.

The balance equation representing Y is taken from (3.12) and let d be the coefficient of diffusion of water vapour in air. Assume also that there is a sufficiently small amount of vapour in the air to allow the density of the air-vapour mixture to be considered as being constant, and that d is also constant. Show that:

$$\frac{Y(x)}{Y_0} = \frac{1 - \exp\left(-\frac{V}{d}(l-x)\right)}{1 - \exp\left(-\frac{Vl}{d}\right)}$$

where V is the velocity with which the vapour leaves the liquid surface. In order to calculate this velocity, show that we must consider as equal the mass fluxes leaving and entering a control volume surrounding the liquid surface, and that it is possible to obtain:

$$V = \frac{d}{l} \ln\left(\frac{1}{1 - Y_0}\right)$$

What is the retreat velocity of the liquid?

■ **3)** The solution of the Fourier series in 1) can be used to calculate the length of Burke and Schumann's diffusion flame (see section 3.4) but in the case where the flame is situated between two plates a distance $2l$ apart (still with a plane of symmetry).

In order to determine $Z_O(x, y)$ which satisfies the heat equation, first develop the function $f(y)$, as a Fourier series, such that $f = 1$ if $-e_j < y < +e_j$ and $f = 0$ if $-l < y < -e_j$ and $e_j < y < l$. Hence obtain $Z_O(x, y)$ in the form of a series (retaining only the first two terms).

If the lines corresponding to $Z_O = 0.05$ and $Z_O = 0.95$ (which can be considered as the "flame limits") are then plotted in the plane (x, y) then the flame is clearly seen to "reclose" only if e_j/l lies below a certain value. Interpret this and show how it leads to a new formula for L_f/e_j a function of $e_j V_j/d$ and also of e_j/l. Discuss.

CHAPTER 4

SELF-IGNITIONS IN CLOSED SYSTEMS

4.1 GENERAL POINTS

Let us now turn our attention to systems in which no matter is exchanged with the surroundings, i.e. which are closed (or at least continue to be so until an explosion occurs). In addition, due to their practical importance, we will also consider systems which are very nearly closed and indeed behave more or less as though they were.

Consider a closed system in which chemical reactions occur which are exothermic when taken as a whole. After a certain period of time, the concentration, pressure and temperature conditions, in addition to the confinement conditions, i.e. the effectiveness with which the heat produced is removed, may result in the rate of reactions increasing abruptly. The mechanism involved in such a process relates to the not sufficiently effective removal of the heat produced; the temperature rises, usually resulting in the rates of reaction increasing sharply, hence an increase in the amount of heat released, and so on. The pressure also increases. In general, the process is accompanied by a short burst of light being emitted. The walls of the chamber containing the reacting medium may not be strong enough to contain the reaction and may break open, causing an "explosion" in the usual sense of the word. This term may also describe the final, violent phase of the reaction being studied. However, the use of this term can lead to ambiguity, since whether the walls give way or not depends on their strength in relation to the violence of the reactions. The potentially dramatic effects observed are thus dependent both on the chemical kinetics and on the strength of the containing vessel. It is also possible for the system to be closed yet not have any confining walls; this is the case for a solid. A further ambiguity is created by the fact that an explosion may occur in the absence of any chemical reaction, for example if the pressure in a cylinder of compressed gas rises above the pressure that the cylinder can withstand. For these reasons, the term "self-ignition" (sometimes known as "spontaneous ignition") is used to describe the phenomenon of abrupt increases in rates of reaction. However, this term implies that the phenomenon is self-induced, and that in some way a flame is produced. Clearly, specific conditions must be achieved, which may simply be a certain temperature, for the phenomenon to be induced. Moreover the word flame is often reserved for situations in which propagation occurs and where the phenomenon is self-sustaining. The term "inflammation" also alludes to the idea of a flame and, worse still, suggests ignition or lighting (by a spark, hot spot, match, compression etc.), which is not always the case (even though, one way or another, sufficient heat must be transferred to the mixture to attain the minimum initial temperature required for ignition). In the following text, indiscriminate use will be made of the words **explosion** and **self-ignition** (for want of more appropriate terms), although the basic mechanism could be called "reactive divergence".

This type of phenomenon may also occur in systems which are very nearly closed; such as a grain silo, coal depot, landfill site, haystack etc., and does not need to be initiated by an external source. For example, if a haystack is insufficiently ventilated (due to being too closely packed or damp), then the heat produced by fermentation reaction cannot be evacuated and a fire can break out. Were the haystack to be uniformly spread out over a field then an external source would be required to create the same effect. This simple example highlights the essential role played by geometry, and hence confinement, in triggering the phenomenon.

In industrial plant, a fire may break out both in confined or unconfined conditions.

In the first case, if the containing vessel is strong enough to withstand the fire, then the damage caused by the explosion will be limited. This was the case in 1979 at Three Miles Island 2 where, as a result of the mishandling of a minor problem, the zirconium alloy ducting around the reactor heated up and started to react with the cooling water to produce hydrogen; part of which was released into the confined area of the reactor. Some ten hours later, a transient pressure peak of approximately 190 kPa was produced, thought to be caused by the combustion of the H_2 which accounted for 8% by volume of the gas in the reactor at that time. Fortunately, the reactor chamber was designed to withstand pressures of up to 415 kPa (although the resistance of a material to a given pressure is less when this pressure is applied suddenly, and 8% of H_2 by volume was roughly the threshold value for producing a detectable pressure peak; in other words, the pressure could have been even greater).

Another cause of damage results from an incident or malfunction which releases a liquid or a "drifting cloud" of combustible material; a tragic example of which was Flixborough in 1974. In such cases, human activity near the site of the accident, even the passing of a car, is the most likely cause of ignition.

One of the principal dangers in coal mines is the spontaneous combustion of methane (firedamp) or also of coaldust. The dangers can be limited by ensuring sufficient ventilation or by reducing the amount of dust present by spraying with water.

Detailed analysis of such events reveals that initially, for a time period which may last for several milliseconds or a few months, nothing apparently happens: flames "smoulder" prior to the fire abruptly "breaking out". This situation is similar to the process involved in the spread of an infectious disease. Below the disease's proliferation threshold (when the situation is "endemic") a relatively small number of people are infected. The outbreak occurs when the disease becomes epidemic, i.e. when it exceeds the threshold and infects a large proportion of the population. The same type of equations can be used to describe the latter phenomenon since, during the incubation period, the micro-organisms proliferate without being sufficiently numerous to produce a detectable effect.

When considering an explosion instead of an epidemic, it is convenient to consider the propagating centres of any radical chain reactions or the heat released, which are both initially small but subsequently become much greater. Thus, at the start of a chain reaction, although propagating centres are created their numbers are so small that their presence has no notable consequences. However, before they disappear these radical centres may in turn generate new centres. If they appear more quickly than they disappear then their concentration will increase and vice-versa. As the concentration increases, a point will be reached when the increase accelerates rapidly, and the reaction changes suddenly from being slow (or even very slow) to fast or even very fast, with inflammation occurring.

Thus, in a hydrogen-oxygen mixture, the extent to which the OH radicals, which are always present, produce this effect depends on the conditions of the system. The same reasoning can be applied to the release of heat, which increases the temperature, hence increasing the rates of reaction, which in turn further increases the heat released and so on.

In all these cases, the resulting effect can itself become the cause and the phenomenon is sometimes described as "autocatalytic". The term is not particularly appropriate since a "real" catalyst should be unchanged at the end of the reaction, which is clearly not the case for propagating radical centres or for the release of heat.

4.2 TEMPERATURE DISTRIBUTION IN A CLOSED CHAMBER CONTAINING A COMBUSTIBLE MIXTURE

At the start of an explosive (or indeed of a slower, non-explosive) reaction, the heat released is not usually removed via the container walls at the same rate as it is produced. There is also a tendency for the temperature to rise; more quickly at the centre than close to the walls.

This effect has been clearly demonstrated in a study of the exothermic, potentially explosive decomposition of diethyl peroxide vapour $C_2H_5—O—O—C_2H_5$. Precise temperature measurements were made during this slow reaction along a vertical diameter through a spherical chamber with a radius of 6.5 cm, yielding temperature profiles and their evolution with time. The measurements are shown in Figures 4.1a and 4.1b. With an initial temperature of 203.7°C, two experiments were conducted with an initial pressure of

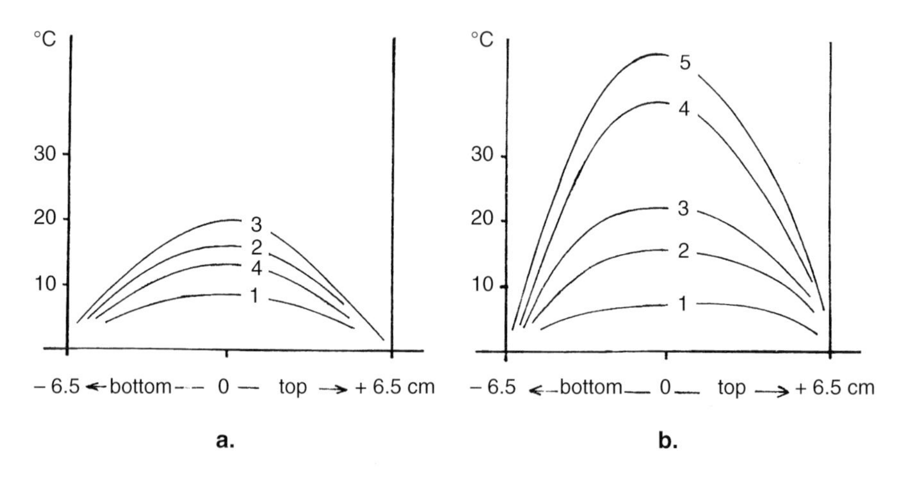

Figure 4.1 Temperature profiles during the exothermic decomposition of diethyl peroxide in a spherical cell of radius 6.5 cm, for an initial temperature of 203.7°C (after Fine, Gray and Mackinven (1968) *12th (International) Symposium on Combustion*, p. 545, The Combustion Institute).

a. Initial pressure: 1.1 torr.
Curve **1**: 0.250 s after the start of the reaction.
Curve **2**: 0.500 s after the start of the reaction.
Curve **3**: 1.250 s after the start of the reaction.
Curve **4**: 3.000 s after the start of the reaction.

The temperature at the centre reaches a maximum at conditions corresponding approximately to curve 3 (slightly afterwards according to the original paper). This maximum is not very high, and no explosion takes place.

b. Initial pressure: 1.4 torr.
Curve **1**: 0.250 s after the start of the reaction.
Curve **2**: 0.340 s after the start of the reaction.
Curve **3**: 0.400 s after the start of the reaction.
Curve **4**: 0.500 s after the start of the reaction.
Curve **5**: 0.537 s after the start of the reaction.

This time, the temperature is not observed to pass through a maximum and, furthermore, it rises much more rapidly. The slow, pre-explosive reaction culminates in an explosion.

1.1 and 1.4 torr respectively. At 1.1 torr, the reaction was non-explosive throughout. In contrast, at 1.4 torr, the reaction was slow initially but then became explosive. The curves in Figure 4.1b show the slow reactions prior to the explosion. The profiles are approximately parabolic in shape, and the temperature rise is almost zero next to the wall. The temperature maxima in Figure 4.1a increase (curves 1, 2, 3) before falling again (curve 4), whilst in Figure 4.1b the maxima are greater and rise continuously although the last curve recorded before explosion (curve 5) does not reach an excessively high temperature. The transition from a regime which is slow and non-explosive throughout, to one which is pre-explosive (Fig. 4.1b), can occur for only a small increase in initial pressure.

The same conclusions would be reached for other initial defining parameters and it seems that the dividing line between the two possibilities is very fine, there being a "catastrophic" aspect which we will consider in the following section.

4.3 EXPLOSION LIMITS

From the above discussion, there appears to be two distinct reaction possibilities, divided by a barrier so thin that it may be considered as having zero thickness, although in practice the situation is not so well defined. Once again a practical criterion is required, a condition which is present for one possibility but not the other. This condition is the appearance, after a so-called "induction delay" of duration τ, of a temperature peak associated with a pressure peak (caused by the rapid expansion) and a peak in the intensity of light emitted (mainly in the visible region, see Fig. 4.2). This emission can only be the result of a thermal population of higher quantum levels (cf. chapter 1, section 1.3) or by chemiluminescent chemical reactions yielding products in excited states (for chemiluminescence in combustion reactions see Caralp and Destriau[1]). Owing to collisional deactivation, which increases with pressure, this latter contribution is therefore always small unless the pressure is sufficiently low.

Based on these points, diagrams can be plotted, for a given reactor geometry, showing the parameter values such as pressure, temperature and initial mixture composition (so-called "limiting" values) which separate the explosive region from that where only slow reaction occurs (plots of pressure/temperature for a given composition, of pressure/composition for a given temperature etc.).

1. Caralp F. and Destriau M. (1984) *J. Chim. Phys.*, 81, p. 285.

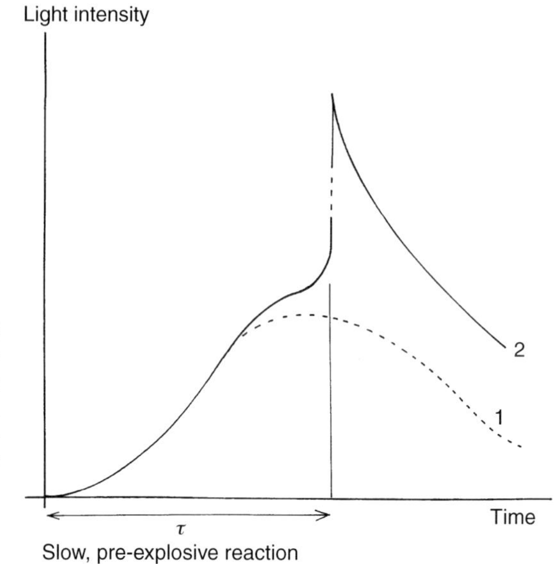

Figure 4.2 Variation in light intensity (arbitrary units) with time. Curve **1** (dashed line): slow reaction throughout. Curve **2** (solid line): slow, pre-explosive reaction followed by explosion.

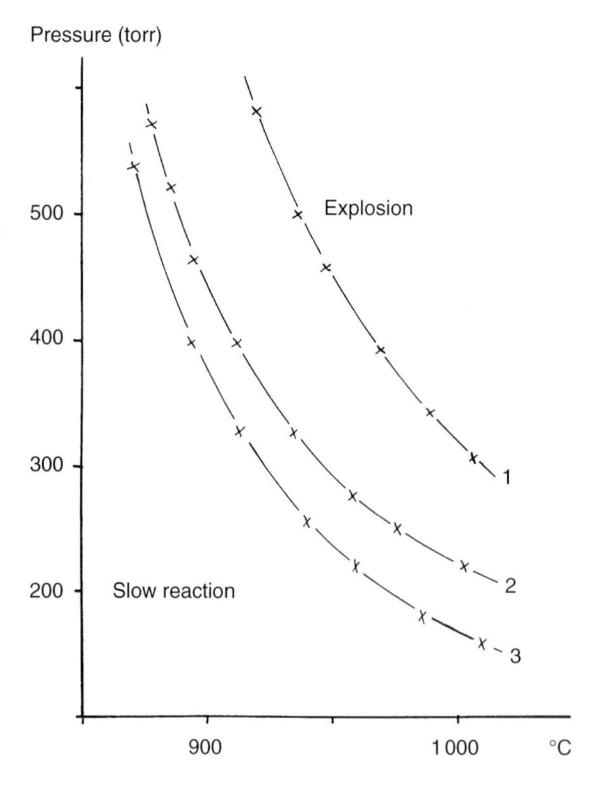

Figure 4.3 Variation of limiting pressure as a function of initial temperature for the decomposition of N_2O into nitrogen and oxygen (exothermic) in a vitreous silica cylindrical cell of internal radius: 9 mm (curve **1**), 15 mm (curve **2**), 20 mm (curve **3**) (after Bonnefois and Destriau (1970) *Bull. Soc. Chim.*, p. 2113).

Often, for a given mixture composition, the limiting pressure decreases as the temperature increases, as is the case for the decomposition (potentially explosive) of pure N_2O (Fig. 4.3). This is, however, far from being a general rule, since for H_2—N_2O mixtures, for example (Fig. 4.4), there is a whole range of temperatures for which the limiting pressure and temperature vary in the same direction.

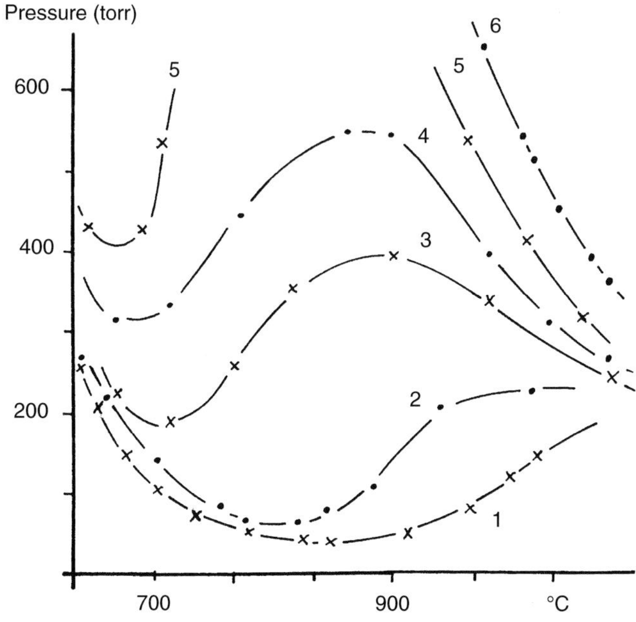

Figure 4.4 Variation of limiting pressure as a function of initial temperature for H_2—N_2O mixtures of different composition, in a cylindrical cell of internal radius 9 mm. Explosion occurs when the pressure exceeds the limiting pressure (after Navailles and Destriau (1968) *Bull. Soc. Chim.*, p. 2295).

Number of moles of H_2 / Number of moles of N_2O	1/7	1/10	1/15	1/18	1/20	Pure N_2O
Curve	1	2	3	4	5	6

For hydrocarbons, “lobes” appear in these plots (Fig. 4.5) such that on the low-temperature side there is a “cool” flame (which releases little heat) whilst on the higher temperature side there is a “long-delay” flame, which has a very long induction delay. Cool flames are involved in the phenomenon known as “knocking” which can occur in spark-ignition engines when, during the compression cycle, a cool flame ignites the fuel, causing “normal” combustion ahead of the ignition-initiated combustion. There is strong evidence to suggest that all akanes can produce cool flames. However, they are more difficult to produce with methane than with the higher members of the same family

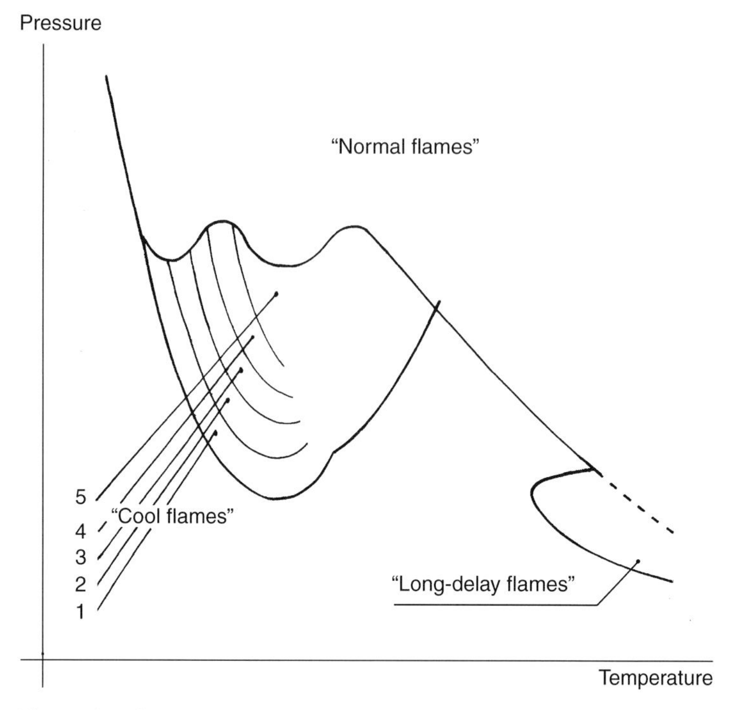

Figure 4.5 General form of the variation of limiting pressure as a function of initial temperature for hydrocarbon-air (or oxygen) mixtures.

1. Region with 1 cool flame.
2. Region with 2 cool flames.
3. Region with 3 cool flames.
4. Region with 4 cool flames.
5. Region with 5 cool flames.

The use of the word "flame" in this figure is not strictly correct (see section 1 of this chapter).

because the flame limits are narrower, and they burn at higher temperatures. As a consequence, which is a positive point, the risk of knocking is less with methane[2]. It is also possible to have three pressure limits for one temperature, as is the case for mixtures of O_2 with H_2, CO, COS, CS_2, H_2S, NH_3, P, and PH_3 (Fig. 4.6). This subject is covered in depth in the book by Lewis and Von Elbe[3].

It can also be shown that for a given pressure and composition, the limiting temperature varies with the diameter of the reaction cell (Figs 4.3 and 4.7), or more generally, with the cell geometry, owing to the heat losses over the cell's surface.

2. Vanpée M. (1993) *Combust. Sci. Tech.*, 93, p. 363.
3. Lewis B. and Von Elbe G. (1987) *Combustion, Flames and Explosion of Gases*, 3rd edition, Academic Press, New York.

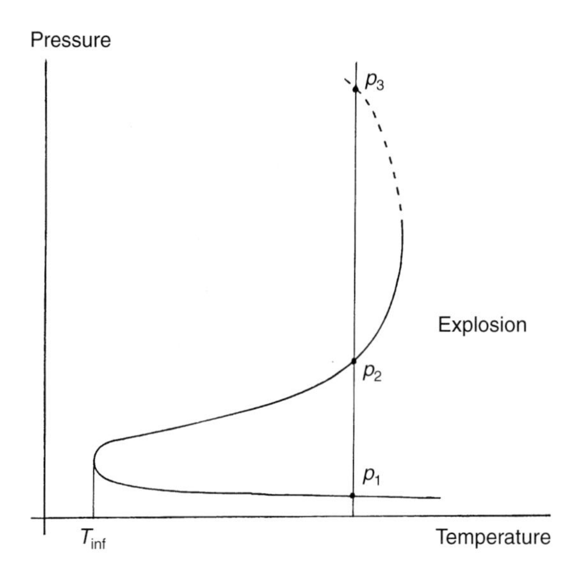

Figure 4.6 General form of a diagram showing three explosion pressure limits p_1, p_2, p_3.

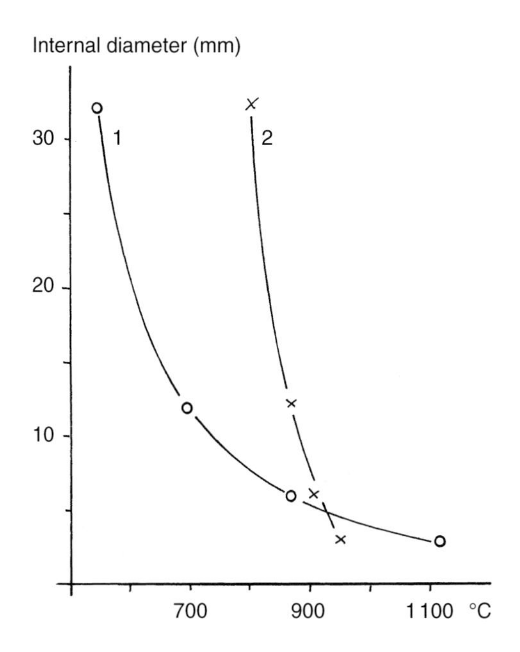

Figure 4.7 Variation of the initial limiting temperature with reactor diameter for a given initial pressure (1 bar).
Curve **1**: Stoichiometric mixture of CH_4—O_2.
Curve **2**: Stoichiometric mixture of H_2—air.

From the pressure/temperature limits for different compositions can be derived the pressure/composition limits at a given temperature (Fig. 4.8). Clearly, these limits will vary depending on the source of ignition, for example the values for ignition by a spark will differ from those by a hot spot source, etc. Moreover, these limits are unrelated to practical concepts such as the "flash point"; the minimum temperature required to ignite the vapour above a liquid at normal atmospheric pressure using an "auxiliary" flame. The flash point may not be hot enough to allow the flame produced to sustain itself and propagate (cf. chapter on deflagration); the temperature at which this occurs is in fact referred to as the "fire point". Usually, the fire point is a few degrees above the flash point.

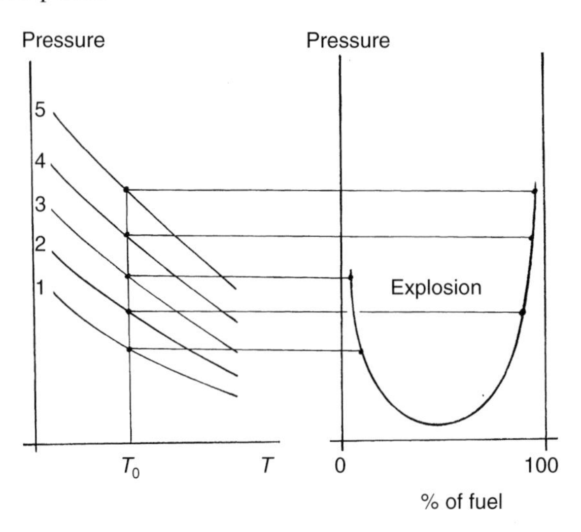

Figure 4.8 Schematic representation of the conversion, for a given cell geometry, from a pressure/temperature diagram for explosion at a given composition, to the corresponding pressure/composition diagram for explosion at the given initial temperature T_0. Curves 1, 2, 3, 4 and 5 correspond to different compositions.

As a general rule, explosion temperatures at atmospheric pressure are slightly higher in air than in pure oxygen, and much higher for ammonia than for hydrazine. This temperature decreases steadily on passing from methane to the decanes and from ethene to 1- and 2-butene, and is greatly lowered on passing from ethene to acetylene. The temperature for benzene, on the contrary, is high. These trends reflect what is already known about the reactivity trends for these types of molecules, the limiting temperatures being lowest for the most reactive compounds.

For flammable liquids such as petrol, kerosene, or benzene, the limiting temperatures for explosion can also, perhaps not surprisingly, depend upon the whether the liquid is introduced by spraying or by vaporisation.

4.4 OCTANE NUMBER

Various ratings are used to define fuel quality. One of the most well-known, the "octane number", is related to the "knocking" (or "pinking") effect. Knock refers to an undesirable pre-ignition effect to which spark-ignition engines are susceptible and involves a fraction of the fuel mixture contained in the cylinder head. The effect is initiated by localised high temperatures during the compression phase prior to "true" ignition by the spark plugs. Knocking can be considered as being an untimely diesel-type ignition which reduces the engine's efficiency and causes excess wear.

Octane number is defined by comparison with two reference compounds, namely a "linear" chain alkane, *n*-heptane CH_3—$(CH_2)_5$—CH_3 a fuel prone to knock, and the highly "branched" alkane, 2,2,4-trimethylpentane (or *n*-octane) a fuel which is resistant to knock. The first fuel, if burnt alone, would score an octane number of 0 and the second would score 100. Octane number is determined in the following way. Under identical conditions, an engine, ideally one with a constant compression ratio, and constructed to precise standards, is run first on a petrol fuel with a complex composition and unknown octane number, and then on an *n*-heptane and iso-octane blend whose percentage composition is known.

The octane number of the petrol being studied is given as the percentage of iso-octane in the *n*-heptane/iso-octane blend which produces the equivalent knocking intensity.

A petrol which produces less knocking than the iso-octane thus has an octane number greater than 100, and would therefore fall outside the 0–100 range. Furthermore, a petrol may have a good octane number without actually containing any octane. Higher octane numbers allow the engine to be designed with a higher compression ratio, resulting in improved engine efficiency.

Octane number can be improved by using additives or "anti-knock" agents, the most well-known being tetraethyl lead, $Pb(C_2H_5)_4$. Anti-knock agents should not be confused with "antioxidants", the latter being additives used to slow down the oxidation of various compounds, and not exclusively alkanes.

It is found that for the same number of atoms and under the same conditions, the more "branched" alkanes produce less knocking than those with less branching and hence have a higher octane number. The kinetic mechanism proposed by Westbrook and Pitz in 1986 to model this effect is mainly based on the thermal decomposition of the fuel (represented as RH) into H atoms and $R^{\bullet}$; the alkyl radicals can combine with O_2 to give $RO_2^{\bullet}$ or in turn decompose to give smaller radicals. The advantage of this model is that it provides a good prediction of the relationship between octane number and the various alkanes with different degrees of branching. However, Baronnet, Simon and Scacchi showed in 1989 that the model predicts, for decomposition in the absence of oxygen, a correlation between octane number and the amount of hydrogen produced. This result has not been verified.

4.5 EXPLOSION THEORIES

In order to be able to construct a mathematical model of an explosion, let us return to the analysis described in section 4.1 above. The chemical mechanism chosen may be a chain reaction mechanism with an unspecified degree of branching and with no temperature variation prior to explosion: the isothermal chain mechanism. For this mechanism to operate, the heat released must be removed continually such that its rate of removal through the walls is equal to its rate of production in the chamber. This can be considered as being a limiting case, which in practice is difficult to achieve, but which can be approximated by diluting the combustible mixture in a bath gas which does not undergo a chemical reaction. Under the conditions produced, the temperature of the bath gas must be controlled to ensure that it is the same as the reactive mixture containing the propagating radical centres, i.e. thermal equilibrium is essential (see chapter 1, section 1.3).

A reaction can also be envisaged which is chain branching in nature but athermal, in which case isothermal conditions can be easily provided. Thus, after a given time, the number of propagating radical centres can be thought of as increasing sharply without any variation in temperature, although it is not easy to see exactly how this is achieved. In this case, the event has the "catastrophic" nature mentioned earlier in this chapter. This case will not therefore be considered in detail here.

Having said this, many explosive (hence exothermic) reactions are chain branching, although there are very probably others which are not chain reactions of any description. In both cases, a convenient approach is to consider the mass and thermal balances. Let us start by considering the case of a globally exothermic reaction (which may or may not be a chain reaction) in order to investigate the thermal theory of explosion, i.e. by analysing heat exchanges in particular.

Unless every point in the reaction mixture is at the same state and hence at the same temperature, (which would appear to be impossible), heat must be exchanged within the mixture. On the other hand, the extent to which the wall can transfer heat depends upon its nature or even thickness. In general, the faster the chemical reactions in the cell, the more the walls can be considered as being adiabatic. In the following analysis, these limiting cases will be considered even though they are not always encountered in practice. The systems will still be considered as closed in the sense that no matter passes between the reacting medium and its surroundings. Moreover, it will be assumed that any light emission need not be taken into account. The first attempt at providing a theoretical description of systems of this type was probably made by Frank-Kamenetskii[4].

4.5.1 Thermal theory of explosions

4.5.1.1 Balance equations and critical conditions for a non-convective system

Assuming that no motion in the reacting medium occurs (and consequently that the density does not vary with temperature) and that heat transport is by diffusion only, the fundamental equation (solved for various geometry by Frank-Kamenetskii) is the approximate thermal balance equation (cf. chapter 3):

$$\rho c_V \frac{\partial T}{\partial t} = \mathrm{div}\,(\lambda\ \overrightarrow{\mathrm{grad}\ T}) + qw \tag{4.1}$$

which for one-dimensional systems is:

$$\rho c_V \frac{\partial T}{\partial t} = \frac{\partial}{\partial x}\left(\lambda \frac{\partial T}{\partial x}\right) + qw \tag{4.1'}$$

- The product qw is the source term which reflects the contribution to the thermal balance of the heat evolved by reaction in the cell per unit volume per unit time.
- In this product, w is the rate of reaction defined in equation (2.3′) of chapter 2, which assumes that only one chemical reaction occurs:

$$a\mathrm{A} + b\mathrm{B} \rightarrow a'\mathrm{A}' + b'\mathrm{B}' + \ldots$$

$$w = -\frac{w_\mathrm{A}}{a} = -\frac{w_\mathrm{B}}{b} = +\frac{w_{\mathrm{A}'}}{a'} = \ldots$$

  where $w_{\mathrm{A}'}$ w_B, $w_{\mathrm{A}'}$ etc. are the molar rates of reaction of A, B, A′, etc. Thus w is a positive quantity.
- The definition of q originates from those of w and of the product qw. Note that if the initial mixture contains, per unit volume, not a but x moles of A

4. Frank-Kamenetskii D.A. (1955) *Diffusion and Heat Exchange in Chemical Kinetics*, Princeton University Press.

and not b but y moles of B, and if y/x is smaller than b/a, then the reactant B (oxidant or fuel) is said to be in "deficit", and then even if the reaction is complete and hence does not lead to equilibrium, A cannot react completely. Under these conditions, the heat released chemically is limited by the total number of moles of B, i.e. y moles per unit volume (y is therefore a molar concentration), and cannot in total exceed, from the beginning to the end of the reaction, the value $(y/b)q$, where q is the quantity produced from the reaction of b moles of B with a moles of A. This corresponds to $-(q/b)(\mathrm{d}y/\mathrm{d}t)$ per unit time, a positive quantity which is in fact equal to qw. This said, even though the reaction occurs in a fixed volume, q is not necessarily equal to Δu of reaction since q is positive while Δu is negative and, furthermore, the final and initial temperatures are not the same (see section 1.2.1 of chapter 1). In fact, q is smaller in magnitude than $-\Delta u$, due to the fact that, under these conditions, the combustion reaction is far from being complete. It goes without saying that this effect is less important during the slow, pre-explosive part of a combustion reaction (see the second worked example at the end of this chapter).

- λ is the thermal conductivity of the medium,
- T its temperature,
- c_v its mean heat capacity at constant volume,
- ρ is either the density of the mixture (if c_v is its specific heat capacity) or the number of moles of all the species present per unit volume (if c_v is the molar heat capacity of the mixture).

ρ and c_v are assumed to be constants. Strictly speaking, each term in the thermal balance equation should be measured in the SI units of $\mathrm{J{\cdot}m^{-3}{\cdot}s^{-1}}$.

Equation (4.1) implies a temperature distribution within the reaction mixture and results from the fact that the heat produced by the reaction over a given period (the product qw) is consumed during the same period, being removed in part through increasing the temperature of the mixture (given by the term $\rho c_v(\partial T/\partial t)$, and also by the diffusion of hotter regions towards cooler regions, as given by the term $\mathrm{div}(\lambda\, \overrightarrow{\mathrm{grad}\, T})$.

Using equation (4.1), it is interesting to consider the conditions under which a steady-state situation can arise, i.e. under which conditions $\partial T/\partial t$ becomes zero for a temperature T which then depends only on x (this of course being impossible during the explosive phase). If, in addition, λ is constant, then equation (4.1) becomes, with $\partial T/\partial t$ zero:

$$\lambda \frac{\mathrm{d}^2 T}{\mathrm{d}x^2} + qw(T) = 0 \tag{4.1''}$$

Take the one-dimensional case described below where the reacting mixture is considered as being contained between two plane, parallel walls. Let $x = 0$ correspond to a hot wall at a temperature T_{hot}, which we are free to choose, and $x = r$ to a cold wall at a temperature T_{cold}, slightly different from T_{hot}. The effect of these walls is to impose the boundary conditions for the mixture. The temperature gradient in the gas mixture is dependent on these boundary conditions and on the reaction. Taking the volume between the internal face of the hot wall and a plane parallel to it, located at a distance x from it, and at which $T = T(x)$, gives upon a first integration of (4.1″):

$$\frac{dT}{dx} - \left(\frac{dT}{dx}\right)_{x=0} = \left(\frac{2q}{\lambda}\int_T^{T_{\text{hot}}} w(T)\,dT\right)^{1/2} \tag{4.2}$$

in which case, if the rate of reaction obeys Arrhenius' law:

$$w(T) = f(c_i)\exp\left(-\frac{E}{RT}\right)$$

where $f(c_i)$, which is dependent on the nature of the mixture, represents the dependence of w on the concentrations c_i.

These concentrations differ little from the initial concentrations when, as is often the case, the consumption of the combustible mixture is small during the slow, pre-explosive reaction. If an explosion occurs, then it should be close to the hot wall, hence in a plane where the difference $(T_{\text{hot}} - T)$ is much smaller than T_{hot}. Based on this assumption, it can be approximately stated that:

$$\exp\left(-\frac{E}{RT}\right) \simeq \exp\left(-\frac{E}{RT_{\text{hot}}}\right)\exp\left[-\left(\frac{E}{RT_{\text{hot}}^2}\right)(T_{\text{hot}} - T)\right] \tag{4.3}$$

With these two approximations, the integral of (4.2) can be calculated analytically and thus the integration of (4.2) will yield $T(x)$. Finally $T(x)$ is known with certainty once the constant $(dT/dx)_{x=0}$ is calculated by expressing the second condition of the problem at the limit where $T(r) = T_{\text{cold}}$.

Once this has been done, the conditions required for an explosion to occur can be investigated at any distance x from the hot wall. This is equivalent to exploring the question of whether a steady-state value $T(x)$ given by the solution to (4.2) can exist (no explosion) or not (explosion). Now the equation yielding $(dT/dx)_{x=0}$ has two different solutions (one stable, the other unstable) if the dimensionless quantity:

$$\delta = \frac{qEr^2 f(c_i)\exp\left(\dfrac{-E}{RT_{\text{hot}}}\right)}{\lambda R T_{\text{hot}}^2} \tag{4.4}$$

is less than 0.88 (or less than 2 for a cylindrical cell of infinite length, or 3.32 for a spherical cell, where r denotes the radius of the cylindrical or spherical cell). If the dimensionless quantity is greater than 0.88, then the equation predicts that there can be no steady state and there is an explosion. The critical condition requires that at a given T_{hot}, if (4.4) gives an r value greater than that of the cell, then an explosion cannot occur at that temperature in the cell.

In order to interpret what actually happens, consider the temperature difference ($T_{hot} - T_{cold}$). If T_{hot} is slightly greater than T_{cold}, then since the temperature of the mixture at the internal face of the hot wall is higher, the rate of reaction will increase in this region, as will the chemically-released heat. This heat will diffuse throughout the entire volume under consideration and gradually the temperature, rate of reaction, and quantity of heat released will all increase throughout this volume. At a specific moment, the temperature of the gas near to the hot wall will be virtually the same as that of the wall, which is the case where $(\mathrm{d}T/\mathrm{d}x)_{x=0} = 0$. If the temperature at the hot wall is just sufficient to allow steady state to be reached, then the chemically-released heat will flow in the direction of decreasing temperature, i.e. towards the cold wall and the exterior beyond. If the temperature increases further still, the heat transfer will not be quick enough to maintain steady state conditions and an explosion will occur. For the limiting case, by setting $(\mathrm{d}T/\mathrm{d}x)_{x=0} = 0$ in equation (4.2) and with approximation (4.3), integration gives:

$$\left(\frac{\mathrm{d}T}{\mathrm{d}x}\right)^2_{\text{critical}} = \frac{2q}{\lambda} f(c_i) \exp\left(-\frac{E}{RT_{hot}}\right) \int_T^{T_{hot}} \exp\left(E\frac{T - T_{hot}}{RT^2_{hot}}\right) \mathrm{d}T$$

$$= \frac{2q}{\lambda} f(c_i) \exp\left(-\frac{E}{RT_{hot}}\right) \frac{RT^2_{hot}}{E} \left(1 - \exp\left(E\frac{T - T_{hot}}{RT^2_{hot}}\right)\right)$$

which by using:

$$\left(\frac{\mathrm{d}T}{\mathrm{d}x}\right)_{\text{critical}} \simeq \left(\frac{T_{hot} - T_{cold}}{r}\right)_{\text{critical}}$$

and:

$$\exp\left(E\frac{T - T_{hot}}{RT^2_{hot}}\right) \simeq 0$$

$$\left(\frac{T_{hot} - T_{cold}}{r}\right)^2_{\text{critical}} \simeq \frac{2q}{\lambda} \frac{w(T_{hot})RT^2_{hot}}{E} \tag{4.5}$$

and incorporating condition (4.4), yields:

$$T_{hot} - T_{cold} \simeq 1.33 \frac{RT^2_{hot}}{E}$$

By applying the same theoretical principles, the phenomenon of pre-explosive heating may also be addressed, i.e. the rise in temperature caused by the slow pre-explosive reaction from the initial temperature T_0 of the mixture up to that at which the explosion is triggered. It is found:

$1.20\ RT_0^2/E$ for a cell with planar, parallel faces,
$1.37\ RT_0^2/E$ for a cylindrical cell,
$1.60\ RT_0^2/E$ for a spherical cell.

4.5.1.2 Balance equations and critical conditions when the cell temperature is equalised by convection (for diathermal walls)

The term – div ($\lambda\ \overrightarrow{\text{grad}\ T}$) in equation (4.1), which is positive, can be replaced by the term $\beta A(T - T_0)/V$ (also positive) so long as the temperature T is substantially constant throughout the reacting medium, since the drop in temperature occurs almost exclusively immediately next to the wall, which is at a temperature T_0. This was not the case for the experiment from which Figure 4.1 was plotted. Hence, T_0 is also approximately equal to the initial temperature, β is a transport coefficient (in cal or $J{\cdot}cm^{-2}{\cdot}K^{-1}{\cdot}s^{-1}$), A is the surface area of the wall, and V the volume of the chamber (for this reason the ratio A/V depends upon the geometry, and is equal to $2/r$ for a cylinder of infinite length, and $3/r$ for a sphere). For these reasons an expression for Newton's rather than Fourier's Law is produced. If the combustible mixture is a solid, then no convection currents can be created, however, if the solid is a good heat conductor (and even if the solid is heterogeneous) the temperature T produced will still be approximately the same throughout. The thermal balance equation becomes:

$$qw = \beta \frac{A}{V}(T - T_0) + \rho c \frac{\partial T}{\partial t} \qquad (4.6)$$

In Figure 4.9 this equation is represented by the equality:

$$\overline{IC} = \overline{IB} + \overline{IA}$$

for any time t.

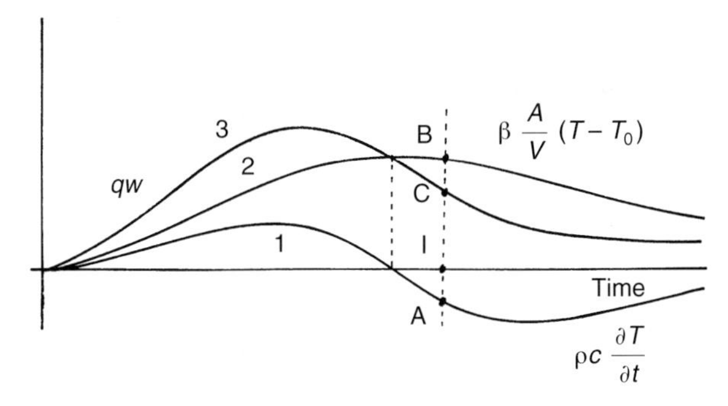

Figure 4.9 Graphical representation of the thermal balance (see equation 4.6).

Now consider the conditions under which T, and hence $\beta A(T - T_0)/V$, passes through the maximum (when $T = T_M$.) At this point the product qw has started to decrease and $\partial T/\partial t$ is zero. If, furthermore, it is assumed that $w = f(c_i) \exp(-E/RT)$, then equation (4.6) reduces to equation (4.7):

$$q f(c_i) \exp\left(-\frac{E}{RT_M}\right) = \frac{\beta A}{V}(T_M - T_0) \tag{4.7}$$

thus:

$$h(T_M) = g(T_M) \tag{4.8}$$

For a given pressure and concentration, h increases with temperature as $\exp(-E/RT)$ (see chapter 2, section 2.5.2 and Figures 2.6a and 2.6b) and g increases linearly. The following three situations can arise (see also Fig. 4.10 and 4.11):

a) h and g intersect at A, B and C (Fig. 4.10): the temperature rises from T_0 to T_A, and decreases beyond T_A, with g becoming greater than h, which leads to a drop in temperature. The representational point for the system becomes A again, which is a stationary point.

b) h and g intersect at only one point, point D, (Fig. 4.10), which is at a very high temperature (see chapter 2, section 2.5.2). This point has virtually no physical significance and there is no real point in attempting to verify if Arrhenius' law still holds at this temperature. All that can be said is that the temperature rises to a very high level, as does the rate of reaction, leading to the mixture exploding without it being possible to determine at what temperature the transition from the pre-explosive to explosive regime occurred. There is apparently a degree of continuity between the two regimes, in contrast to the conclusion of the discussion concerning explosion limits in section 4.3.

c) h is a tangent to g at L (Fig. 4.11). This is the limiting case and a value can be determined for the overheating limit ($T_{M,\,\text{limit}} - T_0$). In fact at L, besides equation (4.8) $h = g$, the gradients are equal, thus:

$$\beta \frac{A}{V} = q f(c_i)\left(\frac{E}{RT^2_{M,\,\text{limit}}}\right) \exp -\left(\frac{E}{RT_{M,\,\text{limit}}}\right) \tag{4.9}$$

Taking the ratio of (4.7) to (4.9) term by term, yields:

$$T_{M,\,\text{limit}} - T_0 = \frac{RT^2_{M,\,\text{limit}}}{E} \tag{4.10}$$

a quadratic equation in $T_{M,\,\text{limit}}$. Since T_M, is often only very slightly higher than T_0, the following approximate solution can be derived:

$$T_{M,\,\text{limit}} - T_0 \simeq \frac{RT_0^2}{E} \tag{4.11}$$

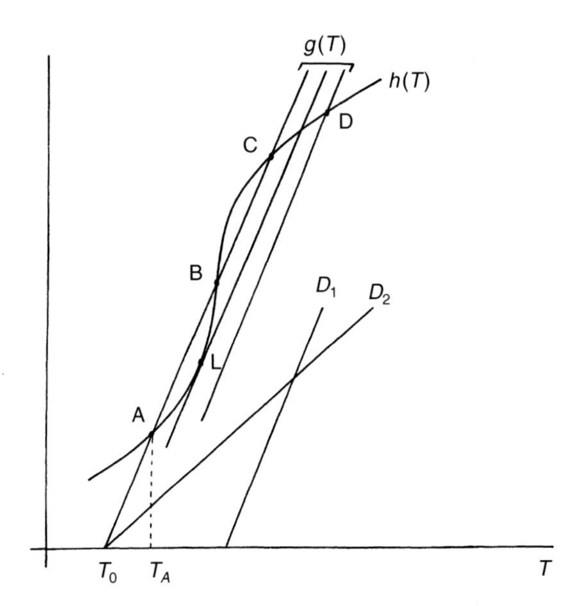

Figure 4.10 Influence of initial temperature on the triggering of an explosion, based on the thermal balance equation (4.6).

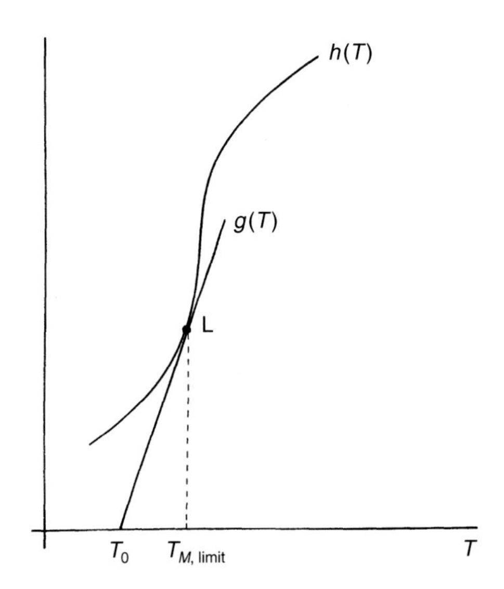

Figure 4.11 Pre-explosive overheating at the explosion limit ($T_{M,\ \mathrm{limit}} - T_0$). Point L in this figure is the same as point L in Figure 10.

giving, for example, 10 K for T_0 = 440 K and E/R = 20 000 K. This overheating does not therefore depend on the reactor geometry, unlike the above result based on equation (4.1). If the initial temperature is only slightly above that corresponding to the situation in Figure 4.11 then an explosion will occur, with the temperature increasing from T_0 to T_M. At $T = T_M$ the change in temperature passes through a point of inflexion, with a horizontal tangent, before rising sharply.

If the wall temperature is raised, then the line ABC in Figure 4.10 is translated to the right, to D_1 for example, which is then in the explosive region. If the slope of the line is reduced, i.e. $\beta A/V$, without changing T_0, it can be moved to D_2, for example, which is also in the explosive region.

It may be that the rate of reaction, w, does not follow an Arrhenius-type relationship but instead varies with temperature in a more complex way, as does the product $qw = h(T)$ which features in the thermal balance equation (4.7). Consider the case represented by Figure 4.12a where, for a temperature range between points M and N, $h(T)$ decreases if the temperature increases, in a situation known as "negative temperature dependence". This is apparently the case for explosions involving H_2—N_2O mixtures and for cool flames. So long as the initial temperature is lower than that at point F then there will be a slow, pre-explosive region represented by the area indicated by a minus (–) sign in Figure 4.12b. At a point such as G, between F and H, a large rise in temperature occurs from that at point G to that at point J, and hence in practice an explosion occurs in a similar way to the explosion at point D in Figure 4.10. This explosion is shown by a plus (+) sign in Figure 4.12b. Beyond a point such as H the situation once again resembles that at point E, i.e. the explosive regime can no longer exist. It is, however, encountered again when the point I is passed through again. Experimentally, it is often quite difficult to determine point H accurately, since the boundary between the non-explosive and explosive regimes is poorly defined. In fact, the transition from point G to point H is quasi-continuous and in Figure 4.12b the poorly-defined point H is represented by a shaded area to signify that the boundary between the two domains is not well defined.

4.5.1.3 Temperature and concentration profiles in the case of a closed system in the absence of convection

Writing equation (4.1′) again, but without a convective term and for a one-dimensional system, gives:

$$qw = -\frac{\partial}{\partial x}\left(\lambda \frac{\partial T}{\partial x}\right) + \rho c \frac{\partial T}{\partial t} \tag{4.1'}$$

and for every species i, a mass balance equation of the form:

$$w_i = -\frac{\partial}{\partial x}\left(\rho d_i \frac{\partial Y_i}{\partial x}\right) + \rho \frac{\partial Y_i}{\partial t} \tag{4.12}$$

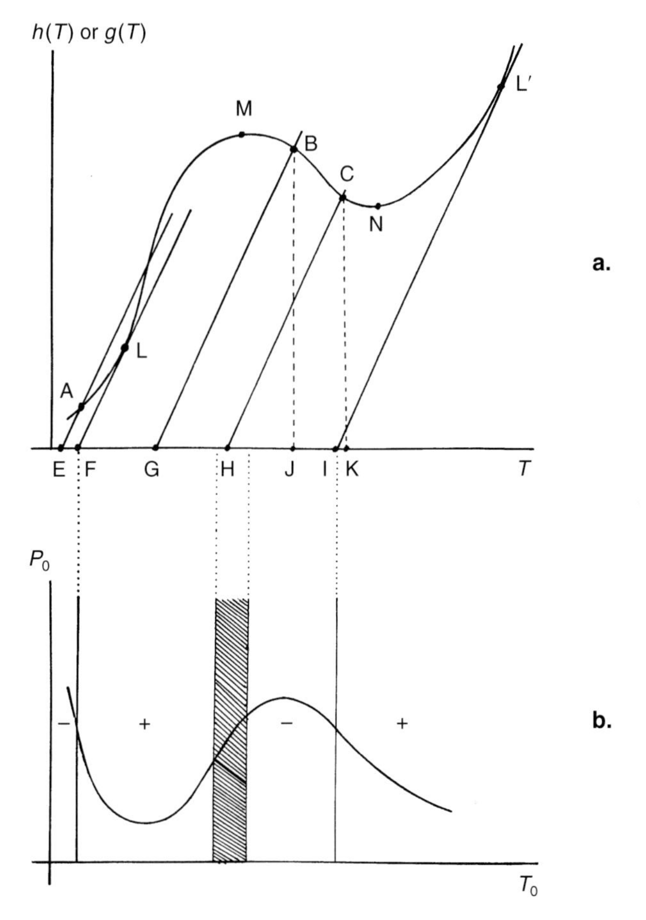

Figure 4.12 Explosive region in the case of "negative temperature dependency".
a. Adaptation of Figure 4.10 for this case.
b. Resulting explosive region. A temperature range appears within which the limiting pressure increases as the initial temperature is increased.

where q, w, λ, c and ρ have the same meaning as before, w_i denotes the rate of reaction for each species i, d_i its coefficient of diffusion and Y_i its mass fraction (for mass-based parameters in (4.1′) et (4.12)). The boundary conditions are now:

- the system is nearly isolated, the walls allow no exchange of matter or heat,
- at $t = 0$ and at the centre (chosen as the origin, $x = 0$), $T = T_0$ and $Y_i = Y_{i,0}$,
- moreover by assuming a single chemical reaction, $w_i = \alpha_i w$ can be written with (according to equation (2.3′) from chapter 2) $\alpha_i = \nu_i$, where ν_i is the

stoichiometric coefficient for constituent i (with a change of sign for molecules on the reactants side). Hence equation (4.1′) can be written:

$$\frac{w}{\rho} = -\frac{\partial}{\partial x}\left(\frac{\lambda}{\rho c}\frac{\partial}{\partial x}\left(\frac{c(T-T_0)}{q}\right)\right) + \frac{\partial}{\partial t}\left(c\left(\frac{T-T_0}{q}\right)\right)$$

and (4.12):

$$\frac{w}{\rho} = -\frac{\partial}{\partial x}\left(d_i\frac{\partial}{\partial x}\left(\frac{Y_i - Y_{i,0}}{\alpha_i}\right)\right) + \frac{\partial}{\partial t}\left(\frac{Y_i - Y_{i,0}}{\alpha_i}\right)$$

A number analogous to the Lewis number is introduced, but with c replacing c_p, hence $\lambda/\rho c d_i$. If it is taken to be equal to 1 also, then the terms:

$$\frac{c\,[T(x,t)-T_0]}{q} \quad \text{and} \quad \frac{a_i}{\alpha_i}[(Y_i(x,t)-Y_{i,0}] + b_i$$

where a_i and b_i are constants, satisfy both the first and second equation. If Z is used to denote the common factor in these two terms, (4.1′) and (4.12) reduce to a single equation:

$$\frac{w}{\rho} = -\frac{\partial}{\partial x}\left(\frac{\lambda}{\rho c}\frac{\partial Z}{\partial x}\right) + \frac{\partial Z}{\partial t} \tag{4.13}$$

The boundary conditions require that b_i must be zero. Therefore, if initially $T(x)$ and $Y_i(x)$, i.e. the temperature and concentration profiles have the same form, then they will continue to have the same form at any subsequent time t, thus defining the constant a_i.

4.5.2 Theory of isothermal chain branching

Thermal theory is not able to predict "peninsular-shaped" domains such as those shown in Figure 4.6. It is therefore necessary to apply the theory of chain reactions.

Let us begin by considering the simple, albeit rather unrealistic case of chain branching under strictly isothermal conditions. The first step is that of initiation, with a rate I_0, giving the first propagating centres X. These then react with the initial molecules I to produce final, stable molecules F and more propagating centres X, in accordance with the overall bimolecular process:

$$\mathrm{I} + \mathrm{X} \rightarrow \mathrm{F} + y\mathrm{X} \quad \text{where } y \text{ is positive} \tag{4.14}$$

This is a branching process if y is greater than 1 and, if a branching step involves the decomposition of a relatively stable product, e.g. in a step of the form P → X + ..., the chain is referred to as having undergone "degenerate-branching" (a term already defined in section 2.8.4 of chapter 2). The net branching factor is given by $\phi = (y - 1)/\delta t$, where δt is the average lifetime of the radical centres, of concentration [X]:

$$\frac{\mathrm{d[X]}}{\mathrm{d}t} = I_0 + \phi[\mathrm{X}] \tag{4.15}$$

giving:

$$[\mathrm{X}] = \frac{I_0(e^{\phi t} - 1)}{\phi} \qquad \text{if} \qquad [\mathrm{X}] = 0 \text{ at } t = 0 \tag{4.16}$$

This relationship predicts, if ϕ is positive, that [X] should increase indefinitely with t and that there will be an explosion if ϕ remains constant, or at least does not decrease, during the period leading up to the triggering of the explosion. On the contrary, if ϕ is negative, [X] tends to the limiting value $-I_0/\phi$ (a positive quantity) and there is no explosion. In this approach the limiting condition is $\phi = 0$, a condition which is independent of I_0.

However the presence of chain-terminating reactions must lead to a reduction in ϕ, which, to a first approximation, can be represented by an equation of the form:

$$\phi = (y - 1)k_b - k_{\mathrm{termination}} \tag{4.17}$$

where k_b (= $1/\delta t$) is the rate constant of the branching reaction, $k_{\mathrm{termination}}$ is an overall correction factor for ϕ, either for termination in the gas phase arising from termolecular recombination:

$$\mathrm{X + X + M \rightarrow X_2 + M}, \tag{4.18}$$

where M is the "third body" (see section 2.6.2 of chapter 2), or from the complex process involving:

- diffusion of the chain-carrying centres towards the wall,
- followed by their deactivation at the wall.

$k_{\mathrm{termination}}$ thus depends in a complex way, in a first approximation, on the smallest of the rate constants k_i of all the processes (uni-, bi- or termolecular), hence on the temperature, T, and on the pressure of the third body and thus on the total pressure, p:

$$k_{\mathrm{termination}} = F(T, p)$$

When expressed in this way, $F(T, p)$ is analogous to a global rate constant for chain termination. However, since the various loss processes for the centres are in some cases termolecular, and in others not, they cannot strictly speaking all be considered together. F does not therefore only depend on T

and p but also on [X]. This effect is neglected insofar as [X] also depends on p and T and since, at a given p and T, d[X]/dt can be considered as being close to zero by the SSA approximation (see section 2.4 of chapter 2). Now let us assume that the rate of formation of propagating centres is given by a product of the form:

$$f([X])(y-1)k_b$$

where $f([X])$ is a function increasing with [X] and the rate of their disappearance is given by a product of the form:

$$g([X])k_{\text{termination}} = g([X])F(T,p)$$

To a first approximation, an explosion should occur if the rate of formation of propagating centres is greater than their rate of disappearance, i.e. for:

$$f([X])(y-1)k_b > g([X])F(T,p) \tag{4.19}$$

At low pressures, where diffusion is faster, termination at the walls predominates over that in the gas phase (and conversely at high pressure). Normally, between the first and second limits, p_1 and p_2, for mixtures which produce the "peninsular" region shown in Figure 4.6, the pressure is high enough to prevent a sufficient number of radicals from recombining at the walls, yet low enough to prevent a sufficient number from recombining in the gas phase. If, at a temperature T, the pressure corresponding to the first condition, p_2, is lower than the pressure p_1 corresponding to the second, then the mixture cannot explode. The temperature at which $p_1 = p_2$ is the temperature T_{inf} in Figure 4.6. As might be expected, the pressure p_1 is highly dependent on the nature of the wall.

The third limit, p_3, is usually considered as being principally governed by thermal theory, although there is insufficient experimental data to enable a more detailed analysis. Furthermore, it is difficult to determine and hence plot p_3 with any accuracy since the duration of the pre-explosive reaction is so short in this case that explosion may occur before the combustible mixture has had time to flow from the storage bulb at room temperature into the reaction cell at the desired temperature. In this case, neither the pressure nor the composition of the gas mixture are accurately known and even if the introduced mixture does not explode, the slow reaction produced is still fast enough by itself to occur significantly before complete introduction into the reaction cell.

Empirical results show that the induction delay τ can often be described approximately by a relationship of the form:

$$\tau p^n \exp\left(-\frac{E}{RT}\right) = \text{Constant} \tag{4.20}$$

where p is the pressure, E and n are parameters (usually positive) which are in practice deduced by adjusting the equation to fit the experimental data.

The values found for τ vary over a wide range extending from fractions of seconds to several minutes or more. The values for two stoichiometric mixtures are given below:

H_2—air	100 ms at 830 K and 1 bar, and 10 ms at 10 bar
	2 ms at 1000 K and 1 bar
CH_4—air	70 ms at 1100 K and 1 bar
	15 ms at 1250 K and 1 bar.

Equation (4.20) can be proven from a theoretical point of view (see Semenov's work), although several assumptions are made, some of which are highly complex. For the purposes of this book, it is preferable to consider the equation simply as a useful relationship to be applied when making approximate predictions of the variations of τ with temperature and pressure. Moreover, the results should be applied with caution, particularly in relation to the chemical mechanics leading to combustion. For this reason, at least initially, it is advisable to consider n and E simply as parameters whose relationship with kinetic mechanisms is complex.

WORKED EXAMPLES

■ **1)** The enthalpy of formation, Δh_f, of ammonium nitrate, a compound used as a fertiliser, is 365 kJ·mol^{-1} which means that its decomposition, given by the reaction:

$$NH_4NO_3 \rightarrow N_2 + 2H_2 + \left(\frac{3}{2}\right)O_2$$

releases 365 kJ·mol^{-1} (or even more if H_2 and O_2 are partially combined as H_2O). The decomposition can thus be explosive, especially if the heat produced by the slow reaction is poorly evacuated. Costly accidents have been caused by the incorrect storage of this product, some of which have involved the loss of human life.

In order to evaluate the potential danger of, for example, a sack of this product which is roughly spherical in shape, the critical radius can be calculated (above which the sack could be considered as dangerous). The solution requires the application of the critical condition described by Frank-Kamenetskii, with $T_{hot} = T_0$. In order to evaluate the consequences of the sack overheating due to poor ventilation, the calculation can be made at two temperatures, e.g. 298 and 373 K.

The rate constant for the decomposition reaction is:

$$k = 6 \times 10^{13} \exp(-20\,450/T)\ \text{s}^{-1}$$

- the density of ammonium nitrate is: $\rho = 1750$ kg·m^{-3},
- its molar mass is: $M = 0.08$ kg·mol^{-1},

- its thermal conductivity is: $\lambda = 0.126\ \text{W}\cdot\text{m}^{-1}\cdot\text{K}^{-1}$,
- $\delta = 3.32$ (the value for a sphere).

The product $qf(c_i)$ in equation (4.4) is taken as being the quantity:

$$\Delta h_f(\rho/M)k \text{ in J}\cdot\text{m}^{-3}\cdot\text{s}^{-1}\text{, in the SI system.}$$

Note that δ is a dimensionless number.

Thus, by writing:

$$r^2 = \frac{3.32\lambda RT_0^2 M}{qEk\rho},$$

where in this case q is taken as being equal to $-\Delta h_f$ for ammonium nitrate (assumed to decompose completely), the following results can be calculated:

$$\text{if } T_0 = 298\text{ K},\quad r = 1530\text{ m},$$
$$\text{if } T_0 = 373\text{ K},\quad r = 1.95\text{ m}.$$

At 298 K, for an isolated sack, the danger is therefore non-existent, at least under the conditions considered here. However, there is a degree of risk at 373 K, especially in consideration of the fact that this is a very approximate calculation.

■ **2)** Consider again the measurements discussed at the start of section 4.2 and shown graphically in Figures 4.1a and 4.1b. In order to check if Frank-Kamenetskii's critical condition holds, knowing that for the reaction in question:

$$k = 10^{14.2} \exp\left(-\frac{17\,200}{T}\right)\text{s}^{-1}$$

$$\lambda = 0.027\text{ SI} \quad \text{and} \quad r = 0.065\text{ m}$$

For the heat released, the choice must be made between taking the value of 195 kJ·mol^{-1} of diethyl peroxide (the value for the non-explosive system), or 159 kJ·mol^{-1} (that for the explosive system, which is less because of the presence of equilibria at higher temperatures which limit decomposition). By referring back to the discussion of the critical condition, it should become clear that the first of these two values should be taken. Furthermore, since $w = kp/RT$, then $\delta = 3.7$, which differs little from the expected value. It can then be shown that $T_M - T_0$ is approximately equal to 13°C, which is less than the value suggested by Figures 4.1a and 4.1b.

■ **3)** For a gas:

- the average molecular velocity is given by:

$$<c> = \left(\frac{8RT}{\pi M}\right)^{1/2}$$

- if n is the number of molecules and σ the molecular diameter, then the average distance between two successive collisions (the mean free path) is:

$$l = \frac{1}{(2^{1/2}\pi n\sigma^2)}$$

- the thermal conductivity is given by the equation:

$$\lambda = \frac{(\rho c_v <c> l)}{3}$$

where c_v is the heat capacity at constant volume,

- and the coefficient of self-diffusion (for molecules of X in a gas X) is given by:

$$d = <c> \frac{l}{3}$$

From this, the variation in thermal conductivity and in the coefficient of self-diffusion with temperature can be deduced.

Writing the ideal gas equation at the molecular level (for n molecules, Boltzmann constant k_B and Avogadro's number N_A):

$$pV = nk_BT \quad \text{which gives} \quad \rho = \frac{M}{V} = M\left(\frac{p}{N_A k_B T}\right) = \frac{pM}{RT}$$

giving:

$$d = \frac{\frac{1}{3}\left(\frac{8RT}{\pi M}\right)^{1/2} k_B T}{2^{1/2}\pi\sigma^2 pV}$$

(see also 3.2.1), which for a given V becomes:

$$d = \text{Constant } T^{3/2} p^{-1}$$

Hence diffusion is favoured by elevated temperatures and low pressures:

$$\lambda = \frac{\frac{pM}{3RT} c_v \left(\frac{8RT}{\pi M}\right)^{1/2} k_B T}{2^{1/2}\pi\sigma^2 pV} = \text{Constant } T^{1/2}$$

Apparently, thermal conductivity is not dependent on pressure. Indeed pressure acts inversely on density and mean free path. However, the relationships given here are only approximations since thermal conductivity must tend to zero as pressure tends to zero.

CHAPTER 5

LAMINAR FLAMES AND DEFLAGRATIONS

5.1 GENERAL POINTS

Phenomena related to self-ignitions (or explosions) were studied in chapter 4. An explosion can only occur at all points simultaneously in the explosive medium under consideration if the composition, temperature and pressure are exactly the same throughout the medium, which can never be the case in practice. Rather, in real situations, the explosion is normally initiated at some central point before sweeping through the entire volume. This volume represents a closed system within which the fuel and oxidant are already mixed. Although the mixture composition, temperature and pressure around this central point may initially be sub-critical, these parameters must be able to change quickly to provide explosion-sustaining conditions. A common definition states that a "flame" is produced when combustion occurs in a limited zone, smaller than the total volume of the system, and from which the flame then propagates.

Traditionally, a differentiation is made between "premixed" and "diffusion" flames. For premixed flames, the fuel and oxidant are already mixed when the flame front arrives (e.g. in the case described above, or in a tube, or a Bunsen burner, etc.). In a tube, the flame sweeps through the tube as a single wave, called a "combustion wave", which progresses from the burnt medium to the unburnt medium. If the unburnt medium is made to move towards the flame, then a flame may be produced which is stable in relation to its physical surroundings, in which case the reaction zone is an open system. Later in this chapter we shall consider in more detail how a premixed flame can be stabilised.

In diffusion flames, the oxidant and fuel mix by diffusing in opposite directions towards the reaction zone. A candle is often used to illustrate this form of flame: volatile substances evaporating off the heated wax diffuse away from the body of the candle and into the flame to mix with ambient air also diffusing towards the flame. A very similar process is involved when an oil well catches fire at the surface. Diffusion flames will be considered later in this chapter.

In general, for flame propagation to be "autonomous", a mechanism must be involved which ensures that the flame is self-sustaining; this mechanism is the exothermic chemical reaction called combustion. If the amount of heat transferred from the burning volume to the neighbouring unburnt gases is sufficient to create the right conditions (of mixture composition, temperature and pressure) then the surrounding volumes will burn and the flame can propagate. We shall see that there are two possible modes of heat transfer; one when the difference in pressure before and after the flame passes is negative, and the other when it is positive. In the first case, known as "deflagration", the chemical reaction can be considered as being coupled (producing heat and variations in concentration) with an exchange of heat and matter (by diffusion) from the flame's reaction zone towards the neighbouring unburnt gas zone. In this chapter, deflagration will be dealt with in detail whilst the second mode, "detonation", in which the chemical reaction is coupled with a shock wave which raises the temperature of the unburnt gas, will be considered in a later chapter.

5.2 THE STABILITY AND PROPAGATION OF A PREMIXED FLAME FRONT IN LAMINAR FLOW

For non-turbulent flow in a premixed medium, the flame is usually very thin and, to a first approximation, its width can be considered as being zero. The flame can thus be represented by a surface of zero thickness, which creates a discontinuity separating the burnt and unburnt medium, known as a "flame front". Its "spatial velocity", $\overrightarrow{V_S}$, at a point O (Fig. 5.1) is the velocity at which a unit area, dA, of this surface propagates (centred at point O in relation to a fixed coordinate system). However, since the fluid mixture may itself be in motion, the propagation velocity of the flame, $\overrightarrow{V_P}$, in relation to the fluid flowing at a velocity, $\vec{V}$, at the point O in question, in the fixed coordinate system, is defined as:

$$\overrightarrow{V_P} = \overrightarrow{V_S} - \vec{V} \tag{5.1}$$

However, as will be shown below, the most representative velocity for the combustible mixture itself is the "fundamental" or "normal" velocity $\overrightarrow{S_L}$ of

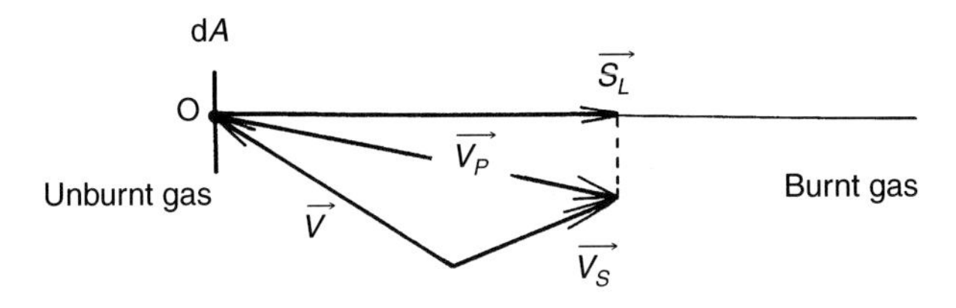

Figure 5.1 Fundamental velocity S_L in a laminar flow (at a point O) for a unit area, dA, of a flame front.

the laminar flame, which is the component of $\overrightarrow{V_P}$ normal to the flame front at O. Ideally, for a planar, stable flame on a burner, V_S is zero and $\overrightarrow{V_P} = \overrightarrow{S_L} = -\overrightarrow{V}$.

A propagation velocity relative to the unburnt fluid can be defined, for a fluid of density ρ_0, flowing with velocity $\overrightarrow{V_0}$:

$$\overrightarrow{V_{P,0}} = \overrightarrow{V_S} - \overrightarrow{V_0}$$

and similarly for a velocity relative to the burnt fluid, of density ρ flowing at a velocity $\overrightarrow{V}$:

$$\overrightarrow{V_P} = \overrightarrow{V_S} - \overrightarrow{V}$$

The conservation of mass perpendicular to the flame front (Fig. 5.2) can be written as:

$$\rho_0 V_{P,0} \cos \alpha = \rho V_P \cos \beta \tag{5.2}$$

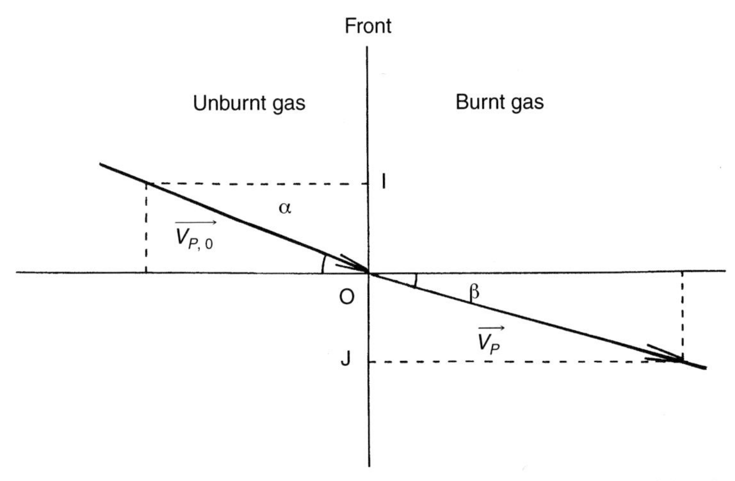

Figure 5.2 Deviation of the propagation velocity V_P on either side of the flame front. According to equation (5.3), $OI = OJ$ in this figure and, from equation (5.4), for α to be greater than β then the density of the burnt gas mixture must be lower than that of the unburnt mixture.

or with the specific volumes $v_0 = 1/\rho_0$ and $v = 1/\rho$:

$$v V_{P,0} \cos\alpha = v_0 V_P \cos\beta \tag{5.2'}$$

Moreover, if friction and long-range forces are neglected, the conservation of momentum equation (see chapter 3, equation 3.19) for a tangential projection at the flame front is given by:

$$\frac{\partial(\rho V_{P,N} V_{P,T})}{\partial N} + \frac{\partial(\rho V_{P,T} V_{P,T})}{\partial T} = -\frac{\partial p}{\partial T}$$

where $V_{P,N}$ and $V_{P,T}$ are projections of the propagation velocity normal to the flame front along the ON axis, and tangentially along the OT axis. If the flame front is planar, and if each point on it is considered as being equal in all respects, then along OT no variable depends on the coordinate T and hence $\partial p/\partial T$ is zero; from which $\rho V_{P,N} V_{P,T}$ must be constant across the flame front. Now since, from equation (5.2), this is also the case for the product $\rho V_{P,N}$, the component $V_{P,T}$ must also be constant, thus giving:

$$V_{P,0} \sin\alpha = V_P \sin\beta \tag{5.3}$$

which when combined with equations (5.2) and (5.3) yields:

$$\rho \operatorname{tg}\alpha = \rho_0 \operatorname{tg}\beta \tag{5.4}$$

where ρ is different from ρ_0, and α is different from β with the result that the burnt gases are deviated. If ρ_0 is greater than ρ, β is smaller than α.

This deviation might destabilise the flame front if its thickness were exactly zero, indeed, even a small distortion might be amplified with important consequences for the flame front. A slight distortion, approximated to a sinusoidal wave, of an initially planar and of zero thickness flame front (such as that shown in Figure 5.3a) must worsen as the amplitude of the disturbance increases. The deviation causes the build up of burnt gases in the regions identified as II in the diagram and this phenomenon would be compensated for by the retreat of the interface towards the unburnt gases, which in turn amplifies the distortion. Similarly, the interface moves towards the burnt gases in regions such as I. Any shift from the stable position thus results in an even greater shift, which is known as the Darrieus-Landau instability.

On the other hand, as we shall see (albeit very qualitatively), if the flame front is of non-zero thickness then stabilisation of the planar flame can occur. On line *b* of Figure 5.3b the concentration deficit of unburnt species causes a lateral diffusion from a point such as D towards the line *b*. This reduces the rate of reaction at D, and hence the propagation rate, which can then no longer compensate for the flow velocity, thus tending to increase further the concavity. On the contrary, the diffusion of heat in the opposite direction raises the temperature and causes the opposite effect. A state of stability tends to be reached when this effect, related to the thermal diffusion, predominates over that of mass diffusion. In fact this mechanism can only stabilise disturbances whose wavelength is fairly small.

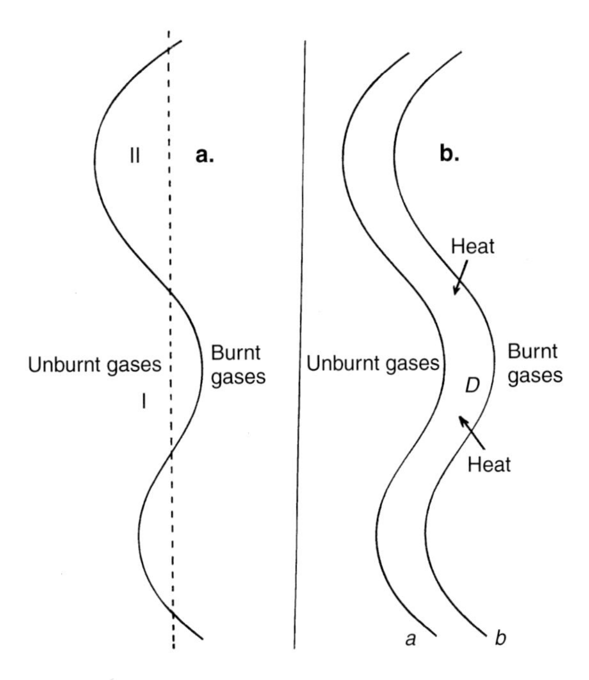

Figure 5.3 Possible stabilisation of a flame front.
a. A flame front of zero thickness is inherently unstable.
b. Stabilisation of a flame front of non-zero thickness.

A typical example of a stabilised premixed laminar-flow flame is the flame above a Bunsen burner, as shown in diagrammatic form in Figure 5.4a. The flame is roughly conical, at an angle θ to the flow velocity V of the unburnt gases such that $V_P = -V$, hence:

$$\sin\theta = \frac{S_L}{V}$$

or r/l, equivalent to A_b/A_f, where A_b is the area at the burner's mouth and A_f that of the flame front. θ is constant for a given flow rate and mixture (hence for a velocity S_L) which explains why the flame is conical. θ increases if V, and hence the flow rate, decreases; the specific case of $\theta = 90°$, hence $r = l$ and $A_b = A_f$, produces what is known as a "flat" flame. The drop in flow rate from the centre to the periphery of the flame, as well as from the mouth to the apex of the flame, and the localised cooling produced by greater heat transfer around the burner's rim are responsible for the modifications shown in Figure 5.4b. Fine particles of refractory material, of appropriate size, can be used to visualise the flow lines and proves that the flow deviates as it crosses the flame front.

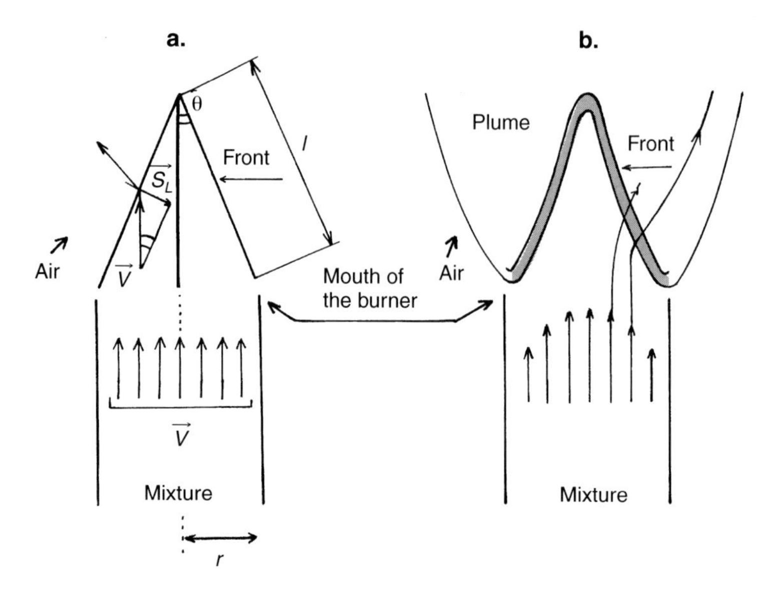

Figure 5.4 Flame stabilisation over a Bunsen-type burner.
a. Ideal case of a laminar flow of a combustible mixture whose flow rate is uniform across a straight section, and giving an infinitely thin, perfectly conical flame front. Flow deviation across the flame front is governed by Figure 5.2, assuming that all the conditions are uniform along the entire front.
b. More realistic case of a parabolic laminar flow, giving a non-conical flame front of finite thickness.
In both situations, the flame "lifts off" the burner mouth, since the mouth is cooler than the flame front.

A flame can also be stabilised in a slightly conical tube (Fig. 5.5). The front locates at a point in the tube where the cross-sectional area satisfies the various conditions required by the continuity equations and for stability.

The flame front thickness can be measured. For a Bunsen burner at atmospheric pressure the flame front is only about 0.5 mm thick. The luminous dark blue region produced when the fuel is natural gas only occupies a fraction of this zone. In the region known as the "plume" (Fig. 5.4b), a yellowish emission may be produced which is simply due to the thermal emission of unburnt solids (in varying quantities depending on the burner setting).

The system's inherent instability may lead to the flame adopting other states with different structures. For example, by varying the composition of a Bunsen burner's supply gas, the flame can be changed (at a critical value) from a roughly conical geometry to another which is more or less polyhedral.

In the same way, for a flame propagating in a tube, it is possible to create a substantially stable cellular flame structure. It is even possible to produce a flame shape which appears to be completely disordered, a phenomenon known as "autoturbulence". Although these types of transition are not easy to analyse, they are not specific to the field of combustion, and are the subject of a great deal of research work and hence our understanding of them is continually increasing.

Since the flame front has a certain thickness, all the physical properties of the medium (temperature, pressure, composition) vary continuously through it, as shown schematically in Figure 5.6 for temperature and concentration. Evidently, the temperature of the unburnt gases is less than that of the burnt gases, and it is relatively easy to predict the trends in concentration variation. For pressure, however, two possible mechanisms may be involved; deflagration, which will be studied in this chapter, and detonation, which will be considered in a later chapter.

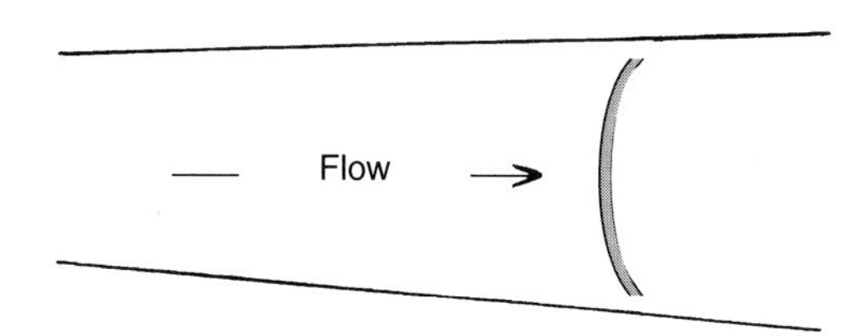

Figure 5.5 Flame stabilisation in a slightly conical tube, for laminar flow. For a constant flow, the flow rate is greater on the left than on the right and the flame will stabilise in the tube at the point where its burning velocity S_L is equal and opposite in sign to that of the gas flow.

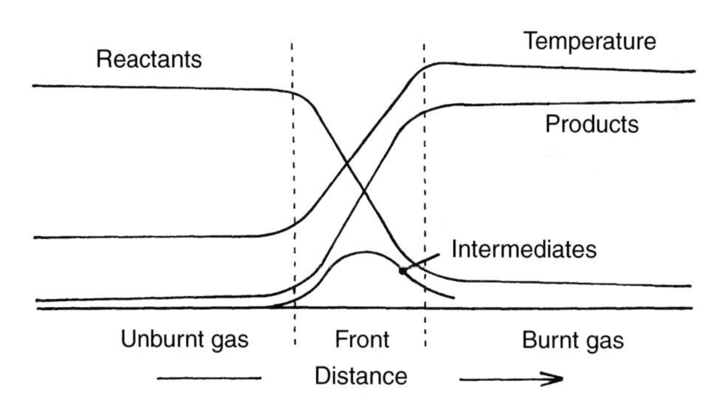

Figure 5.6 Typical temperature and concentration profiles across a flame front.

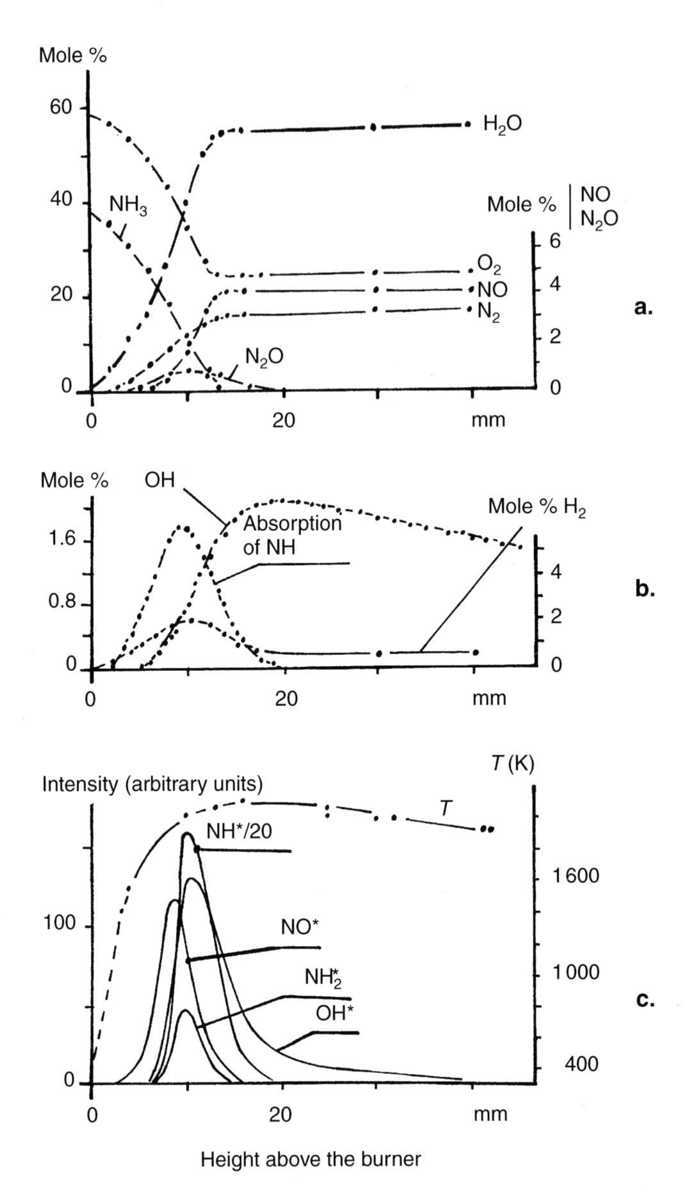

Figure 5.7 Profiles of concentration, temperature and light intensity (from several emitters) for an NH_3—O_2 flame (containing 40% NH_3 by volume) at 20 torr and for a flow rate of 80 $cm^3 \cdot s^{-1}$ (the burner mouth has a diameter of 8 cm) (After Maclean and Wagner (1967), *11th Symposium (International) on Combustion,* p. 871, The Combustion Institute).

a. Figure 5.7a shows that the amounts of NH_3 and O_2 reactants decrease progressively with distance away from the burner mouth whilst those of the final species (N_2, H_2O and NO) increase, which is as expected. On the other hand, the transient appearance of N_2O is not so easy to predict and should be taken into account when considering the kinetic mechanism.

b. Figure 5.7b shows that NH appears before OH, which would seem logical.

c. Comparison of Figures 5.7b and 5.7c shows that OH* (OH excited into a radiative state) disappears well before ground state OH.

Measurements can also be made of temperature and concentration at various heights above a burner, preferably with a low supply pressure and a flat flame (where $\theta = 90°$ as discussed above; see also chapter 2) in order to minimise the lateral diffusion of heat and mass. Figures 5.7a,b,c show measurements taken in this way for an NH_3—O_2 flame with 40% NH_3 at 20 torr. The concentrations of NO, H_2O, N_2, NH_3, H_2 and N_2O were measured by mass spectrometry on extracted mixture samples, the concentrations of ground state OH and NH radicals by absorption spectroscopy, and those of the excited radicals NH_2^*, NO^*, NH^* and OH^* by emission spectroscopy.

The less the flame front is flat, the greater is the need to consider the radial variations of all the variables, including temperature and concentration.

5.3 QUENCHING

In practice, the flame front can never be completely flat, nor can it touch the walls (of the tube for propagation along a tube, or of the burner mouth for a Bunsen flame, etc.). These walls in fact act as "inhibitors" either chemically, through the removal of active species at the walls, or physically, by facilitating heat transfer from the reacting medium to the surrounding medium (since solids are much better heat conductors than gases).

These effects reduce the burning velocity next to the walls with the result that the flame front becomes curved (the effect is increasingly apparent as the tube narrows), in addition to creating a zone where reactions occur slowly immediately next to the walls (generally 0.1 to 5 mm thick). A flame cannot propagate in a cylindrical tube if the tube diameter is less than twice this thickness; this minimum diameter, d_m, is termed the "quenching diameter". An analogous parameter is defined for propagation between parallel planes, the "quenching distance" d. These two parameters are related by the equation $d_m = 1.54\ d$. If λ is the thermal conductivity, ρ the density and c_p the heat capacity of the medium, then the ratio $a_T = \lambda/\rho c_p$ is then its thermal diffusivity. A dimensionless number relative to the unburnt gases, $S_L d/a_T$, is known as a Peclet number. Theory shows that this number is constant for a given mixture.

The quenching phenomenon has been usefully applied in safety systems. Before the advent of electric light bulbs, miners' lamps were surrounded by a fine wire mesh (the Davy lamp). Each hole in the mesh effectively produced a channel small enough to prevent the propagation of deflagration. More recently, gauzes have been used to surround spark-producing systems, such as electric motors and circuit-breakers, when this equipment is used in an atmosphere which may contain combustible gas mixtures. The mesh size is selected as a function of the nature and composition of the combustible mixture under the worst imaginable conditions in which the device would be used.

5.4 PREMIXED FLAMES: THE HUGONIOT-RANKINE EQUATION

Consider a reactive or unreactive fluid (gas or liquid) flowing continuously and in the absence of friction in a tube of uniform cross-sectional area, and in which a phenomenon such as a shock without reaction, a deflagration, or a detonation occurs over an almost negligible length. The phenomenon causes a sudden, quasi-discontinuous change in the properties of the fluid. The pressure, density, flow velocity and internal energy (per unit mass) of the fluid will all change radically across this discontinuity. In this chapter, which deals with deflagration, the subscript 0 will refer to the values of these parameters before deflagration, i.e. to unburnt gases, whilst no subscript will be used to refer to parameters relating to burnt gases. In a later chapter, when considering shocks and detonation, the subscript 1 will refer to parameters before shock, 2 to post-shock parameters and no subscript for the final state after shock **and** reaction).

In a coordinate system moving with the discontinuity, V_0 denotes the flow rate of the unburnt gases and V that of the burnt gases. The conservation of mass condition gives, as derived in chapter 3, the relationship:

$$\rho V = \rho_0 V_0 \tag{5.5}$$

and the relationship for momentum normal to the flame:

$$p + \rho V^2 = p_0 + \rho_0 V_0^2 \tag{5.6}$$

Combining equations (5.5) and (5.6) gives:

$$p - p_0 = \left(\frac{1}{\rho_0} - \frac{1}{\rho}\right)(\rho_0 V_0)^2 \tag{5.7}$$

i.e. a linear relationship between p and $1/\rho$ which is a straight-line equation, known as the Rayleigh's line, whose slope is equal to $-(\rho_0 V_0)^2$. The following relationships can then be deduced directly:

$$V_0 = \frac{1}{\rho_0}\left[\frac{p - p_0}{\dfrac{1}{\rho_0} - \dfrac{1}{\rho}}\right]^{\frac{1}{2}} \tag{5.8}$$

$$V = \frac{1}{\rho}\left[\frac{p - p_0}{\dfrac{1}{\rho_0} - \dfrac{1}{\rho}}\right]^{\frac{1}{2}} \tag{5.8'}$$

and thus there are two possibilities:

- either $p > p_0$, in which case $1/\rho_0 > 1/\rho$
- or $p < p_0$, in which case $1/\rho_0 < 1/\rho$.

Moreover, by applying the concept of conservation of energy, as discussed in chapter 3, using the equation:

$$u + \frac{V^2}{2} + \frac{p}{\rho} = u_0 + \frac{V_0^2}{2} + \frac{p_0}{\rho_0} \tag{5.9}$$

and combining equations (5.5), (5.6), (5.8), (5.8') and (5.9) gives:

$$2(u - u_0) = (p - p_0)\left(\frac{1}{\rho_0} + \frac{1}{\rho}\right) + \frac{p_0}{\rho_0} - \frac{p}{\rho} = (p + p_0)\left(\frac{1}{\rho_0} - \frac{1}{\rho}\right) = f(p, T) \tag{5.10}$$

which is the Hugoniot-Rankine's equation for the curve of the same name (also called the "dynamic adiabatic"). A similar equation for the enthalpy per unit mass, $h = u + p/\rho$, gives:

$$2(h - h_0) = (p - p_0)\left(\frac{1}{\rho_0} + \frac{1}{\rho}\right) = g(p, T) \tag{5.11}$$

in which u and u_0 from equation (5.10), as well as h and h_0 from equation (5.11), vary depending on the inherent chemical characteristics of the medium and hence are different for reactants and products. The extent of the variations in these parameters due to a chemical reaction can, however, be determined from tabulated values of the enthalpies of formation. There is, therefore, a Hugoniot-Rankine curve, H_0, for the fluid medium before combustion and another, H, for the medium after combustion. The effect of the heat released means that the second curve is situated above the first (Fig. 5.8). Indeed, because of this release

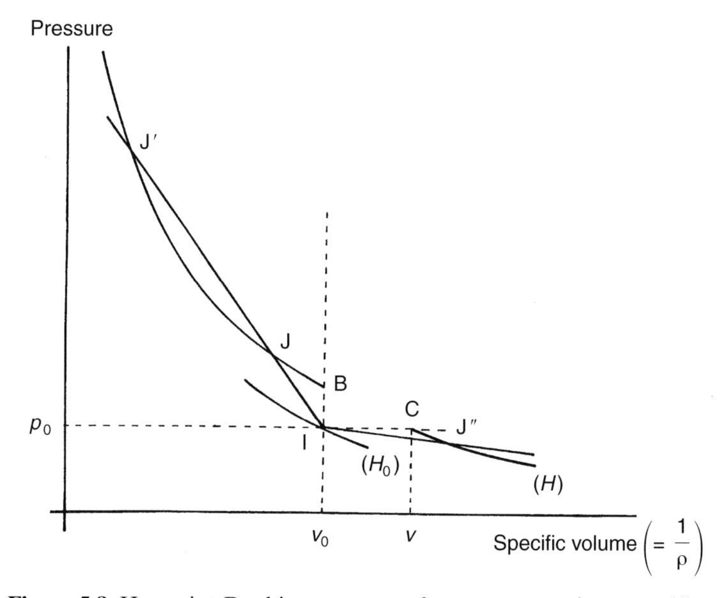

Figure 5.8 Hugoniot-Rankine curves of pressure against specific volume (specific volume = 1/density). H_0 is the adiabatic dynamic for the unburnt mixture and H that for the burnt mixture.

of heat within the reacting medium, the temperature of the products is greater than that of the reactants, as are also the ratio p/ρ and the pressure p on curve H for the same volume (point B is directly above point I). From point I on the first Hugoniot curve, a combustion reaction is represented by a line to a point such as J, J′ or J″ on the second curve:

- either above point B, hence with $(p - p_0)$ positive and $(1/\rho - 1/\rho_0)$ negative, i.e. a detonation (studied in a later chapter),
- or below point C, hence with $(p - p_0)$ negative and $(1/\rho - 1/\rho_0)$ positive, i.e. a deflagration (studied in this chapter).

Note that the transition from point I to a point such as J but between B and C would require both $(p - p_0)$ and $(1/\rho - 1/\rho_0)$ to be positive, which is impossible.

Figure 5.8 also shows that the Rayleigh line, described by equation (5.7), has a slope equal to $(\rho_0 V_0)^2$ (once the sign has been changed) and is steeper for detonation than for deflagration. Hence, for a given ρ_0, V_0 is greater for a detonation than for a deflagration.

Consider a given initial point (p_0, ρ_0) through which passes a Hugoniot curve H_0, an isotherm ($p/\rho = nRT$ for an ideal gas) and an isentrope (p/ρ^γ = constant for an ideal gas). From a point (p_0, ρ_0) on the Hugoniot curve for the reactants, the points (p, ρ) can be reached which are situated on the curve for the products, with ρ, ρ_0, p and p_0 obeying equations (5.7-5.11). All the straight lines linking the first points to the second are Rayleigh lines (5.7) whose slope, $(\rho_0 V_0)^2$, (with changed sign) is thus dependent on the initial flow rate V_0.

5.5 PREMIXED FLAMES: DEFLAGRATION PROPAGATION MECHANISM

5.5.1 Thermal balance

Equations (5.7-5.11) do not take heat and mass transfer processes into consideration. Later we will see that these can be neglected for detonation but not for deflagration, since the latter phenomenon propagates mainly by heat and mass transfer.

For the ideal case of a flat, stationary flame front in a one-dimensional flow and for which $S_L = -V$, the variation of temperature with x is as shown schematically in Figure 5.9. T_0 is the initial temperature of the unburnt gases and T_M the maximum temperature after combustion (assumed to be constant). Neglecting heat losses and losses due to light emission, the energy balance equation reduces to that of the thermal balance (i.e. in SI units of $\mathrm{J{\cdot}m^{-3}{\cdot}s^{-1}}$):

$$qw = -\frac{\mathrm{d}}{\mathrm{d}x}\left(\lambda\frac{\mathrm{d}T}{\mathrm{d}x}\right) + \rho c_p V\frac{\mathrm{d}T}{\mathrm{d}x} \tag{5.12}$$

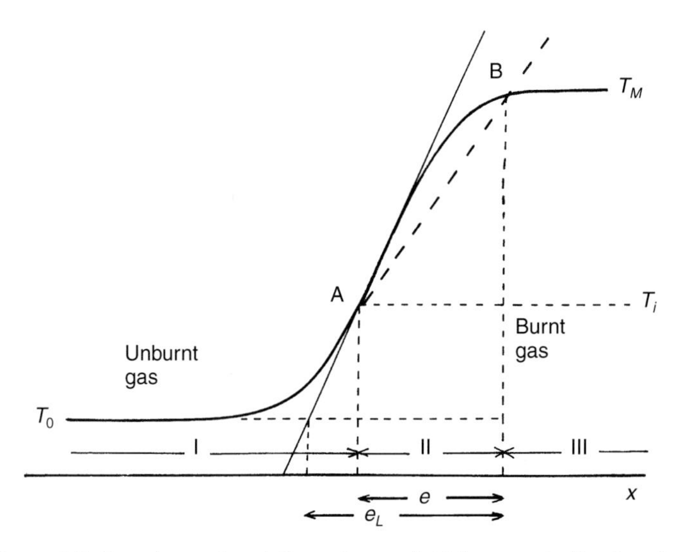

Figure 5.9 One-dimensional flame front of thickness *e*, indicating the three regions I, II and III considered in the analytical solution. The flame front is stationary with respect to a fixed reference if the flow velocity towards the right is equal and opposite to the front's velocity.

- The product qw represents the contribution of the heat released by the reaction (per unit volume per unit time) to the thermal balance (as shown by equation (4.1) in chapter 4).
- In this product, the rate of reaction, w, is defined by expression (2.3′) given in chapter 2, assuming that a single chemical reaction is occurring. If this reaction is then written as:

$$aA + bB \rightarrow a'A' + b'B' + \ldots$$

then w is given by:

$$w = -\left(\frac{w_A}{a}\right) = -\left(\frac{w_B}{b}\right) = +\left(\frac{w_{A'}}{a'}\right) = \ldots$$

where w_A, w_B, $w_{A'}$, etc. are the rates of reaction of A, B, A′, etc.
- The definition of q results from the definitions of w and of the product qw. Since the initial mixture contains (per unit volume) not a but x moles of A and not b but y moles of B then, if (y/x) is smaller than (b/a), constituent B (oxidant or fuel) is said to be the minor component. If the reaction is complete, i.e. it does not attain equilibrium, B can react completely, although A cannot. Under such conditions, the heat released chemically is dependent on the number of moles of B but not on the number of moles of A. If q is the total heat released between the beginning and end of the reaction of A with b moles of B, then the heat released is $(y/b)q$ per unit volume, or:

$$-\frac{qw_B}{b} = +qw$$

per unit time. Hence the product qw indeed has the required meaning. This treatment can of course also be extended to q and w as specific (i.e. per unit mass) rather than molar parameters.
Unlike the case considered in chapter 4, the system does not react at a given volume, nor even truly at a given pressure. Either way, q is not equal (even after taking into account a possible sign change due to the convention used) either to the Δu or Δh of reaction simply because the initial and the final temperature after combustion are different. Moreover, the temperature is sufficiently high for the reaction to be limited by the attainment of chemical equilibria.

- λ is the thermal conductivity of the medium.
- ρ is the total number of moles (or the mass) of the medium per unit volume (depending on whether molar or specific parameters are used.
- c_p is the heat capacity at constant pressure (molar or specific), assumed to be independent of the temperature and of the composition of the medium, and
- V is the flow rate.

The second term in equation (5.12) refers to thermal diffusion (Fourier's Law) and the third term to convection.

The rate of the global combustion reaction depends on the chemical processes occurring in the flame. It is therefore a function of pressure, temperature, and concentration, although especially the latter two due to the fact that pressure changes only very slightly in a deflagration flame. Since temperature and concentration vary with x, the global reaction rate at a given point can be assumed to be dependent only on the temperature at this point. If T_0 is used to denote the initial temperature of the unburnt gases and T_M the maximum temperature reached, then the global reaction rate must be approximately zero at these two temperatures. The reason for this is that at $T = T_M$ the reaction is over, and the system is at chemical equilibrium, whereas at $T = T_0$ the temperature is too low for the rate to be much above zero. The same reasoning applies at temperatures above T_0, for practically the entire temperature range extending from T_0 to the temperature marked T_i in Figure 5.9. In other words, the reaction mainly occurs between the temperatures T_i and T_M.

Equation (5.12), despite being extremely approximate, is nonetheless difficult to solve. The normal approach is to calculate approximate solutions for each of the regions marked I and II in Figure 5.9, based on the assumptions made above for these two regions.

Region I, between T_0 and T_i, where the reaction rate is close to zero, is known as the preheating zone. Assuming that the thermal conductivity is roughly constant gives:

$$\lambda \frac{\mathrm{d}T}{\mathrm{d}x} = \rho c_p V(T - T_0) \text{ which is positive} \tag{5.13}$$

with $T = T_0$ and $\mathrm{d}T/\mathrm{d}x = 0$ as x tends to minus infinity. A second integration yields:

$$T = T_0 + (T_i - T_0) \exp\left(\rho c_p \frac{Vx}{\lambda}\right) \tag{5.14}$$

Region II is the reaction zone with a thickness e. If $\mathrm{d}T/\mathrm{d}x$ is considered as being constant in this region and that the tangent at A is similar to the line AB (with the aim being to replace, in Figure 5.9, the real temperature variation (solid line) with the linear variation (dashed line)), then the following can be derived:

$$\frac{\mathrm{d}T}{\mathrm{d}x} = + \frac{T_M - T_i}{e}$$

where T_i, the temperature at A, is the temperature at which ignition occurs under the given conditions, which are different from the conditions required for explosion in a closed system. At A the reaction rate is zero and hence equation (5.13) holds if $T = T_i$, giving:

$$\frac{\mathrm{d}T}{\mathrm{d}x} = + \frac{T_M - T_i}{e} = \frac{\rho c_p V (T_i - T_0)}{\lambda}$$

If the reaction begins at A, and not before, and ends at B, and not after, the reaction time is equal to e/V, as a consequence of which the mean reaction rate defined by:

$$<w> \text{ between } T_0 \text{ and } T_M = \frac{1}{T_0 - T_M} \int_{T_0}^{T_M} w(T)\,\mathrm{d}T$$

is such that:

$$<w> = k\left(\frac{S_L}{e}\right) \tag{5.15}$$

where k is a constant (in kg·m^{-3} if $<w>$ is in kg·m^{-3}·s^{-1}). Setting $(T_M - T_i)/e$ equal to $\mathrm{d}T/\mathrm{d}x$ as calculated at A in region I, then from equation (5.13), we have:

$$V^2 = S_L^2 = \frac{\lambda}{\rho c_p} \frac{T_M - T_i}{T_i - T_0} \frac{<w>}{k} \tag{5.16}$$

Despite the various assumptions made on the way to obtaining expression (5.16), the equation does provide predictions which are quite well verified by experiment. For example, S_L and the rate of reaction, w, vary in the same direction, even if S_L is found not to vary exactly with $w^{1/2}$, and S_L reduces as the ignition temperature increases etc. Furthermore, the equation gives S_L as a function of the parameters λ , ρ , c_p, T_0, T_i and T_M (T_M via the heat released q) for the combustible mixture.

The flame thickness can also be calculated since it is the sum of the thicknesses of zones I and II. Strictly speaking, the flame thickness in preheating zone I is infinite since, according to (5.14), T tends to T_0 as x tends to minus infinity. However, a characteristic flame thickness can be defined, e_L, as indicated in Figure 5.9, by taking the tangent to the profile at $T = T_i$, giving:

$$e_L \approx e + \frac{e(T_i - T_0)}{T_M - T_i} = \frac{e(T_M - T_0)}{T_M - T_i}$$

which, combined with equation (5.15), gives:

$$e_L = \frac{kS_L}{<w>} \frac{T_M - T_0}{T_M - T_i} \tag{5.17}$$

and, finally, by combining with equation (5.16):

$$e_L = \frac{T_M - T_0}{T_M - T_i} \left[\frac{(T_M - T_i)k\lambda}{(T_i - T_0) <w> \rho c_p} \right]^{1/2} \tag{5.18}$$

Experimental measurements have shown (Fig. 5.10) that S_L is a maximum close to the stoichiometric mixture, i.e. an equivalence ratio of unity.

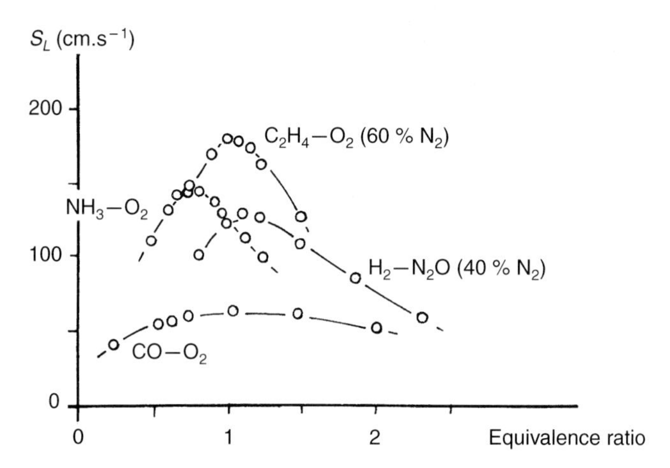

Figure 5.10 Influence of equivalence ratio on burning velocity for laminar flow, S_L (After Van Tiggelen *et al.* (1968) *Oxydations et combustions*, volume 1, p. 361. Éditions Technip, Paris).

5.5.2 Mass balances

For steady-state conditions the same as those described above, but by additionally neglecting any diffusion of matter arising from the temperature gradient, the following mass balance equation can be applied for each species *i*:

$$w_i = \frac{\mathrm{d}}{\mathrm{d}x}\left(d_i \frac{\mathrm{d}c_i}{\mathrm{d}x}\right) + \frac{\mathrm{d}}{\mathrm{d}x}(c_i V) \tag{5.19}$$

where d_i is the diffusion coefficient of species *i*, and w_i is the rate of change of its concentration c_i (moles per unit volume).

The subscript *i* may designate a reactant, a product, or an intermediate. When considering intermediates, the steady-state approximation assumes that

w_i is close to zero. If this approximation is extended to all active species, then the rate of disappearance of a reactant is similar to that of the appearance of a product, which effectively means that a single global reaction rate can be defined, irrespective of whether a reactant or a product is being considered (cf. section 5.5.1 above and section 2.4 of chapter 2). Moreover, if all the values of d_i are equal for all the reactants and products, then the mass balance equations are equal in value for all species (allowing for any constant factor throughout). Also, equation (5.19) for the mass balance becomes equal to that of (5.12) for the thermal balance if:

$$\lambda = d_i \rho c_p \quad \text{i.e. if the Lewis number,} \quad \mathrm{Le}_i = \frac{\lambda}{d_i \rho c_p} = 1$$

for all i. In this case, the temperature profile, $T = T(x)$ and the concentration profiles, $c_i = c_i(x)$ of the reactants and products are the same (again allowing for any constant factor throughout).

As discussed in chapter 3, if radiation and kinetic energy losses are disregarded and if the Lewis number is equal to one for all species, then the balance equation for energy can be written:

$$\frac{\mathrm{d}}{\mathrm{d}x}\left(\frac{\lambda}{c_p}\frac{\mathrm{d}h}{\mathrm{d}x}\right) = \frac{\mathrm{d}}{\mathrm{d}x}(\rho V h) = \rho V \frac{\mathrm{d}h}{\mathrm{d}x} \qquad \left(\text{since } \frac{\mathrm{d}(\rho V)}{\mathrm{d}x} = 0\right)$$

if ρ is the density of the mixture and hence h its specific enthalpy. Since $\mathrm{d}h/\mathrm{d}x$ tends to zero as x tends to infinity, the solution of this balance equation is:

$$h = \text{Constant}$$

or, if j is used to denote one section of gas, which may be unburnt, and k another section, which may be burnt, then:

$$h_j = h_k \tag{5.20}$$

Now, based on equation (1.2) from chapter 1, where the enthalpies are the "absolute" enthalpies for the system under consideration (a mixture or consisting of a single component):

$$(h, \text{specific, at } T) = \int_0^T (c_p, \text{specific})\, \mathrm{d}T + \frac{1}{\rho}\sum_i c_i (h_{i,0} \text{ molar})$$

hence:

$$c_p(T_j - T_k) + \left(\frac{1}{\rho}\sum_i c_i h_{i,0}\right)_j - \left(\frac{1}{\rho}\sum_i c_i h_{i,0}\right)_k = 0 \tag{5.20'}$$

so long as the specific c_p, of the mixture is assumed to be constant and c_i is the **molar** concentration of constituent i (equally specific $h_{i,0}$ and specific c_i could have been taken, i.e. the mass of **constituent** i per unit volume). Along the direction of flow, the concentrations of reactants, c_i vary in the opposite sense to both the c_i for the products and the temperature. Since combustion releases heat, the sum of the $c_i h_i$ for the products must be less than the sum for the reactants. Progressively as one moves across the flame front, from the unburnt gases towards burnt gases, the contribution of the difference in the two $c_i h_{i,0}$ terms in equation (5.20′) decreases as the term $c_p(T_j - T_k)$ increases, or vice-versa.

5.6 CHAIN-BRANCHING THEORY

There is a great deal of evidence to suggest that chain mechanisms play a fundamental role in flame phenomena, particularly:

- that burning velocity is highly dependent on the presence of a variety of species, which may be present in very small quantities, which either act as promoters (e.g. hydrogen and water in CO/O_2 flame mixtures) or as inhibitors (e.g. halogenated compounds in certain situations);
- the presence of high concentrations of free radicals in the flame front.

The chain-branching theory has been expressed in the form of equations which have themselves been used to calculate theoretical values for the burning velocity S_L which vary depending on the assumptions made and which can be quite complex. However, it is always found, as shown by equation (5.16) derived from thermal theory and described above, that S_L values vary, at least approximately, as the square root of the product of a diffusion coefficient (or of thermal diffusivity) and a reaction rate (or a combination of rates). A large amount of research work has been devoted to this subject. Essentially, the assumption is made that the concentrations of free radicals in the flame front do not vary (or at least vary little) because they are being continually regenerated by branching reactions. Hence it can be shown that :

$$S_L^2 = \frac{2d_X (w_r - w_t)}{c_X} \tag{5.21}$$

where:

c_X denotes the concentration of the chain propagating free radicals (noted as X),
d_X is their diffusion coefficient,
w_r is the rate of branching, and
w_t the rate of termination.

This relationship, like equation (5.16) above, indicates that the fundamental velocity S_L is a direct function only of the properties of the combustible mixture and does not depend on the geometry or flow conditions. However, since equation (5.21) contains quantities with unknown numerical values, its verification can only ever be qualitative.

5.7 FLAMMABILITY LIMITS

Thermal theory, as considered in section 5.5, and chain-branching theory, briefly summarised in section 5.6, both assume that flames propagate only if the burning velocity S_L is positive. In terms of chain-branching theory, this means that the rate of branching w_r, is greater than the rate of termination w_t.

It is possible to plot the various flammability limits in a manner similar to how values were plotted in chapter 4 for explosive limits. The limits enclose regions which include the initial values of the variables which allow the flame, once it has been created in the medium, to propagate through it. The variables include composition as a function of pressure at a given temperature, or composition as a function of temperature at a given pressure, etc. In this way it is possible to specify, for example at a given temperature and pressure, the lower limit for the amount of fuel or oxidant which will allow a flame to propagate through the medium; extremely useful information when assessing potential safety problems. Moreover, the dilution limit can be measured for increasingly dilute mixtures which nonetheless have the same equivalence ratio, as shown in Figure 5.11 for ternary methane-oxygen mixtures diluted to varying degrees in N_2 or CO_2.

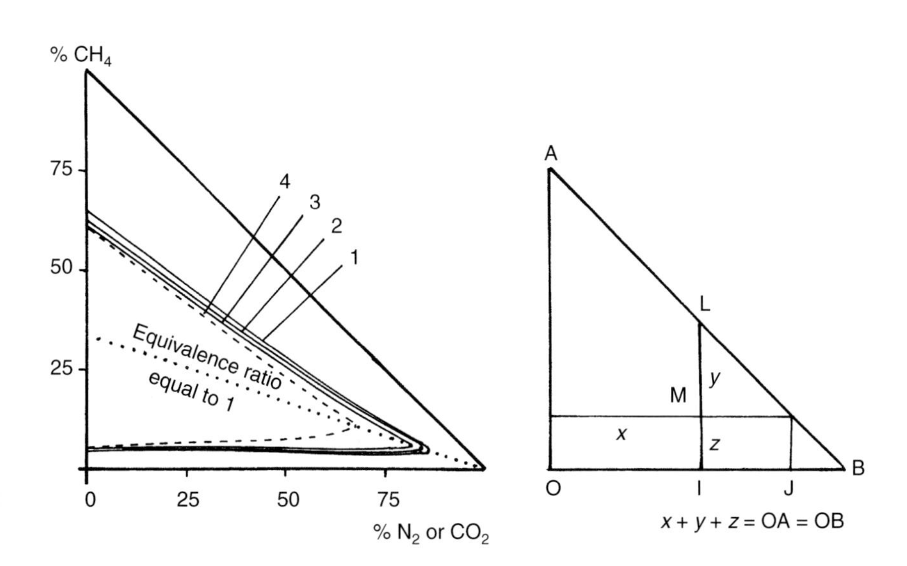

Figure 5.11 Flammability limits:
- for CH_4—O_2—N_2 at T_0 = 375°C (curve 1)
- for CH_4—O_2—N_2 at T_0 = 160°C (curve 2)
- for CH_4—O_2—N_2 at T_0 = 20°C (curve 3)
- for CH_4—O_2—CO_2 at T_0 = 20°C (curve 4).

(After P. Remmerie and A. Van Tiggelen, *Thèse*, Louvain).

For any point M in Figure 5.11:

OI $= x$ is the percentage of N_2 or CO_2 diluent,
ML $= y$ is that of O_2,
IM $= z$ is that of CH_4,

with: $$x + y + z = OI + IJ + JB = OA = OB = 100.$$

Stoichiometric mixtures are those of the form $CH_4 + 2\,O_2 + n\,N_2$ or $n\,CO_2$, i.e.:

- in the absence of N_2 or CO_2, there are three moles in total and therefore 33% CH_4,
- for $n = 1$ (for N_2 or CO_2), and with one mole of CH_4, giving 4 moles in total, thus: 25% CH_4 and 25% N_2 or CO_2,
- for air, containing about eight times as much N_2 as CH_4 giving 11 moles in total, thus: 9% CH_4, 73% N_2 and 18% O_2.

The line passing through B, such that $z/y = 1/2$ with x variable, thus relates to the stoichiometric mixture (an equivalence ratio of unity). For all other lines passing through B, the ratio (z/y), corresponding to (CH_4/O_2), is constant.

The region in which deflagration occurs enlarges when the initial temperature is increased.

However, in order to plot the region within which S_L is non-zero and positive, i.e. where propagation occurs, it is necessary to ignite the mixture externally with an auxiliary flame or, more often, with an electric spark. For the flame ignited in this way to then be able to propagate, a sufficient amount of heat and an adequate number of propagating radical centres must be created in the ignited volume just to ensure that the surrounding, unburnt gases are brought into a combustible state (cf. chapter 8). It is thought that this condition cannot be satisfied if the ignition energy supplied is too low, or on the contrary, if it is too high, in which case the burning velocity is increased (at least initially). Ultimately, a flame may appear to be unable to propagate through a medium where the conditions are theoretically favourable, yet be able to propagate over a fairly long distance through a medium in which these same conditions are not favourable. Unless strict criteria are applied, the regions within which deflagration occurs may vary as a function of the ignition conditions. Researchers have not always considered ignition-related aspects in their calculations, which is why different published articles may give significantly different limits for the same mixture. Such discrepancies are particularly apparent in the two most sensitive zones; those near to the upper and lower limits, which is the result predicted in the rather rudimentary analysis outlined above.

So long as the initial composition, pressure, and temperature are the same at all points in a gas mixture, this source of error can be avoided. This is achieved by eliminating the measured values of the burning velocity which would otherwise vary depending on which point was being considered. In other words, the values measured near to the ignition volume are taken and extrapolated to the conditions for which the burning velocity is zero. It is also possible, although not easy, to prepare a mixture in which a concentration gradient can be measured, and then to note the composition at which propagation ceases.

5.8 LAMINAR DIFFUSION FLAMES

"Diffusion" flames are characterised by the fact that combustion occurs at the point where the fuel and oxidising gases meet and mix. In its simplest form, a diffusion flame is typified by a cigarette lighter, in fact this example was used in chapter 3 to investigate and explain convection and diffusion phenomena. A candle also produces a diffusion flame, although it is slightly more complex since the wax must first liquefy, then moisten the wick, before vaporising at its surface to form the combustible current of gas which is carried upwards and diffuses laterally towards the reaction zone. At the same time, the surrounding air is entrained by the hot, burnt gases moving away from the reaction zone. Although some of this air serves only to mix with and hence cool these hot gases, the rest diffuses towards the reaction zone where it reacts with the fuel vapours vaporising off the wick, to produce hot, burnt gases which rise in accordance with Archimede's principles of buoyancy.

This system, shown schematically in Figure 1 in the Introduction, can only be differentiated from that of a gas cigarette lighter (Figure 9a of the same chapter) by the fact that it is the heat transferred (mainly by radiation) from the flame towards the wick which vaporises the fuel (after it is liquidised) and also due to the key role played by buoyancy in carrying the gas upwards away from the wick. The principal characteristic of diffusion flames is therefore common to both examples; namely the existence of a reaction zone which separates one medium, where fuel mixes with burnt gases, from another zone where oxidant mixes with burnt gases. When a diffusion flame is established in a jet-type flow, as in the above examples, it is often called a "jet flame". Jet flames are used commercially and in industry, in all combustion systems where it is potentially dangerous to pre-mix the fuel and oxidant, i.e. in almost all combustion applications. Very often, however, the fuel is injected as a liquid in the form of a spray of droplets, and the gas velocity is sufficiently high to create turbulent flow. Since these two factors greatly complicate the study of diffusion flames, this chapter will consider only gaseous, laminar-flow flames, leaving turbulent diffusion flames to be dealt with in chapter 6, and the combustion of sprays in chapter 9.

A limiting case often considered when studying diffusion flames is that encountered when the chemical reactions occur infinitely quickly, in which case it is the physical processes of diffusion and convection which dictate the flame characteristics. Sometimes the use of the term "diffusion flame" is restricted to this limiting case, with the term "non-premixed flame" being used for more general systems.

5.9 THEORETICAL RESULTS FOR LAMINAR DIFFUSION FLAMES

In section 3.4 of chapter 3, we saw how to calculate the diffusion and convection phenomena associated with laminar jet flames. The simple case envisaged by Burke and Schumann in 1928 was used; which was the flame produced between a planar (i.e. non-cylindrical) gas jet and a parallel flow of air. The two flows are taken to have the same unit flow ρV at all points and a single, infinitely fast, irreversible chemical reaction is assumed to occur between the fuel and oxidant.

The resulting flame structure is very similar to that shown in Figure 1 of the Introduction for a candle. The chemical reactions occur over a surface (a zone of infinitely small thickness since the reactions are assumed to occur infinitely quickly) which separates a gaseous medium rich in fuel from one rich in oxidant. Consider the case of a single reaction:

$$\text{fuel} + \nu \text{ oxidant} \rightarrow \text{products}$$

or, in shortened form:

$$\mathrm{K} + \nu\,\mathrm{Ox} \rightarrow \mathrm{P}$$

where ν is the molar stoichiometric coefficient, related to the specific coefficient ν_S, which is the product of ν and the ratio $M_{\mathrm{Ox}}/M_{\mathrm{K}}$ of the molar masses of oxidant and fuel. If α is the quantity d_j/V_j (where d_j is the coefficient of diffusion per unit mass or per mole since these are assumed to be equal, and V_j is the gas velocity, both being taken at the exit section, and e_j the half-thickness of the planar gas jet) then a plot of the surface separating the fuel and oxidant in a plane (x, y) normal to the fuel jet, is given analytically, yet implicitly, by the formula:

$$\frac{\dfrac{0.23}{\nu_s}}{1+\dfrac{0.23}{\nu_s}} = \frac{1}{2\sqrt{\pi\alpha x}} \int_{-e_j}^{+e_j} \exp\left[\frac{-(y-y')^2}{4\alpha x}\right] \mathrm{d}y' \qquad (5.22)$$

which is obtained by setting the variable Z_{O}, representing $(Y_{\mathrm{K}} - Y_{\mathrm{Ox}}/\nu_s)$, to zero at this point (where Y_{K} is the fuel mass fraction and Y_{Ox} the mass fraction of the oxidant). 0.23 is the mass fraction of oxygen in air. Figure 5.12a is a schematic representation of this plot, which can bc referred to as the reaction front.

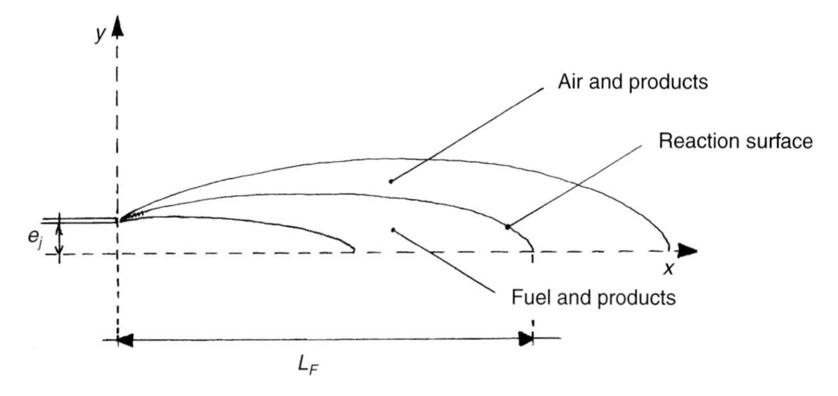

Figure 5.12a Schematic representation of a diffusion jet flame.

At the reaction front, the temperature is maximum and constant. According to the calculations outlined in section 3.4.2 of chapter 3, the temperature is such that $Z_O = 0 = \alpha_0 h + \beta_0$, which gives, for $h = \overline{c_p}\, T - q$, $h_{jet} = \overline{c_p}\, T_j$ and $h_{ext} = \overline{c_p}\, T_e$:

$$T = \frac{q}{\overline{c_p}} + \frac{T_e + \dfrac{0.23 T_j}{\nu_s}}{1 + \dfrac{0.23}{\nu_s}} \tag{5.23}$$

where T_e is the temperature of the surrounding air, T_j that of the gas jet and $\overline{c_p}$ a mean heat capacity. The symbol q has the same meaning here as it does in the thermal balance (5.12), except that in this case the mixture is stoichiometric.

In section 3.4.4, the flame length, L_F, was defined as being the distance, in the plane of symmetry indicated in Figure 5.12a, between the fuel-supply tube outlet and the surface corresponding to the maximum temperature (beyond which no fuel is found in the gaseous mixture). With this definition, the flame length was determined in section 3.4.4 to be equal to:

$$\frac{L_F}{e_j} = \frac{V_j e_j \left(\dfrac{0.23 + \nu_s}{0.23}\right)^2}{\pi d_j} \tag{5.24}$$

where d_j is the coefficient of diffusion.

A similar approach for a round jet gives, instead of (5.24):

$$\frac{L_F}{r_j} = \frac{\dfrac{V_j r_j}{d_j}}{\ln\left(\dfrac{0.23 + \nu_s}{\nu_s}\right)} \tag{5.24'}$$

where r_j is the radius of the fuel-supply tube.

In fact, the assumption that ρV does not change with radial distance on which these formulae are based, is fairly restrictive. Even so, it provides a good prediction of the variation of L_F/e_j (or of L_F/r_j) with Peclet number for the combustible gas jet ($Pe_j = V_j e_j/d_j$ or $V_j r_j/d_j$), and even its variation as a function of stoichiometric ratio, which is involved in the ratio $0.23/\nu_s$. Experimental results have confirmed formulae (5.24) and (5.24′) to within a multiplicative constant; the value of which depends slightly on the condition of the boundary layers in the fuel-supply tube, and on the precise design of its mouth.

However, one parameter which is not well accounted for in these formulae is the ratio of the density of the combustible gas, ρ_j, to that of the flame, ρ_F, since the assumption that ρV = constant does not allow this factor to be taken into consideration. Experiment has shown that, in general, the above formulae may be applied if ρ_j is very different from ρ_F, but with V_j replaced by $V_j(\rho_j/\rho_F)^{1/2}$.

5.10 DIFFUSION FLAMES IN MORE REALISTIC CHEMICAL APPLICATIONS

Real laminar jet flames, in which the chemistry does not reduce to a single, infinitely fast, irreversible reaction, do not in fact differ greatly from the basic case considered above. The additional complexity resulting from applying more realistic chemistry may come from two sources: the possibility that there may be numerous reversible reactions occurring, and the fact that their rates may not be great enough to ensure that all reactions reach equilibrium. These two factors may combine.

The main consequence of this is that the reaction zone can no longer be considered as being a surface but rather a zone of finite thickness. Its position continues to be that predicted accurately by the calculation of the position of the reaction front, assuming a unique, very fast and irreversible reaction. However, the zone extends on either side of this idealised reaction front (but still over a much smaller thickness than the mixing zones shown in Figure 5.12a).

If we assume that the chemical reactions occurring, which may be numerous, are close to equilibrium at all points, then the previous theory can be generalised, so long as the assumption is maintained that all the coefficients of diffusion are equal. In this case, the Zeldovitch variable, Z, can be defined as being the mass fraction of a given chemical element, for any molecule containing the element. For example, for the element hydrogen, if the fuel is methane and the combustion products are CO, CO_2, H_2O, OH, H, H_2, then by definition:

$$Z_H = Y_H + 2\left(\frac{M_H}{M_{H_2}}\right) Y_{H_2} + 2\left(\frac{M_H}{M_{H_2O}}\right) Y_{H_2O} + \left(\frac{M_H}{M_{OH}}\right) Y_{OH}$$

where M_i are the molar masses. If necessary, the products HO_2 and H_2O_2 can also be taken into account.

This variable, insofar as the coefficients of diffusion are assumed to be equal, verifies closely a balance equation for energy without a reaction term, such as Z_{Ox} and Z_P as described in chapter 3. Furthermore, the enthalpy h still satisfies the same partial differential equation given in chapter 3. This is also the case for all the variables Z_i corresponding to all the other chemical elements in the mixture (O, C, N, and occasionally others). It follows that all Z_i, as well as h, can be expressed as functions of Z_H, using linear relationships which depend only on the limiting conditions, as was found for Z_{Ox} and Z_P:

$$Z_i = Z_i(Z_H) \text{ for any } i \quad \text{and} \quad h = h(Z_H) = \sum\nolimits_1^n Y_j h_j(T)$$

If we now assume that chemical equilibrium is reached, or almost reached, throughout, then the local flow composition at any point can be calculated as a function of elemental composition, temperature and pressure, i.e. for any species j:

$$Y_j = Y_j(Z_i, T, p)_{\text{equil}}$$

However if $Z_i = Z_i(Z_H)$ and $h = h(Z_H)$, it must follow that:

$$Y_j = Y_j(Z_H, p)_{\text{equil}} \quad \text{and} \quad T = T(Z_H, p)_{\text{equil}} \tag{5.25}$$

In practice, the variable f (known as the mixture fraction) is used, which makes Z_H dimensionless:

$$f = \frac{(Z_H - Z_{H,\,\text{surr}})}{(Z_{H,j} - Z_{H,\,\text{surr}})}$$

where $Z_{H,j}$ is the value of Z_H in the fuel jet and $Z_{H,\,\text{surr}}$ is that in the surrounding air (zero for pure air !).

Once the equations (5.25) have been calculated and tabulated, all that remains is to solve the equation for f, which is again:

$$\frac{\partial f}{\partial x} = \alpha \frac{\partial^2 f}{\partial y^2}$$

in the Burke and Schumann assumption (as in section 3.4.4), to obtain the solution to the entire problem. Equations (5.25) are shown schematically in Figure 5.12b, where they are compared with the results given by assuming that a single chemical reaction occurs, as in chapter 3. Figure 5.12c gives more accurate numerical values for the example of a hydrogen-air flame at atmospheric pressure and with a pre-flame gas temperature of 300 K. In fact, the assumption of a single, infinitely-fast reaction seems to be a first approximation to local chemical equilibrium, which allows the profiles of the major species and of the temperature to be calculated reasonably correctly. Furthermore, by assuming chemical equilibrium at each point, it is then possible to calculate the amounts of minor species present (O, OH, H, and even HO_2 and H_2O_2), as well as to improve the accuracy of the calculation of the temperature profiles and major species.

Taking into consideration chemical kinetics away from equilibrium is only theoretically possible if numerical calculations can be made. Care must be taken when performing these calculations, for the following reasons. When the flame is well-attached, the results of these calculations are not significantly different from those obtained with the assumption of chemical equilibrium, at least in terms of the length and extent of the flame. Figures 5.13a and 5.13b provide an example of one such calculation, showing the isotherms for an H_2—air jet flame. However, although the maximum temperature is still located on a line resembling the reaction front as previously defined, it is not constant throughout. Moreover, the calculation can give differences of 50 to 100% when considering the overall quantities of the minor species in the flame, depending on whether the assumption of local chemical equilibrium or that of the inclusion of chemical kinetics is made, with a mechanism which appears to be suitable (see chapter 2 for aspects related to chemical mechanisms).

If the rates of chemical reaction are not very high, if the gas velocity is too high, or if the gases are diluted with an inert constituent, then the results of such calculations will deviate significantly from those found on the basis of an assumed chemical equilibrium, and potentially the flame may no longer be able to self-sustain. This situation will now be considered.

5.11 STABILISATION OF LAMINAR FLAMES

5.11.1 Problem definition

At the beginning of this chapter, a description was given of the stabilisation of a premixed, laminar conical flame on a Bunsen burner. The structure of a diffusion flame, established around a combustible jet of gas and maintained within it was also discussed in detail. Both types of flame cannot simply

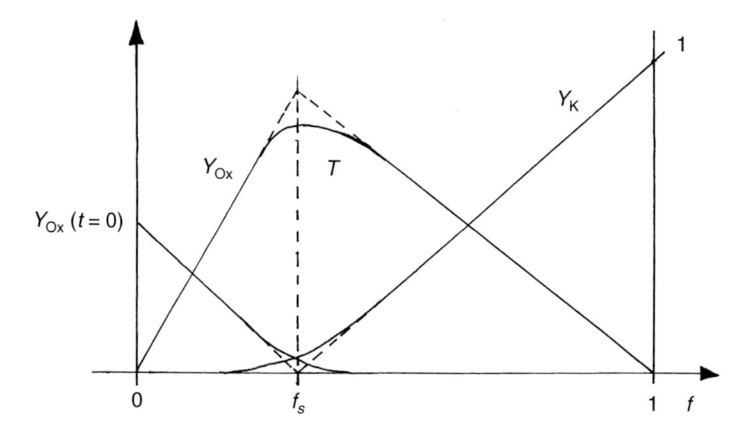

Figure 5.12b Internal structure of a diffusion flame showing the temperature and mass fractions of the various species as a function of the "mixture fraction" through the flame. The solid curves can be approximated, using the assumption of a single irreversible reaction, by a series of straight lines. f_S is the value of f which corresponds to the stoichiometric ratio, and is placed arbitrarily between 0 and 1 to make the diagram easy to read.

f	0.005	0.01	0.02	0.025	0.0275	0.03	0.038	0.05
T(K)	847	1301	2015	2275	2365	2399	2308	2131
M(g)	28.096	27.278	25.753	24.980	24.540	24.008	22.114	19.692
ρ (kg/m^3)	0.40445	0.25545	0.15578	0.13381	0.12647	0.12195	0.11678	0.11260
Ar	0.00900	0.00869	0.00812	0.00784	0.00768	0.00750	0.00685	0.00602
CO				0.0002	0.00004	0.00008	0.00015	0.00015
CO_2	0.00029	0.00028	0.00026	0.00024	0.00021	0.00016	0.00007	0.00004
H			0.00003	0.00053	0.00140	0.00261	0.00330	0.00175
H_2			0.00036	0.00392	0.01105	0.02749	0.10836	0.21692
H_2O	0.06968	0.13530	0.25391	0.30245	0.31947	0.32566	0.30608	0.27048
NO			0.00392	0.00434	0.00320	0.00166	0.00019	0.00002
N_2	0.75367	0.72788	0.67847	0.65445	0.64181	0.62702	0.57343	0.50435
O			0.00017	0.00059	0.00061	0.00036	0.00003	
OH		0.00001	0.00242	0.00628	0.00706	0.00564	0.00150	0.00027
O_2	0.16735	0.12750	0.05233	0.01933	0.00746	0.00181	0.00004	

Figure 5.12c Table of the curves relating the "mixture fraction" to the mole fractions (for various species), T, ρ, and M (molar mass of the mixture), for an H_2—air flame at chemical equilibrium. Here, f_S = 0.02795. (After R.W. Bilger (1976) *Progress in Energy and Combustion Sciences*, Vol. 1, p. 87.)

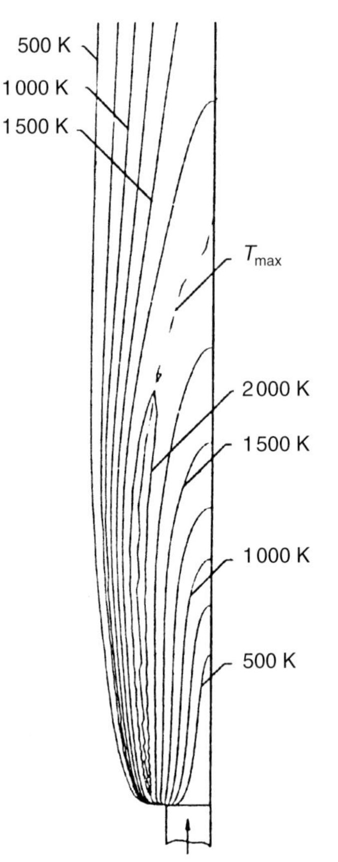

Figure 5.13a Example of a calculation of the isotherms in a laminar H2—air diffusion flame not at chemical equilibrium. The hydrogen jet has a velocity of 10 m/s and a radius of 0.32 cm. (Results of one of the first calculations of this type, performed by Kee and Miller (1978) *AIAA Journal*, 16, 2).

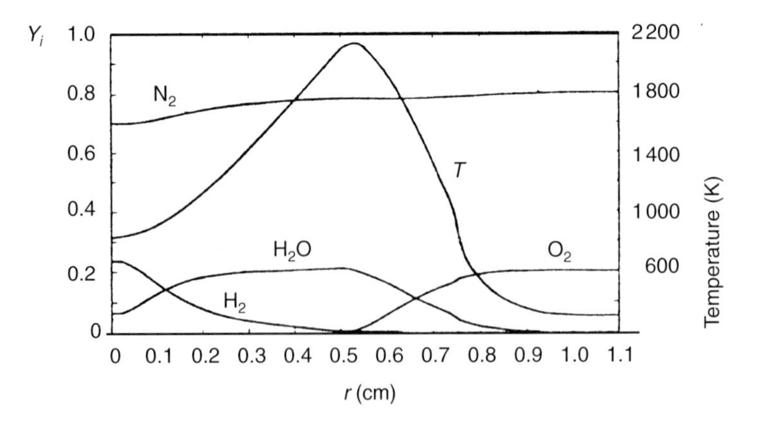

Figure 5.13b Calculated transversal profiles, at 1.5 cm from the mouth, of the mass fractions of the species and the temperature. (Results of one of the first calculations of this type, performed by Kee and Miller (1978) *AIAA Journal*, 16, 2).

"appear" spontaneously; they must be ignited by an external source of heat such as a spark or small auxiliary flame (e.g. a match). Ignition is only required for a limited period of time since, once lit, both types of flame should remain stable with no further external energy source needed. Is this always the case ? Does this phenomenon depend on the nature of the gas mixture, the gas velocity, etc.? The answer is of course "yes", although the theory is more complex, as we shall see below.

5.11.2 Stabilisation of a premixed, Bunsen burner flame

Deflagration theory explains, to a first approximation, why a premixed conical flame seats itself on top of the mouth of a Bunsen burner (see Figure 5.4a). The flame remains stable with respect to the burner since the speed with which it advances into the premixed flow is the same as the flow of unburnt gases in the opposite direction. However, the velocity profile of the gas at the mouth of the gas-supply tube, V_g, is not uniform. The velocity is zero or very small next to the wall and rises to a maximum at the central axis, as indicated in Figure 5.14a. Furthermore, across the same section, the burning velocity of the premixed flame in a laminar flow, V_F, in the gas mixture, is not uniform, and is also greatest on the tube axis (equal to S_L and less than the gas velocity) but decreases towards the tube's wall. The reasons for this are two-fold; firstly because the tube's wall absorbs heat from the flame (producing a cooler region) and secondly because the equivalence ratio of the mixture and hence the basic burning velocity are reduced by the effect of air mixing with the gas, since the section in question is slightly above the mouth of the tube.

As Figure 5.14a demonstrates, there are two possible points at which $V_F = V_g$, so long as the central gas velocity V_j, is not too high compared to S_L. However, these are points of instability from which the flame can advance, since $V_F > V_g$ between S_1 and S_2. These two points S_1 and S_2 can be found at several straight sections near the mouth of the Bunsen tube and it is at the section closest to the burner, corresponding to where the two curves $V_g(r)$ and $V_F(r)$ are tangential to the point S, that the flame stabilises, since it is unable to propagate further.

If the velocity V_j is too high, or the velocity S_L too low (i.e. if the mixture is too fuel-lean), stabilisation becomes impossible since the curves $V_g(r)$ and $V_F(r)$ can no longer intersect at a section and a lit flame will extinguish (causing "blow-off").

Conversely, if the velocity V_j is too low, the stabilisation point S will be too close to the burner mouth. Depending on the rim characteristics (geometry, thermal conductivity), the rim may gradually heat up until eventually point *S* falls within the tube. The flame can then propagate against the flow, through attaching to the walls, a phenomenon known as "flash-back".

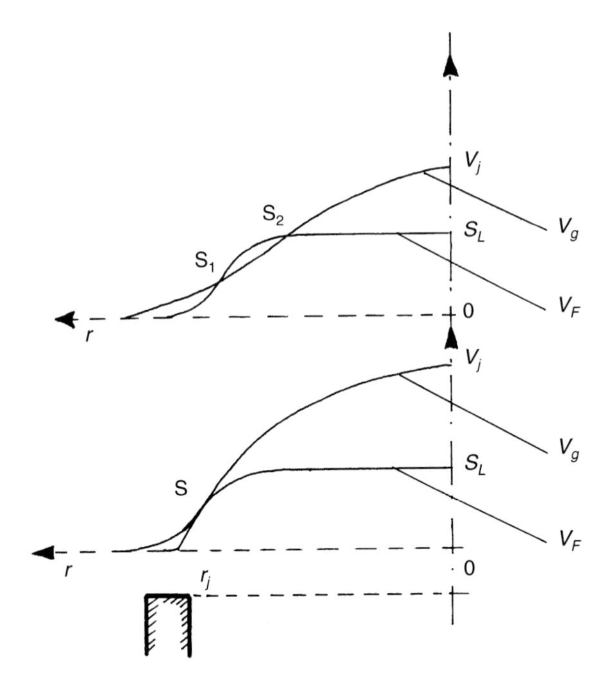

Figure 5.14a A premixed flame stabilises at a point where its own burning velocity V_F is equal to the flow velocity V_g. Point S is the closest point to the burner where this can happen and it is here that stabilisation occurs.

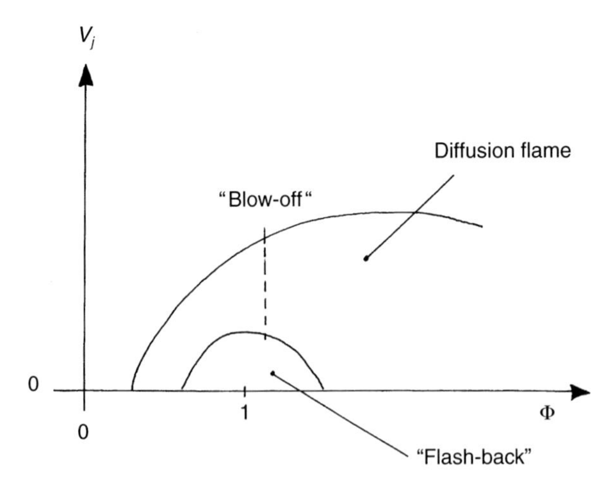

Figure 5.14b The velocity-equivalence ratio plane for a Bunsen burner, representing the region in which the flame is stable.

Thus, as shown qualitatively in Figure 5.14b, a plot can be drawn of the equivalence ratio of a gas jet (Φ) against its velocity (V_j), and hence of the region in which a stable flame exists. When the equivalence ratio exceeds unity, the flame attaches itself to the burner and becomes a diffusion flame where the gases burn in the external air, and hence it is no longer strictly a premixed gas burner.

5.11.3 Stabilisation of diffusion jet flames

The ideal jet flame in which chemical reactions occur infinitely rapidly is only encountered in practice if the gases react with each other spontaneously at the temperature at which they are injected, (an example of which is a fluorine-hydrogen mixture at room temperature). Under these conditions, the reaction front can attach immediately to the rim, thus separating the two reactants, even if the rim is cold (see Figure 5.15a). The flame becomes stable instantaneously, does not require ignition from an external source, and maintains itself irrespective of the flow velocities. If these velocities become too high, the reaction front may thicken and although the assumption of local chemical equilibrium may then cease to be true, the flame does not blow out.

On the contrary, for the more widely-used propellants (including even H_2—O_2) combustion is not spontaneous and the flame must be ignited by an external energy source which, when removed, may or may not result in flame stabilisation at a certain distance from the mouth of the gas-supply tube.

Let us now consider the situation where the shape of the burner tube is designed such as to avoid the formation of a recirculation zone behind it. In this case, the form of the attachment zone for the diffusion flame is as shown in Figure 5.15a. As soon as the flows of fuel and oxidant meet, a mixing zone develops in which the reactants diffuse into each other. At the centre of this zone is a line where the proportions of reactants correspond to the stoichiometric ratio of the global reaction, and which can provide an approximate representation of their combustion. Ignition using an external source can lead to the formation of a premixed flame in this mixing zone, even though the composition of the premixed medium is not homogeneous at all points. This flame propagates upstream, i.e. towards the gas-supply tube, and propagates most rapidly where the equivalence ratio of the mixture is very close to unity. Hence, if the gas velocity profile is such that the gas velocity is less than S_L, for an equivalence ratio of unity and in a given region containing the line of equivalence ratio equal to unity, then the premixed flame can be stabilised at the upstream limit of this region, in the same way as for a gas burner.

The premixed flame develops two branches, one on either side of the line corresponding to an equivalence ratio of unity. One is rich in fuel (and leaves in its wake hot reducing gases) and the other is rich in oxidant (and leaves in its wake hot, oxidising gases). Under these conditions, the premixed flame is

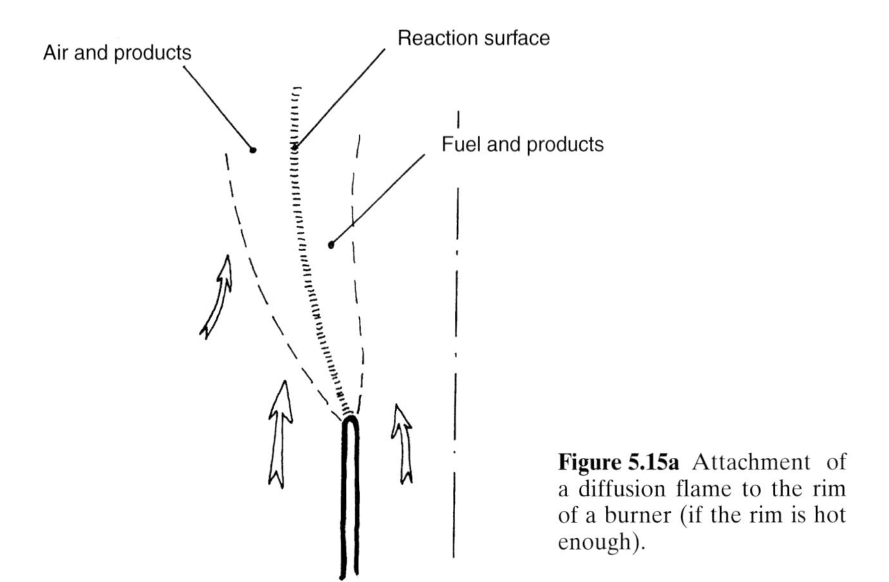

Figure 5.15a Attachment of a diffusion flame to the rim of a burner (if the rim is hot enough).

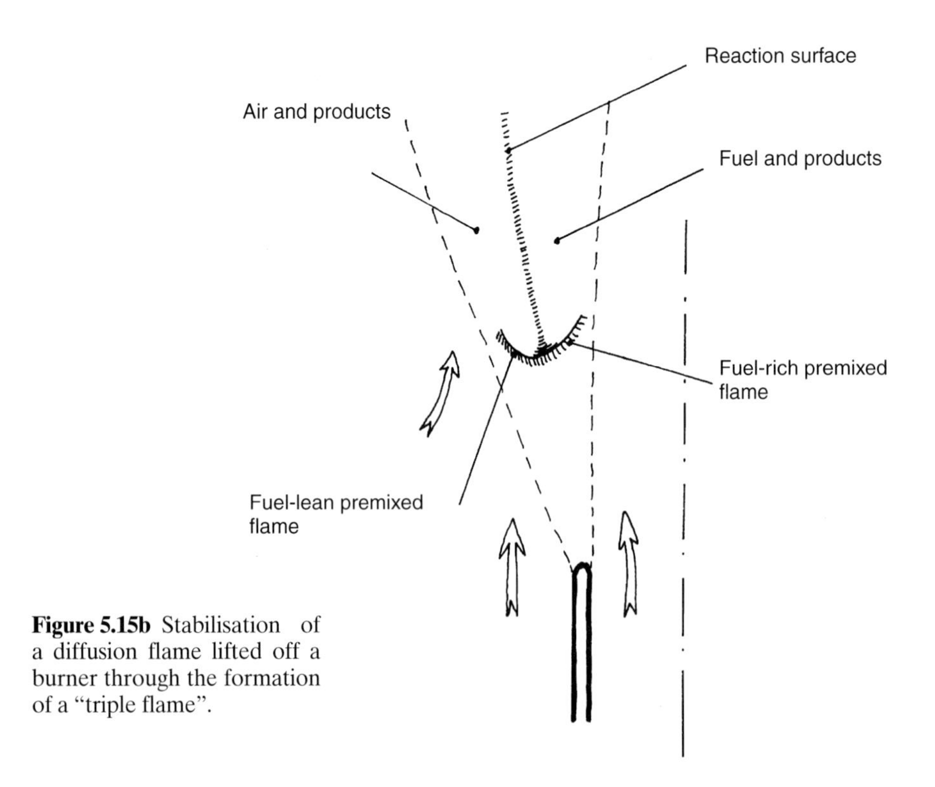

Figure 5.15b Stabilisation of a diffusion flame lifted off a burner through the formation of a "triple flame".

followed by a reaction front of the type found in diffusion flames, where the gases which were not burnt in the premixed flame (i.e. fuel or oxidant) now burn together.

This particular arrangement is often called a "triple flame" (see Fig. 5.15b), and can only arise if a zone with an equivalence ratio of unity intersects a zone where the gas velocity is equal to S_L. This is not always possible, and when this condition is not satisfied the flame cannot stabilise at a finite distance from the burner. This phenomenon is observed for certain fuels (e.g. methane, ethane, etc.) if the Schmidt number is between 0.5 and 1. Apart from this criterion, the region in which a diffusion flame can exist depends only on the approach velocity of the gases and not on the equivalence ratio of the mixture, since there will always be a zone somewhere in the flow where the equivalence ratio is equal to unity.

5.11.4 Stabilisation in a recirculation zone

The theoretical explanations presented in the preceding sections are only approximate insofar as they neglect the effects of the two-dimensional structure of the premixed flames near their attachment zone. For this reason it is not certain that they are totally valid, although when numerical calculations are performed which take these two-dimensional effects into consideration, the resulting data are, in general, found to be qualitatively similar. However, there is one situation in which such quasi-one-dimensional representations of flame fronts are no longer valid, namely when stable vortices of recirculating gas develop in the flow or in the wake of propellant-supply tubes.

A recirculation zone can be created in a premixed gas flow by placing an obstacle in the flow (e.g. a cylindrical wire as in Figure 5.16a, or a sphere, or even a disk). In practice, this requires the condition $V_j d/\nu > 50$ to be satisfied, where V_j is the flow velocity, d the diameter of the obstacle, and ν the kinematic viscosity of the fluid. The effect of this recirculation zone is to greatly

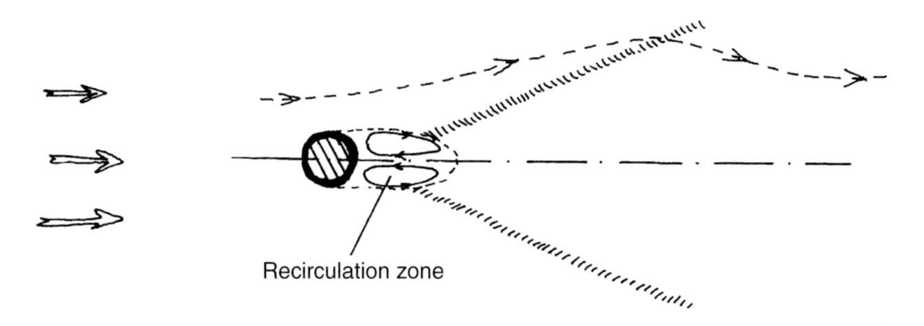

Figure 5.16a A premixed flame attached behind a cylindrical obstacle. The flame develops in the form of a "V" whose base is the recirculation zone.

increase the residence time for the gases such that, once ignited, combustion can be sustained and an oblique flame can develop into the flow downstream of this zone.

The same effect may be obtained by allowing the flow to pass through an abrupt increase in cross-sectional area. Figure 5.16b shows a "pilot" flame which is designed into some gas burners to stabilise the main flame. The pilot flame is produced by a circular trough, located around the rim of the gas burner, and whose base is drilled with holes at regular intervals.

Diffusion flames can be stabilised between two flows whose velocities are both greater than S_L simply by the blunt end of the tube separating the flows. So long as the recirculation zones produced are large enough, then combustion will be initiated within these zones (see Fig. 5.16c).

These flame-holding devices are only effective over a specific operating range, the limits of which can be determined, at least qualitatively, by applying the relevant theory. This determination is based on the concept of a continuously-stirred reactor (as mentioned in chapter 2).

The recirculation zone, whose structure is complex both in terms of gas dynamics and thermochemistry, can to a first approximation be likened to a given volume of gas fed by a certain flow rate of gas (containing fuel and oxidant) and from which escapes the same flow rate, but of partially burnt gases, whose composition is the same as the actual composition within the zone. Mixing in this volume is assumed to occur, by vortex recirculation, over a very short period of time compared to that required for chemical reactions to take place, and hence combustion occurs at uniform composition.

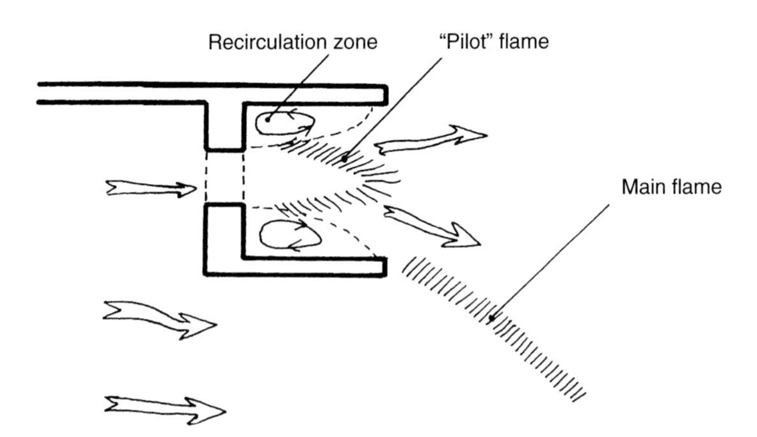

Figure 5.16b Stabilisation of a premixed flame on a burner by a small "pilot" flame. Part of the gas mixture is taken out of the main flow and slowed down in a small trough around the periphery of the rim. The flame stabilised in this trough can develop obliquely into the main flow.

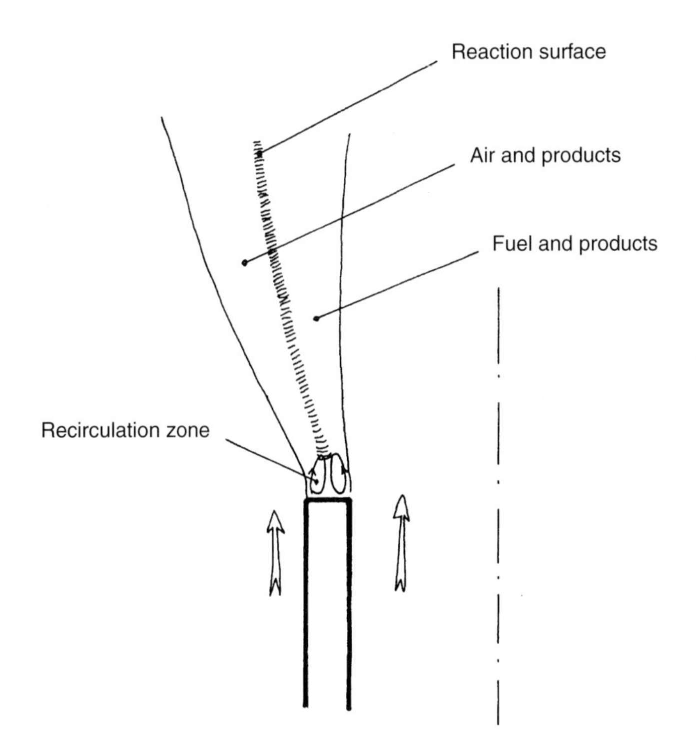

Figure 5.16c Stabilisation of a diffusion flame by recirculation zones on the blunt end of the rim of the burner.

If T_e denotes the inlet temperature and A_e the cross-sectional area of the inlet, the thermal balance, at least to the first approximation applied in this chapter (see section 5.5.1), can be written simply as:

$$\rho c_p V(T - T_e)A_e = + qw\, V_{\text{reactor}} \tag{5.26}$$

where ρV is the flow rate per unit straight section passing through the reactor, whose volume is V_{reactor}, w is the rate of reaction per unit volume per unit time, q is what has been termed the heat released and c_p the specific heat capacity of the system (assumed to be constant in this approximate approach). Equation (5.26) is in fact equation (5.12) of section 5.5.1, integrated over the reactor volume, taking into account the fact that the composition and temperature are constant throughout this volume.

The mass balance for the reactants can be written in a similar way, by assuming that a single chemical reaction occurs, at the specific stoichiometric ratio ν_s, of the fuel K and oxidant Ox, hence:

$$\rho V(Y_{\text{Ox}} - Y_{\text{Ox, inlet}})A_e = -\nu_s w\, V_{\text{reactor}} \tag{5.27}$$

$$\rho V(Y_{\text{K}} - Y_{\text{K, inlet}})\text{A}_e = -w\, V_{\text{reactor}} \tag{5.28}$$

which, by comparison with equations (5.26), (5.27) and (5.28), leads to:

$$Y_{\text{Ox}} - Y_{\text{Ox, inlet}} = \nu_s (Y_{\text{K}} - Y_{\text{K, inlet}}) \tag{5.29}$$

and:

$$Y_{\text{K}} - Y_{\text{K, inlet}} = -\frac{c_p (T - T_e)}{q} \tag{5.30}$$

Once again a linear relationship has been determined between Y_{Ox} and Y_{K} and between T and Y_{K}, which results from assuming that a single chemical reaction occurs between the fuel and oxidant.

As discussed previously, the rate of reaction, which depends on the chemical processes involved in combustion, is a function of Y_{Ox}, Y_{K} and T. Based on equations (5.29) and (5.30), the rate of reaction can be considered to depend ultimately only on T. Hence the solution to equation (5.26), which is an equation involving T explicitly and implicitly, should yield the temperature in the reactor (knowing $\rho V A_e$, V_{reactor} and the inlet temperature and mass fractions).

Equation (5.26) can be rewritten, by defining a "mean residence time" $t_S = \rho V_{\text{reactor}} / (\rho V A_e)$, as:

$$\frac{(T - T_e)}{t_S} = w(T) \frac{q}{\rho c_p} \tag{5.31}$$

which can be solved graphically once the form of the variation of w with T is known (see Figure 5.17a). This variation is the same as that in the premixed flame, within the limitations of our approximate approach, and thus yields the same function as that mentioned in paragraph 5.5.1! The reaction rate w is zero at $T = T_e$ and at $T = T_f$ (when combustion is complete), i.e. when all the fuel or all the oxidant has been consumed (or both!).

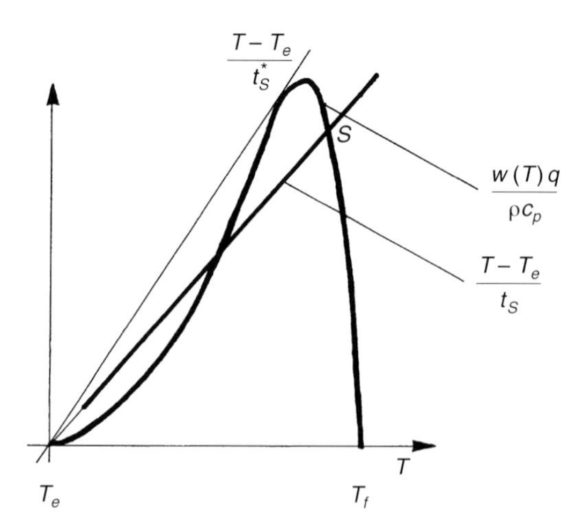

Figure 5.17a Determination of the operating point of a continuously-stirred reactor. t_S^* corresponds to the residence time at the extinction limit.

When t_S is greater than the value t_S^*, which corresponds to the gradient of the tangent at $w(T)q/\rho c_p$ passing through the origin, then each of the three points common to $w(T)q/\rho c_p$ and $(T - T_e)/t_S$ may offer a solution to the problem. The point $T = T_e$ gives the null solution that the zone was not lit. It can be shown that for the two remaining points only point S is the real solution since it is stable.

On the contrary, when t_S is less than t_S^*, the only possible solution is $T = T_e$ and thus at the centre of the zone a temperature corresponding to a certain extent of advancement of the combustion reactions cannot be maintained. Stabilisation is then only possible if the mean residence time of the gases is long enough; with this time being dependent on the chemistry and supply conditions, since it is related to $w(T)$.

Applying this theory to the practical examples mentioned above is complicated since ρV, V_{reactor} and the inlet values for Y_K, Y_{Ox} and T must be correctly identified, which is not always easy. Even so, irrespective of the possible errors associated with these quantities, it is possible to predict qualitatively the general shape of the stabilisation regions.

In terms of the stabilisation of a premixed flame, the inlet values of Y_K, Y_{Ox} and T are well-known, since they are those of the upstream flow. $w(T)q/c_p$ increases as the equivalence ratio approaches unity, irrespectively of whether it approaches from the rich or the lean side. Furthermore, the residence time in the zone lengthens as the zone volume increases, (and is consequently dependent on the diameter of the obstacle), and shortens as the supply flow rate increases (i.e. as the flow velocity increases). Regions of stabilisation may be plotted on the equivalence ratio/velocity plane of the form shown in Figure 5.17b. Above a certain velocity, stabilisation is no longer possible, even at the most favourable equivalence ratio value (unity), unless the size of the obstacle is increased.

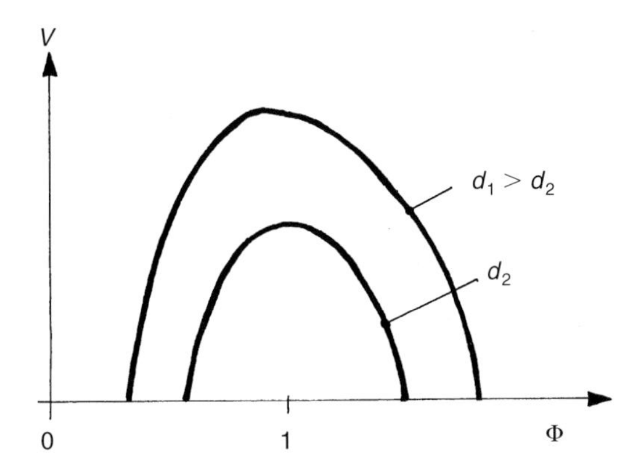

Figure 5.17b Stable operating regions for a flame holder whose operating principle is based on the recirculating zone effect. In the velocity-equivalence ratio plane, d_1 and d_2 correspond to two different sizes of flame holder.

In the case of a diffusion flame, the only problem lies in knowing the equivalence ratio of the mixture within the zone, which should be a function of the ratio of the velocities on the oxidant and fuel side respectively. Once the equivalence ratio has been estimated, the problem returns to being one of a premixed flame, since it is assumed that there is very rapid mixing of the gases within the zone.

WORKED EXAMPLES

In the following examples, both specific (i.e. per unit mass) and molar variables are given, since both are used in practice.

■ **1)** For combustible hydrocarbon-air mixtures, the Peclet number, Pe (defined in section 5.3 of this chapter) is approximately 50. The quenching distances can thus be determined, at 298 K and 1 bar (= 10^5 Pa), for stoichiometric mixtures of CH_4 in air and C_2H_2 in air using the data given in the table below.

	CH_4	C_2H_2	air
S_L ($m \cdot s^{-1}$)	0.45	1.58	
λ ($W \cdot m^{-1} \cdot K^{-1}$)	0.0342	0.0213	0.0261
c_p ($J \cdot mol^{-1} \cdot K^{-1}$)	35.6	44.1	29.2

The molar stoichiometric mixtures are:

$$CH_4 + 2\ (\text{air}) \quad \text{and} \quad C_2H_2 + 2.5\ (\text{air})$$

where air is assumed to consist of $O_2 + 4N_2$. Approximate thermal conductivities and heat capacities of these mixtures can be estimated by assuming that their values result from the linear combinations of their constituent gases, allowing for their respective molar fractions of the gases in the mixture. Thus, the heat capacity of the CH_4 + 2 (air) mixture can be calculated, knowing that there are 11 moles in total:

$$c_p = (1/11)\, c_p(CH_4) + (2/11)\, c_p(O_2) + (8/11)\, c_p(N_2)$$

In this way the following values are found:

	for CH_4 + 2 (air)	for C_2H_2 + 2.5 (air)
c_p ($J \cdot mol^{-1} \cdot K^{-1}$)	31.3	33.5
λ ($W \cdot m^{-1} \cdot K^{-1}$)	0.0288	0.0247

ρ is known from the state equation:

$$\rho\ (mol \cdot m^{-3}) = \frac{p}{RT}\ (p \text{ in Pa and } RT \text{ in } J \cdot mol^{-1})$$

hence, since $d = Pe\, a_T / S_L$: $d = 2.5$ mm for CH_4 + 2 (air)
$d = 0.58$ mm for C_2H_2 + 2.5 (air).

■ **2)** For the stoichiometric methane-air mixture:

$$CH_4 + 2O_2 + 8N_2$$

if the initial temperature, T_0, is approximately 300 K, the temperature, T_i, around 2000 K, and the maximum temperature, T_M, around 2200 K, then S_L is calculated to have a value of 0.45 $m{\cdot}s^{-1}$, and the following values for the flame front mixture: around 2100 K, $\lambda = 0.12\ W{\cdot}m^{-1}{\cdot}K^{-1}$ and $c_p = 36\ J{\cdot}mol^{-1}{\cdot}K^{-1}$. Whilst ρ is the density of the mixture if c_p is the specific heat capacity, if, however, as is the case here, c_p is the molar heat capacity, then ρ is the ratio $n/V = p/RT$, which at 2100 K is equal to 5.7 $mol{\cdot}m^{-3}$ at 10^5 Pa.

The thickness of the flame front e can be evaluated by combining equations (5.16) and (5.15) to give the average reaction rate $\langle w \rangle$, which leads to:

$$e = \frac{\lambda}{\rho c_p S_L} \frac{T_M - T_i}{T_i - T_0} = 1.5 \times 10^{-4}\ \text{m}$$

If the sweeping approximation is made that the methane concentration on the unburnt gas side of the flame front is the initial concentration, c_0, and that its concentration is zero on the burnt gases side, then an average value of the rate of disappearance of CH_4 can be calculated (in fact this will be an upper limit since reaction begins before the flame front and continues beyond it). Since, of the 11 moles of gas in the initial mixture (1 of CH_4, 2 of O_2 and 8 of N_2), only one is methane:

$$c_0 = \frac{p}{11RT}$$

and if c_0 moles disappear during a time t equal to e/V ($= e/S_L$), then the average rate of reaction is:

$$\langle w \rangle = \frac{p}{11RT} \frac{V}{e}, \text{ i.e. approximately } 1600\ mol{\cdot}m^{-3}{\cdot}s^{-1}$$

which is much higher than the value of 84 (same units) from section 2.2.2.2 of chapter 2, but for a different mixture (9.5% CH_4 + 90.5% O_2) and at a lower pressure (40 torr).

■ **3)** Let us again take the example of a stoichiometric methane-air mixture, but this time consider how the basic burning velocity would change if the nitrogen in the air is replaced by helium, a much lighter gas and a much better conductor of heat. It can be assumed that the rate of reaction will be the same in the presence of either bath gas and that the values of T_0, T_i and T_M are the same as above. For the conditions under consideration, the thermal conductivity of the combustible mixture is approximately 13.5 times greater with helium and its heat capacity is a factor of 1.4 less. On the contrary, the quantity $\rho = p/RT$ (see above), does not change if T and p remain the same. It can thus be deduced that:

$$\frac{S_{L,\text{ with He}}}{S_{L,\text{ with air}}} = \frac{(\lambda/c_p)^{1/2}{}_{\text{with He}}}{(\lambda/c_p)^{1/2}{}_{\text{with air}}} = 4.3$$

■ **4)** Let us briefly study how to adjust a Bunsen-type burner to suit the combustible gas being used, ignoring for the sake of simplicity pressure and temperature gradients. We know that the diameter of the burner mouth must be greater than the quenching diameter, whereas the fuel inlet must be smaller than this same quenching diameter. However these are not the only constraints which must be considered.

Let ρ_M denote the density of the mixture and ρ_K that of the fuel. V_M is the flow velocity of the mixture at the burner mouth (assumed to be uniform at all points over its area A_B) and V_K the velocity of the fuel at all points at the inlet orifice, whose area is A_K. Applying the principle of conservation of momentum for the entrainment of air by the fuel gives the condition:

$$(\rho_M A_B V_M)V_M = (\rho_K A_K V_K)V_K$$

The molar fraction of the fuel in the mixture, X_K, is roughly equal to the ratio of the volumes $A_K V_K / A_B V_M$ over unit time which, combined with the preceding relationship, leads to:

$$X_K = \left(\frac{\rho_M A_K}{\rho_K A_B}\right)^{1/2}$$

This molar fraction is close to 1/11 for a stoichiometric methane/air mixture which burns in accordance with the reaction:

$$CH_4 + 2O_2\ (+\ 8N_2) \rightarrow CO_2 + 2H_2O\ (+\ 8N_2)$$

and to 1/26 for a stoichiometric propane/air mixture burning in accordance with the reaction:

$$C_3H_8 + 5O_2\ (+\ 20N_2) \rightarrow 3CO_2 + 4H_2O\ (+\ 20N_2)$$

The densities are known from the equation of state:

$$\rho = \frac{pM}{RT}$$

For the molar mass, M, let us take an "average" value obtained by a linear combination of the individual molar masses (16 g for CH_4, 44 g for C_3H_8 and 29 g for air, i.e. roughly $O_2 + 4N_2$) and weighted by the molar fractions in the two mixtures being studied. Thus, for the first mixture:

$$M_1 = \left(\frac{16}{11}\right) + \left(10 \times \frac{29}{11}\right) = \frac{306}{11}$$

and for the second:

$$M_2 = \left(\frac{44}{26}\right) + \left(25 \times \frac{29}{26}\right) = \frac{769}{26}$$

from which, for the first:

$$\left(\frac{\rho_M}{\rho_K}\right)_1 = \frac{M_1}{M \text{ of } CH_4} = \frac{306}{11 \times 16}$$

and for the second:

$$\left(\frac{\rho_M}{\rho_K}\right)_2 = \frac{M_2}{M \text{ of } C_3H_8} = \frac{769}{26 \times 44}$$

Hence if, on a burner of fixed area A_B, a stoichiometric mixture of methane/air is burned rather than a stoichiometric propane/air mixture, the cross-sectional area of the fuel inlet must be changed such that:

$$\frac{\left(\dfrac{\rho_M A_K}{\rho_K A_B}\right)_1^{1/2}}{\left(\dfrac{\rho_M A_K}{\rho_K A_B}\right)_2^{1/2}} = \frac{Y_{K,1}}{Y_{K,2}} = \frac{26}{11}$$

and, since A_B is equal in both cases, we must have:

$$\left(\frac{A_{K,1}}{A_{K,2}}\right)^{1/2} = \frac{d_1}{d_2} = \frac{26}{11}\left[\frac{\left(\dfrac{\rho_M}{\rho_K}\right)_2}{\left(\dfrac{\rho_M}{\rho_K}\right)_1}\right]^{1/2}$$

giving 1.47. Having said this, the pressure of the combustible gas must also be adjusted and hence its flow rate, if the heat produced during a given period is to be approximately equal for the two cases, and knowing that the gross calorific value (GCV) of one mole of propane is about 2.5 times that of one mole of methane. Of course, this entire analysis is assumed to occur within the flame stabilisation region.

■ **5)** Calculating the stability range of a flame holder.

Consider the single reaction:

$$K + \nu_s Ox \rightarrow P \quad \text{with} \quad \rho w_K = -\rho^2 k Y_{Ox} Y_K \exp\left(-\frac{T_a}{T}\right)$$

where K is propane and Ox the oxygen in air.

It follows that:

$$Y_K \nu_s - Y_{Ox} = \text{Constant} = Y_{K,0}\,\nu_s - Y_{Ox,0}$$
$$c_p T + q Y_K = \text{Constant} = c_p T_0 + q Y_{K,0}$$

and the equivalence ratio Φ is defined such that:

$$\Phi = (Y_{K,0}/Y_{Ox,0})/(1/\nu_s), \quad \text{with} \quad Y_{K,0} + Y_{Ox,0} + (4 \times 28/32)Y_{Ox,0} = 1.$$

a) At $T_a = 15000$ K, $q/c_p = 2000$ K, $T_0 = 300$ K and $p = 1$ atm, plot $w_K(Y_K)$ for different values of Φ for $0 < \Phi < 2$. Take $k = 1$ (SI) arbitrarily.

b) Find, using graphical methods, the critical extinction value, t_S^*, of a recirculation zone behind a flame holder as a function of Φ. What value of k gives a t_S^* value of 10^{-3} s for $\Phi = 1$? This value will be used in c) and d).

c) Consider a cylindrical flame holder of diameter D positioned perpendicularly to the flow, which creates a recirculation zone in its wake whose section roughly corresponds to an isosceles triangle of base D (see Fig. 5.16a). Assume that the volumetric flow rate per unit surface area which enters the recirculation zone due to turbulent mixing is $V_0/10$, where V_0 is the velocity of the premixed flow ahead of the flame holder. The parameter Ω can be defined, equal to $(D/V_0)(p/p$ atmospheric). Plot the stability range for the flame holder in the plane (Ω, Φ).

d) With the previous value of w_K and the value of k from question b, calculate $\tau_c(\Phi)$ such that:

$$\frac{1}{\tau_c} = \frac{1}{Y_{K,0} - Y_{K,\text{end}}} \int_{\text{end}}^{0} w_K(Y_K)\,\mathrm{d}Y_K$$

where $Y_{K,\text{end}}$ is the value of Y_K at the end of combustion, and is a function of Φ. Plot the stability region in the plane $(\Omega/\tau_c, \Phi)$. Comment on the results.

CHAPTER 6

TURBULENT FLAMES AND DEFLAGRATIONS

6.1 WHAT IS A TURBULENT FLAME?

One of the simplest applications for turbulent flames is the ramjet, which burns fuel in a continuous, high-velocity flow of air (at approximately 50 $m \cdot s^{-1}$). The air is usually sufficiently hot and the fuel (gas or liquid) is usually injected sufficiently far upstream from the combustion zone for the combustion process to be considered as taking place in a gaseous medium where the fuel and oxidant almost perfectly mixed.

Figure 5a in the introductory chapter shows the most commonly-used apparatus. The flame-holders, in the form of ring-shaped gutters (troughs), create recirculation zones in their wake. Once ignited, flames stabilise in these recirculation vortices and are the source from which combustion then develops throughout the whole flow in the form of oblique "flames". Within this flame, which is about ten centimetres thick, the temperature rises gradually from the temperature of the inlet gases to a value approaching the combustion adiabatic temperature.

Premixed laminar flames were studied in depth in chapter 5. The aim of this chapter is to consider turbulent flames in the same way; highlighting the similarities and differences, and explaining the advantages offered by turbulent flames and how to estimate their properties.

The flow in a ramjet's combustion chamber is turbulent, which means that the gas velocity at any point, and even in a well-established flow regime, varies continuously in an apparently random way, as indeed does the temperature or any other characteristic parameter of the medium. This turbulence is due to the high inlet flow velocity, and develops in the nozzle upstream of the chamber due to the effects of shear flows at the walls and eddy formation on obstacles (particularly injectors). Reynolds showed that turbulent flow in a tube can be produced simply by increasing the flow velocity. In a similar way, a turbulent gas burner can be created simply by increasing the supply flow rate (hence velocity); this turbulence in the flame can be measured, and its effects observed. With the naked eye, the flame has a wide shape similar to that of a ramjet but, when photographed with a very short exposure time, the flame appears to be thinner and wrinkled, oscillating and moving continuously in a chaotic manner, a phenomenon which will be explained later (Fig. 6.1a and 6.1b).

From a practical point of view, turbulence can be a positive phenomenon for a system. If combustion in a ramjet or over a laminar gas burner is com-

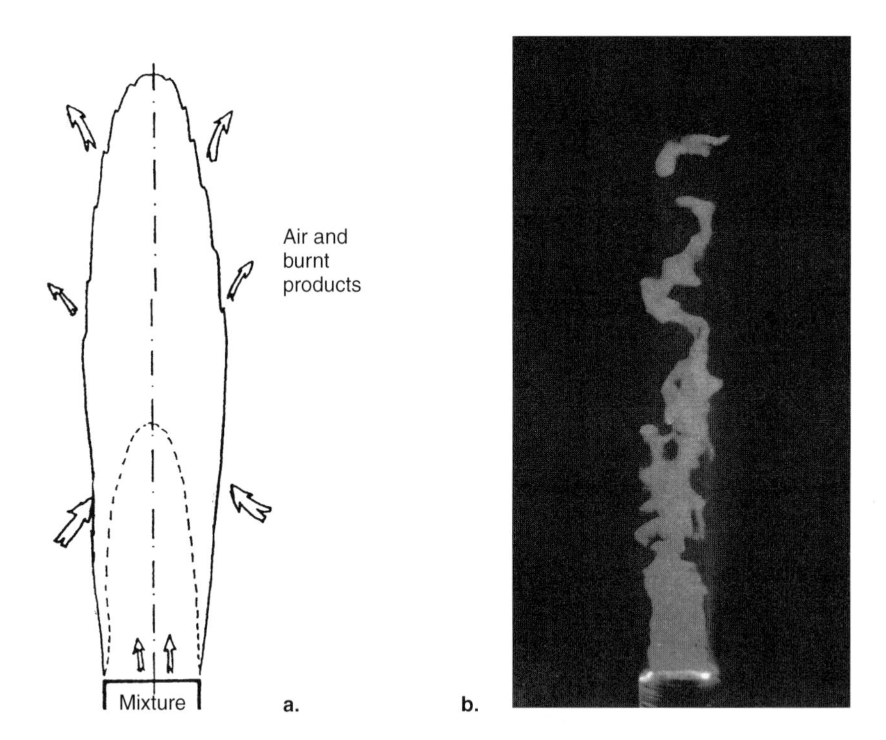

Figure 6.1 Turbulent flame on a gas burner.
a. Sketch of the flame as seen with the naked eye.
b. Instantaneous photograph of a cross-section of the flame (the mixture is seeded with a spray of oil droplets which vaporise above about 300°C). The flame is illuminated by a planar light beam lasting 20 nanoseconds produced by a YAG laser. Only the turbulent part of the flame comprising the unburnt mixture is visible (in green, the colour of the laser beam). The wrinkled profile of the front between the unburnt gases and the reaction zone can be clearly seen which, in this case, is very thin.

pared, the average angle of divergence of the flame is found to be virtually the same, which means that the ratio of flame length to its diameter is approximately equal for both. Now, since the inlet flow velocity is greater for the turbulent case, the flow rate of burnt fuel per unit volume is proportionately greater. Thus turbulence considerably increases the energy released per unit volume. The fact that the "visible" thickness is greater for a turbulent than for a laminar flame is an unfavourable factor when considering the intensity per unit volume of turbulent combustion since it limits the above effect but without compensating for it.

From a physical point of view, as we have seen, the existence of premixed flames is explained by the combined effect of two processes, chemical reactions and heat diffusion. In laminar flames, combustion reactions occur predominantly in a very thin gaseous layer on the downstream (hottest) side of the flame, whereas the molecular diffusion of heat and chemical species occurs throughout the upstream zone (thus introducing heat and radicals at this point which will later initiate the combustion reactions). The importance of these diffusion processes is enhanced by turbulence since it considerably accelerates this phenomenon.

Other types of turbulent flame are also used in industry. In spark-ignition internal combustion engines, a premixed turbulent flame propagates through the gaseous medium following ignition by the spark plug. The extensive movements of gases in the cylinder head immediately prior to sparking (caused by the gas entering via the inlet valve and compression by the rising piston) produce small scale turbulence which results in average burning velocities of several metres per second. This phenomenon is termed turbulent deflagration, and propagates much more quickly than laminar deflagration.

In industrial gas burners, on the other hand, a jet of combustible gas is often injected surrounded by an annular stream of air travelling in the same direction (see Figure 11 in the Introduction). After ignition, combustion is then established as a turbulent diffusion flame. Once again, turbulence plays a beneficial role in accelerating the mixing of the gases present (oxidant, fuel and also the burnt gases). This improvement is largely due to the high velocity gradients (causing shearing stresses) which exist between the inlet flows of the air and the fuel. The practical advantages of this process soon became apparent, since experiment shows that the total length of a turbulent diffusion flame is independent of the inlet velocity of the gases (once stabilisation has been achieved) whereas the length of a laminar diffusion flame is proportional to this velocity, as shown in Figure 6.2 (taken from the work of Hottel and Hawthorne in 1949). In this way, the length of the turbulent flame does not increase even for high rates of gas flow through the burner.

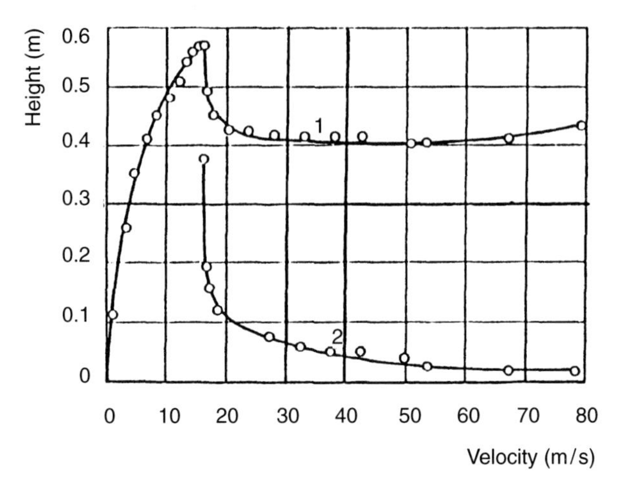

Figure 6.2 Characteristic heights of a diffusion jet flame as a function of velocity and of the inlet jet of fuel (After Hottel and Hawthorne, *Third Symposium on Combustion*, Williams and Wilkins, Baltimore, p. 254, 1949).

1. Total height; 2. Height above which the flame becomes turbulent.

The structure of turbulent flames is, however, much more complex than that of laminar flames and defies a simple description. Indeed, their structure is not yet fully understood, and this leads to the fact that turbulent flame structures are much more difficult to describe mathematically.

6.2 FUNDAMENTAL CONCEPTS AND CHARACTERISTICS OF TURBULENCE

The transport properties relating to mass, energy and momentum for turbulent systems were introduced in chapter 3. A detailed explanation of the structure and calculation of turbulent flames now requires an analysis of other properties of turbulence.

It would be impossible here to provide a full explanation of the physical nature of turbulence and its detailed characteristics since such an analysis would require a whole book in itself; and indeed several already exist (refer, for example, to Tennekes and Lumley[1], and then to Lesieur[2] for an even more advanced analysis). Suffice it to say here that turbulence must be characterised using a minimum of two critical quantities, namely its kinetic energy and its length scale. These quantities will be defined below, followed by a discussion of their relevance.

6.2.1 The kinetic energy of turbulence

One of the characteristics of any turbulent flow is that its velocity, at a given point, has random fluctuations. In chapter 3, the statistical average of each velocity component was defined as:

$$\overline{v_\beta} = \lim_{N \to \infty} \frac{1}{N} \sum_{i=1}^{N} v_\beta^{(i)} \tag{6.1}$$

where the subscript β denotes the component of the velocity vector ($\beta = 1, 2, 3$) for measurement number i. In practice, a time-average rather than a statistical average is often used, and is defined as:

$$\overline{v_\beta^t} = \lim_{T \to \infty} \frac{1}{T} \int_0^T v_\beta(t)\,\mathrm{d}t \tag{6.2}$$

This integral only converges to a finite value if the flow is statistically steady.

These two averages are assumed to be exactly equivalent, in this case, (where $\overline{v_\beta}$ does not vary with time), although the statistical definition of equation (6.1) is the only one of the two which is still valid when $\overline{v_\beta}$ is a function of time and the flow is not statistically steady.

The variance of the fluctuation in v_β can be defined as:

$$\overline{(v_\beta - \overline{v_\beta})^2} = \lim_{N \to \infty} \frac{1}{N} \sum_{i=1}^{N} (v_\beta^{(i)} - \overline{v_\beta})^2 \tag{6.3}$$

where this variance is directly related to the turbulence (and zero for laminar flow), the average value being characteristic of the mean flow.

The first important characteristic quantity of turbulence, called the kinetic energy of turbulence, is the half sum of the three variances for the three velocity components. It is often written:

$$v'_\beta = v_\beta - \overline{v_\beta}$$

1. Tennekes and Lumley (1974) *A First Course in Turbulence,* 3rd ed., MIT Press.
2. Lesieur M. (1990) *Turbulence in Fluids: Stochastic and Numerical Modelling,* 2nd ed., Kluwer Academic Publishers.

and hence the kinetic energy of turbulence, per unit mass, k, is defined:

$$k = \frac{1}{2}\left[\overline{v'^2_1} + \overline{v'^2_2} + \overline{v'^2_3}\right] \tag{6.4}$$

The variance in the velocity fluctuations is only given exactly by k if the turbulence is isotropic. If this is not the case the value of k is meaningless in relation to the respective magnitudes of $\overline{v'^2_1}$, $\overline{v'^2_2}$, $\overline{v'^2_3}$. Moreover, it is the probability density function of the velocity fluctuations, and not simply the variance, which provides information about the fluctuations occurring at a given point in the flow. Although the kinetic energy of turbulence is not the only characteristic of interest, it is the first and most important characteristic to consider when analysing velocity fluctuations.

When the flow is not incompressible, i.e. when the density ρ cannot be assumed to be constant, which is particularly the case for flames, it is often convenient to define average quantities with a weighting by ρ. Hence the mean velocity is defined as:

$$\tilde{v}_\beta = \frac{\overline{\rho v_\beta}}{\overline{\rho}} = \frac{\lim\limits_{N \to \infty} \frac{1}{N} \sum\limits_{i=1}^{N} \rho^{(i)} v_\beta^{(i)}}{\lim\limits_{N \to \infty} \frac{1}{N} \sum\limits_{i=1}^{N} \rho^{(i)}} \tag{6.5}$$

noting that: $\quad v'_\beta = v_\beta - \tilde{v}_\beta \quad$ and $\quad \widetilde{v'^2_\beta} = \dfrac{\overline{\rho v'^2_\beta}}{\overline{\rho}}$,

and

$$\tilde{k} = \frac{1}{2}\left(\widetilde{v'^2_1} + \widetilde{v'^2_2} + \widetilde{v'^2_3}\right) \tag{6.6}$$

These means are known as mass-weighted means or Favre means, since they were originally proposed by A. Favre. Chapter 3 discussed the relevance of these means in relation to the "Reynolds tensor" in section 3.5.2.

6.2.2 The integral turbulent length scale

The second important characteristic of turbulence (and as important as the first) is the characteristic length scale for the vortices or turbulent "eddies". This characteristic scale idealises the mean size of the turbulent eddies and is classically defined by considering the correlation between the velocity fluctuations at two neighbouring points. For a turbulent medium which is homogeneous (i.e. where $\overline{v_\beta}$ is independent of $\vec{x}$) and isotropic, the correlation is defined as:

$$C(|\Delta x|) = \overline{v'_\beta(\vec{x})\, v'_\beta(\vec{x} + \overrightarrow{\Delta x})} \qquad \forall \beta \tag{6.7}$$

When the two points are very close together (i.e. when Δx is very small), the variation in the velocities at these points is closely related, and hence C is very similar to $\overline{v'^2_\beta}$. On the other hand, for two points far apart, there should be no relationship whatsoever between the fluctuations in their velocities, and thus C is expected to tend to zero as Δx tends to infinity.

The integral scale for turbulence is a mean correlation length, defined as :

$$\overline{v'^2_\beta} \cdot l_t = \int_0^\infty C(|\Delta x|)\, \mathrm{d}|\Delta x| \qquad (6.8)$$

Only when $C(|\Delta x|)$ tends to zero (i.e. when Δx tends to infinity and the integral l_t converges) can truly random, turbulent flow exist. Otherwise, the flow consists of a collection of waves, some of which may be identical (allowing for translation) and widely separated, which is not turbulent flow. For a non-isotropic medium, expression (6.8) continues to be valid when several scales are defined (depending on β and the direction of $\overrightarrow{\Delta x}$). In a non-homogeneous medium the integral (6.8) will not automatically converge if the characteristic scale for the spatial variation of $\overline{v'_\beta(x)\, v'_\beta(x)}$ is small compared to l_t itself. This last condition is always implicitly assumed to be satisfied whenever l_t is used.

6.2.3 The “spectrum” of turbulent length scales

In reality, not all the turbulent eddies are of size l_t since a wide range of eddy sizes are created and l_t only gives their mean size. Instead of size, the term length scale is often used; a physical but vague term which becomes clearer if a plot of $C(|\Delta x|)$ is considered instead of l_t alone. If the associated spectral distribution is defined (the three-dimensional “spectrum”) based on the Fourier transform of $C(|\Delta x|)$, then $E(n)$ is obtained, where n is the wave number associated with $|\Delta x|$. It can be shown that:

$$k = \int_0^\infty E(n)\, \mathrm{d}n \qquad (6.9)$$

which indicates that the kinetic energy of turbulence is distributed over various wave numbers between zero and infinity.

Figures 6.3a et 6.3b show the profile of the function $E(n)$, both for linear and logarithmic coordinates. Various experiments have shown (as have certain theories) that this profile of the $E(n)$ spectrum is nearly always the same, provided that care is taken to use, for the x- and y-axis scales, reduced, i.e. dimensionless coordinates conveniently defined. In fact, if n is reduced by $1/l_t$ and $E(n)$ by kl_t, then different experiments yield the family of plots shown in Figure 6.3b, which deviate from each other only for sizes which are very large (small nl_t) or very small (large nl_t). Since at such very small or very large sizes

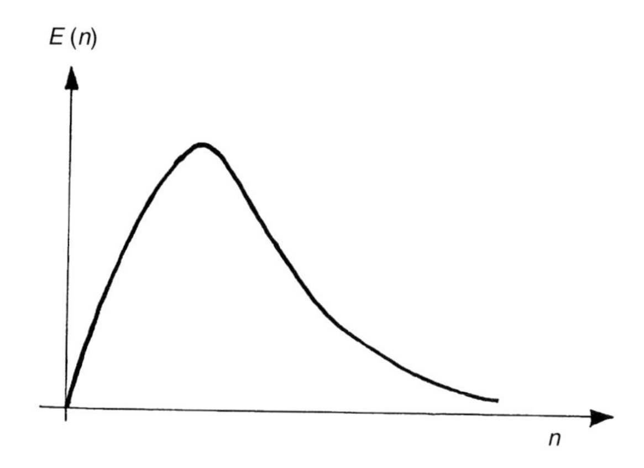

Figure 6.3a Profile of the velocity fluctuation "spectrum" for homogeneous, isotropic turbulence.

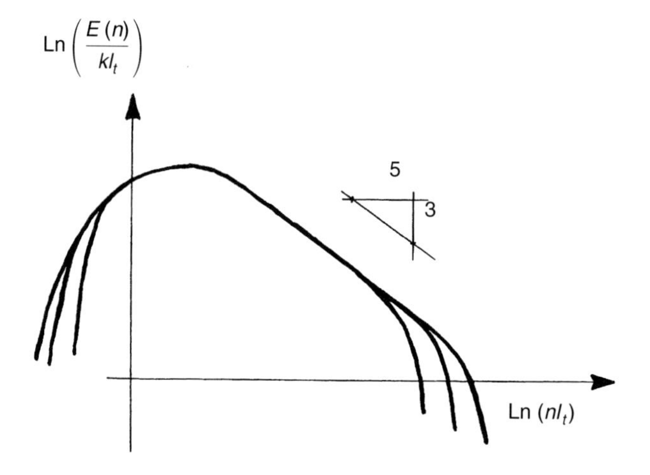

Figure 6.3b The "spectrum" with reduced (i.e. dimensionless) log-log coordinates.

$E(n)$ is very low compared to its maximum value (the scale is logarithmic), these differences can be assumed to be insignificant, and thus turbulence (and $E(n)$) can be characterised simply by using the two quantities k and l_t for all cases.

In a later section we shall consider another characteristic length scale for turbulence. This scale is used for the range of sizes at which the $E(n)$ spectrum falls off sharply at large n (see Fig. 6.3b), and is known as the Kolmogorov length scale. This drop in $E(n)$ is explained physically by the fact that for large n, the associated "eddy sizes" are very small, whereas the velocity gradients related to these eddies are very large. The result is that fluid viscosity becomes a significant factor in dissipating the eddies through friction (since friction is proportional to viscosity and to the velocity gradients). According to Kolmogorov, this scale can be calculated as a function of k and l_t using:

$$\eta = \alpha l_t Re_T^{-3/4} \tag{6.10}$$

where Re_T is the Reynolds number for the turbulence:

$$Re_T = \frac{k^{1/2} l_t}{\nu} \tag{6.11}$$

and α is a constant which depends on the precise determination of l_t (usually 1.35) and ν is the kinematic viscosity, ($\nu = \mu/\rho$), of the fluid in question.

In simplified terms, η represents the size of the smallest eddies in the turbulence, or the distance separating two points in the fluid below which the fluctuations in the fluid velocities at these points are perfectly correlated.

Still in relation to homogeneous and isotropic turbulence, other scales of turbulence can be defined (based on time or on length). For example, one time scale which appears to be very important physically is the integral time scale, τ_L, defined according to the correlation of the velocity fluctuations along the path followed by a fluid particle (known as the "Lagrangian" approach):

$$\tau_L = \int_0^\infty \frac{\overline{v'_\beta(t)\, v'_\beta(t+\tau)}}{\overline{v'_\beta(t)\, v'_\beta(t)}}\, dt \qquad \forall \beta \tag{6.12}$$

This parameter is extremely difficult to measure since it requires a measuring device which can track the turbulent path of one fluid particle. However, it is of great physical importance in turbulence theories and assumes an even greater importance when the fluid particle in question involves a chemical reaction which proceeds with time.

By assuming that turbulence is fully modelled by its kinetic energy and integral scale, as defined previously, and that its spectrum is always similar, then one consequence is that τ_L can be expressed only as a function of k and l_t and, simply for the purposes of dimensional analysis, we have:

$$\tau_L = \alpha' \frac{l_t}{k^{1/2}} \tag{6.13}$$

where α' is a dimensionless multiplying constant.

This equation will hold for all turbulent time and length scales. All integral lengths and times will be proportional to l_t and $l_t/k^{1/2}$ respectively, which from now on will be known as τ_t. The lengths and times relating to small scales of turbulence (such as η) will not only be proportional to l_t and τ_t but also to the Reynolds number for turbulence raised to a certain power.

If homogeneous, non-isotropic turbulence is now considered, or even non-homogeneous, non-isotropic turbulence, then other length or time scales may be involved in addition to k and l_t. In non-isotropic turbulence, the three components of the kinetic energy have been shown to be involved ($\overline{v'^2_1}/2$, $\overline{v'^2_2}/2$, $\overline{v'^2_3}/2$). Equally, scales can be introduced related to different spatial directions. Also involved are the correlation of $\overline{v'_1 v'_2}$ ..., etc. and the length scales associated with these correlations. However, it is usually advisable to limit as much as possible the number of characteristic quantities of turbulence taken into account in the calculations. Particularly, experiments have shown that all characteristics associated with small-scale turbulence (high wave numbers) are related to the isotropic component of the turbulence, even for non-isotropic (and non-homogeneous) media, and that the characteristics k and l_t alone are therefore sufficient to represent this type of turbulence, with Re_T defined as in equation (6.11).

6.2.4 General overview of the "$k - \varepsilon$" turbulence model

As part of the general assumption which considers turbulence as being chiefly characterised by the two quantities l_t and k, studies into turbulence over the last fifty years have led to the proposal of equations governing k and l_t under all turbulent flow conditions, even non-homogeneous and non-isotropic systems. These equations constitute a "mathematical model for turbulence".

The simplest situations for which these equations were initially proposed were jets, boundary layers and mixing layers. For example, Figure 6.4 shows the mixing layer between two planar, parallel flows of different velocity, downstream from a wall.

In practice, it is not l_t but another parameter which is used in the most well-known turbulence model. In fact, based on Kolmogorov's work, it was recognised that the rate at which the turbulence was dissipated, defined as:

$$\varepsilon = \nu \sum_{\alpha,\beta} \overline{\frac{\partial v'_\alpha}{\partial x_\beta}} \, \overline{\frac{\partial v'_\alpha}{\partial x_\beta}} \tag{6.14}$$

played a crucial role. ε has the dimensions L^2T^{-3} and therefore, based on the reasoning that it must be possible to express all parameters in terms of k and l_t, the following can be written:

$$\varepsilon = C_D \frac{k^{3/2}}{l_t} \tag{6.15}$$

where C_D is a dimensionless constant (usually taken to be $(0.09)^{3/4}$).

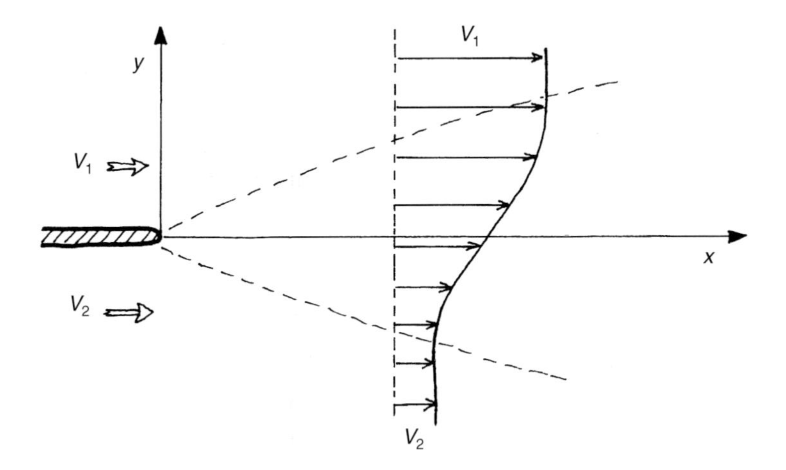

Figure 6.4 The "mixing layer" between two parallel flows of different velocity.

The most classical model uses k and ε instead of k and l_t. For the simple flow shown in Figure 6.4, and using the notation given, the "$k - \varepsilon$" model equations are:

$$\left.\begin{aligned} \overline{v_x}\frac{\partial k}{\partial x} + \overline{v_y}\frac{\partial k}{\partial y} &= \frac{\partial}{\partial y}\left(\frac{\nu_t}{S_{ce}}\frac{\partial k}{\partial y}\right) + \nu_t\left(\frac{\partial \overline{v_x}}{\partial y}\right)^2 - \varepsilon \\ \overline{v_x}\frac{\partial \varepsilon}{\partial x} + \overline{v_y}\frac{\partial \varepsilon}{\partial y} &= \frac{\partial}{\partial y}\left(\frac{\nu_t}{S_{c\varepsilon}}\frac{\partial \varepsilon}{\partial y}\right) + C_1\frac{\varepsilon}{k}\nu_t\left(\frac{\partial \overline{v_x}}{\partial y}\right)^2 - C_2\frac{\varepsilon^2}{k} \end{aligned}\right\} \tag{6.16}$$

where $\overline{v_x}$ and $\overline{v_y}$ are the average longitudinal and transversal velocities (in the x and y directions respectively), S_{ce}, $S_{c\varepsilon}$, C_1 and C_2 are constants (equal to 1, 1.3, 1.45 and 1.95 respectively, from comparison with experiment) and ν_t is the coefficient of "turbulent viscosity", again written as a function of k and ε and given by:

$$\nu_t = C_\mu \frac{k^2}{\varepsilon} \tag{6.17}$$

with the constant C_μ equal to 0.09. Using (6.15), equation $\nu_t = \beta k^{1/2} l_t$ can also be derived where the constant β is found using C_μ and C_D.

A further description of this model can be found in the book by Tennekes and Lumley; of particular interest are the explanations of the form of the equations and the values of the constants.

In the context of our analysis, we will simply note that the term $\nu_t(\partial\, \overline{v_x}/\partial y)^2$ represents the generation of turbulence caused by mean velocity gradients and ε its dissipation due to local velocity gradients (as indicated by equation (6.14)).

This "$k - \varepsilon$" model, with all its variations, is a classical model which has been validated by experiments for certain conditions and which offers a useful, and occasionally very precise way of approximating turbulence in a variety of flows. It is one of the many advances achieved during the one hundred years of study into turbulence, and can be used to advantage in the study of turbulent combustion.

Section 6.8 will include a description of the modifications which must be made to the theory when considering combustion processes.

6.3 THE STRUCTURE OF TURBULENT PREMIXED FLAMES

Before tackling turbulent flames quantitatively, we will discuss what is already known, or suspected, about their qualitative structure. This discussion will begin by considering premixed flames as distinct from diffusion (non-premixed) flames which will be treated in section 6.6.

One of the first questions to be asked when considering turbulent flames is what makes a flame laminar or turbulent? In the majority of cases the answer is simple: it depends on whether or not the flow upstream of the flame is turbulent, i.e. is the flow characterised by fluctuations in velocity and pressure caused by the high flow velocities and steep velocity gradients. This simple answer is complicated by the fact that under certain conditions the flame can change the turbulent nature of the flow, indeed, in general, the expansion of the gases resulting from combustion causes gas motions which interact in a complex manner with the structure of the turbulence, even affecting conditions upstream of the flame itself. Furthermore, the presence of the flame induces variations in the temperature, concentration of chemical species, and density which would not normally be found in the upstream flow.

Moreover, in certain cases the flame can become turbulent even when the flow upstream is not, e.g. for a flame propagating in a premixed medium at rest containing obstacles. However, even here, the velocity field is modified by the flame and the flow becomes turbulent upstream of the flame. The problem is further complicated by the fact that the turbulence is not as well established as it would have been if it had been created further upstream, but is not significantly dissimilar from the case where turbulence is pre-established before the flame.

6.3.1 The three types of turbulent premixed flame

Several different types of turbulent flame can be distinguished, differentiated by the value of two characteristic numbers, $k^{1/2}/S_L$ and l_t/e_L where $k^{1/2}$ and l_t are the characteristic velocity and length for the turbulence at the point considered in the flame, and S_L and e_L the fundamental burning velocity and the planar thickness of a laminar flame which would propagate through the mixture in question (as defined in chapter 5).

Figure 6.5 shows the three principal types of flame: "wrinkled", "thickened", and an intermediate type called "wrinkled-thickened". A wrinkled turbulent flame consists of thin flames or "flamelets" which are almost laminar and wrinkled by the turbulence. The length scales for the turbulence in this case are all greater than e_L. In theory, this type of flame is restricted to the range where the Kolmogorov scale length $\eta > e_L$ (see (6.10)). However, this constraint should not be applied too strictly since the flame structure starts to be modified when η drops below the laminar flame thickness, but only becomes significantly different when the former parameter becomes several times less than the latter. If an instantaneous photograph is taken of a wrinkled-type flame then a long flamelet of uniform thickness e_L is observed, twisted and wrinkled in such a way as to occupy a "width" e_T, which may be considered as being the average thickness of the turbulent flame. (shown schematically in Figure 6.6a). The photograph in Figure 6.1b shows that; since the flow is seeded with a spray of oil droplets which vaporise when the temperature rises to above approximately 300°C, then the line defining the luminous zone is an instantaneous isotherm of the flamelet.

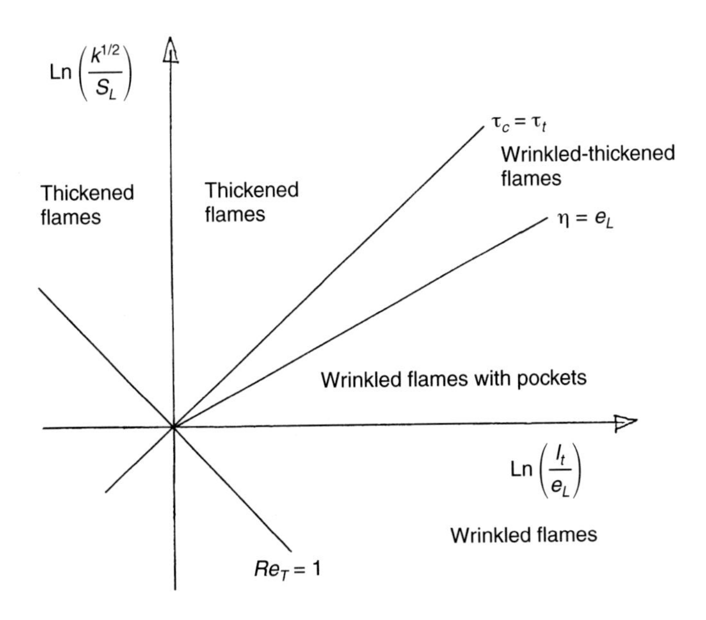

Figure 6.5 Different types of premixed turbulent flames.

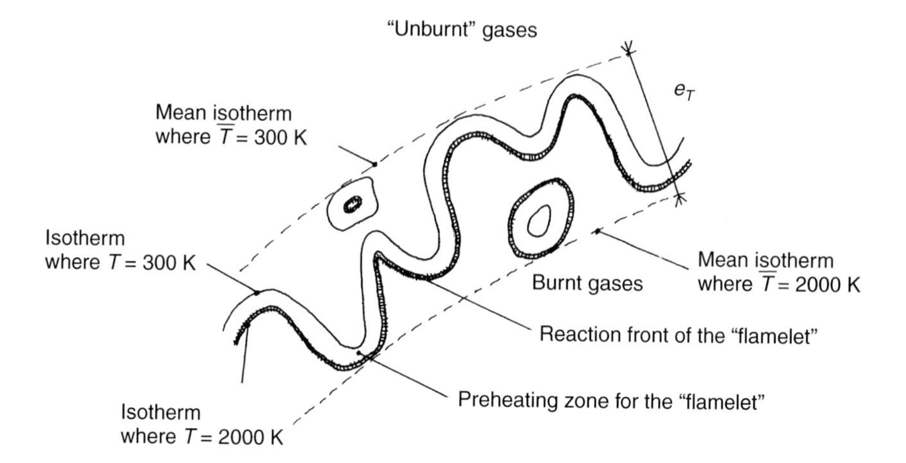

Figure 6.6a Sketch of a "wrinkled" flame at one instant in time.

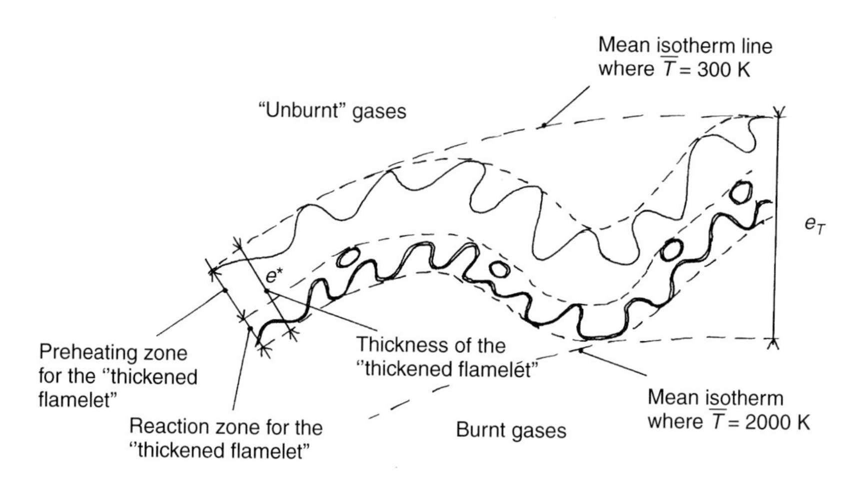

Figure 6.6b Sketch of a "wrinkled-thickened" flame at one instant in time.

If the turbulent energy k is increased, with l_t kept constant, such that $\eta < e_L$, then interactions between neighbouring flamelets occur with greater frequency since the radius of curvature of the flamelets may then, at certain points, be of the same order of magnitude as e_L. The flame structure gradually adopts the profile shown in Figure 6.6b. Note the zones of interaction between flamelets, which locally increase the thickness of the flamelets above e_L. At other points the opposite is true, with this thickness reduced as a result of the flamelets being stretched by the turbulence. The local stretch rate can be approximated to τ_K^{-1}, where τ_K is the Kolmogorov time for the turbulence, defined as:

$$\tau_K = \left(\frac{\nu l_t}{k^{3/2}} \right)^{1/2} \tag{6.18}$$

and it increases as k increases. When we observe the presence of reaction zone loops, separated from the main, continuous reaction zone, this implies that two flamelets have been pushed "back-to-back" by the turbulence and that their interaction results in the extinction of the two reaction zones which came into contact.

If the stretch rate of a flamelet is so great that τ_K becomes less than approximately τ_c (a chemical duration which can be defined by e_L/S_L), then it may be extinguished at the point where the stretching occurs. It can be shown that this phenomenon is governed by Lewis's number, which is a measure of the ratio of the rates of heat diffusion and the rates of diffusion of the species inside the flamelet. In this case, local extinctions of non-curved flamelets may be observed; although these flamelets may subsequently reignite if the stretch rate is reduced, to be replaced by other extinguishing events elsewhere. The overall result is a reduction in the average rate of combustion. A simple calculation shows that the equality $\tau_K = \tau_c$ is equivalent (so long as a constant factor of multiplication is applied) to the equality of laminar flame thickness and the Kolmogorov scale, and hence that these local extinctions occur in the domain of "wrinkled-thickened" flames, at the same time as flamelet interactions.

Turbulent flames of this intermediate type have been termed wrinkled-thickened flames because they contain "averagely-thickened flamelets" which resemble the thickened flame shown in Figure 6.6b. However, since this local thickening is due only to smaller-scale turbulence, the thickened flame which results will remain wrinkled by larger-scale turbulence.

When k is increased further still (for a fixed l_t), then eventually there will only be one non-curved, but very thick flame (much thicker than e_L). The structure of this thick flame is not as simple as that of a laminar flame. Consider a snapshot of a section through the flame. The more intense turbulent perturbations will have initiated numerous flamelet interactions and within the average flame thickness there would probably be variations in the thickness of the preheating zone from one point to another, as well as wrinkling of the reaction zone and the appearance of small-scale "subsidiary" reaction zones (see the schematic representation in Figure 6.6c).

The exact boundaries between the various domains shown in Figure 6.5 are fairly arbitrary and certainly much less well defined than a simple line. The reason why $\eta = e_L$ has been chosen as an upper limit for the thickened flame region is that it is certain that if $\eta > e_L$ then the fluctuations in the velocity due to the turbulence will not modify the internal structure of the laminar

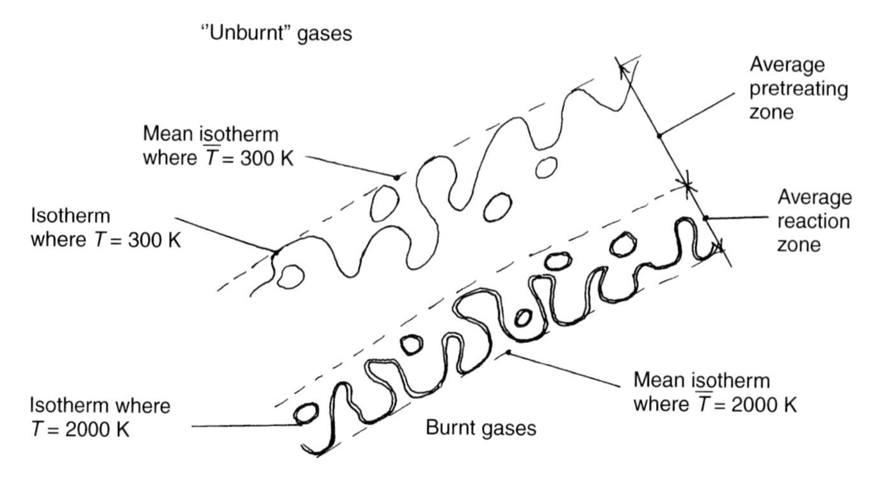

Figure 6.6c Instantaneous view of a "thickened" flame.

flamelets. However the "amount" of velocity fluctuations at an intermediate scale greater than η will still be very small, owing to the significant fall in the spectrum of k (see Fig. 6.3b) for logarithmic coordinates. This definition is therefore quite restrictive.

The lower limit $\tau_t = \tau_c$ for "thickened" flames, where $\tau_t = l_t/k^{1/2}$ and $\tau_c = e_L/S_L$, has been defined as being the upper limit where "wrinkled-thickened" flames are no longer wrinkled, i.e. where the average thickness of the thickened flamelet becomes equal to the integral length scale of the turbulence l_t. This approach was proposed for the first time by V. Zimont in 1979 and requires an estimation of the thickness of the thickened flamelets, for which Zimont also proposed a calculation method. Once again, however, this proposed limit should not be taken as a very precise boundary since the calculations of the turbulence scales often use rather arbitrary values for the constants of proportionality involved.

6.3.2 Experimental evidence

At the present time, not all the theoretical discussions described above have been completely supported by experiment. The experimental methods required to provide the necessary proof have not been fully developed, especially to a level which would provide the detail shown in Figures 6.6a, b, and c.

There is, however, a relatively straightforward way of showing, in various experiments, the presence of the three types of turbulent flame mentioned above. The method involves measuring the temperature fluctuations and the probability density function of these fluctuations in the middle of an adiabatic, turbulent flame. For a wrinkled-type flame, the instantaneous and local measurement (so long as the resolution is high enough for both) will give either the temperature of the unburnt or of the burnt gases, or an intermediate temperature corresponding to an instant when the flame front (of thickness e_L) passes over the measuring probe. If the time taken by the flamelets to pass the probe is very small compared with that taken by the unburnt or burnt gases, then the probability density of the temperatures will have two peaks, as indicated by Figures 6.7a and b.

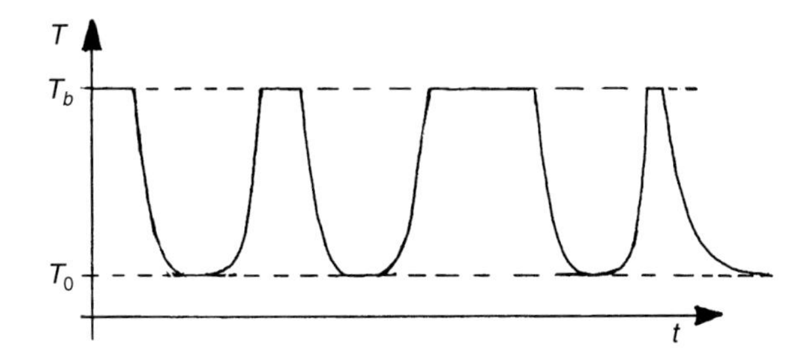

Figure 6.7a Instantaneous temperature signal for a probe crossing a "wrinkled" flame (T_0 = temperature of unburnt gases; T_b = temperature of burnt gases).

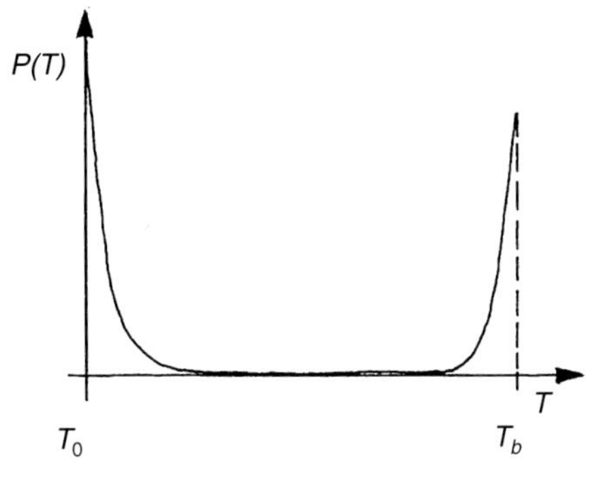

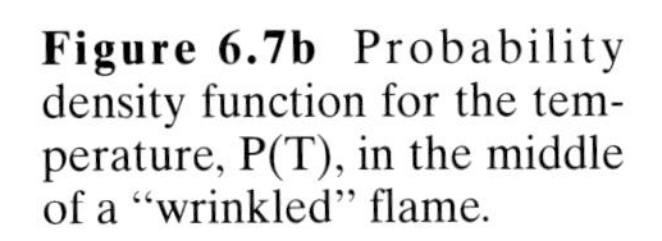

Figure 6.7b Probability density function for the temperature, P(T), in the middle of a "wrinkled" flame.

However, if a thickened-type flame is studied, the temperature fluctuations will not correspond at all to those of the unburnt or burnt gases at the same point for the average flame. The probability density will therefore be much narrower and probably similar to a Gaussian distribution if the measurement were made in the middle of the average flame. These profiles are shown in Figures 6.8a and b.

Between these two cases, wrinkled-thickened flames will simultaneously show the presence of peaks (caused by the wrinkles) and of intermediate situations corresponding to the passing of a "thickened flamelet", for which the probability of finding temperatures other than those of the unburnt or burnt gases will be significant, as shown in Figures 6.9a and b.

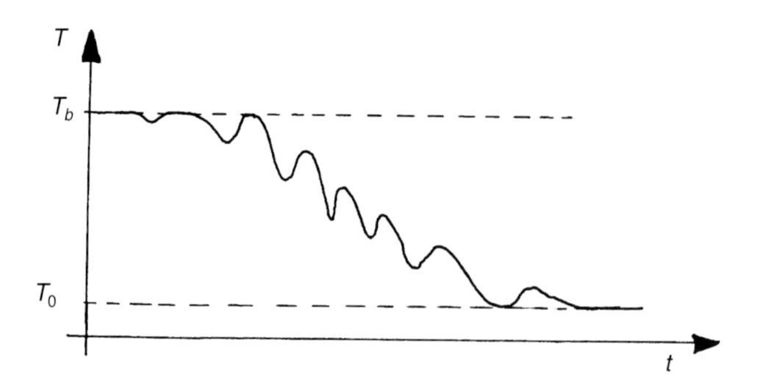

Figure 6.8a Instantaneous temperature signal for a probe crossing a "thickened" flame.

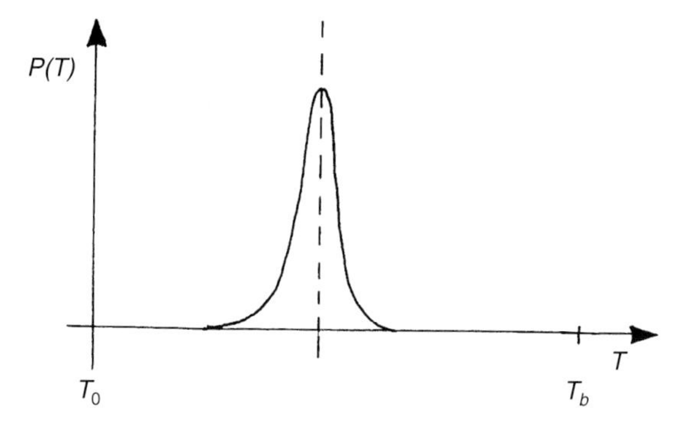

Figure 6.8b Probability density function for the temperature, P(T), in the middle of a "thickened" flame.

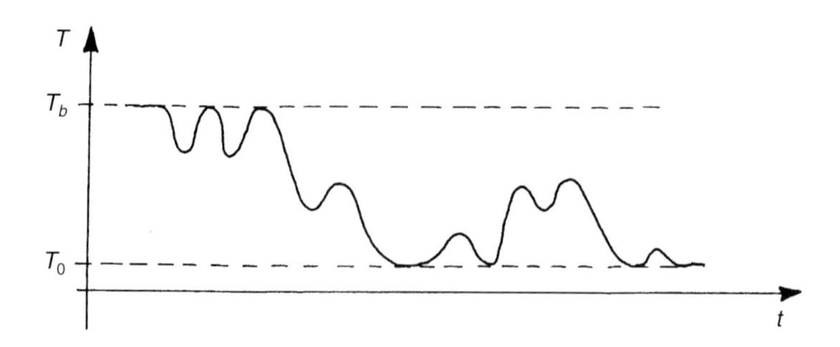

Figure 6.9a Instantaneous temperature signal for a probe crossing a "wrinkled-thickened" flame.

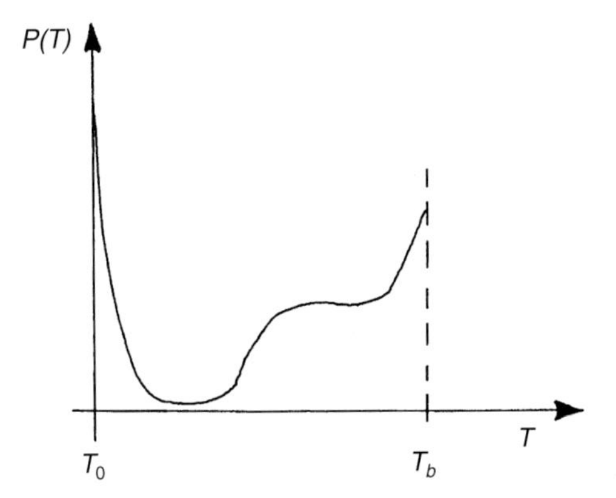

Figure 6.9b Probability density function for the temperature, $P(T)$, in the middle of a "wrinkled-thickened" flame.

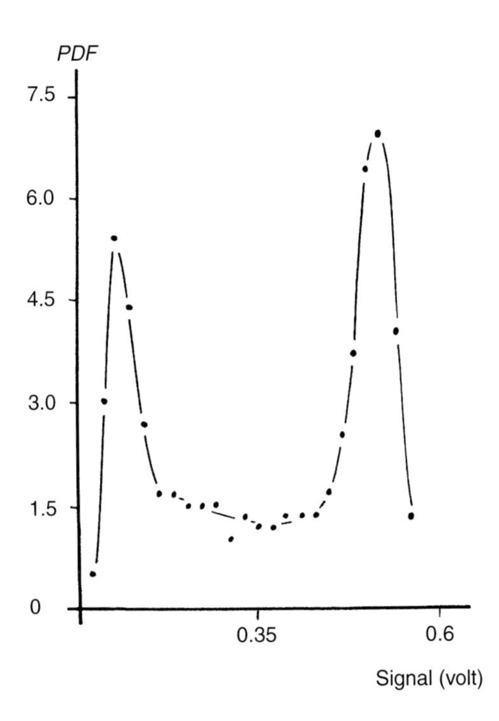

Figure 6.10 Probability density function (PDF) obtained for a wrinkled turbulent flame. The signal is assumed to be approximately proportional to the temperature (After Bill et al. (1982) *Combustion and Flame*, 44, p. 227).

Experimental results for wrinkled flames are easily obtained from turbulent gas burners or other turbulent laboratory flames. For thickened flames, some results have been obtained from more specific experimental set-ups, such as stirred turbulent reactors (known as "Longwell reactors") or from reactors in which very small-scale turbulence can be produced by arrays of small jets (see section 2.2.2 of chapter 2). Results have also been found for the intermediate case, obtained using reactors resembling the combustion chambers of ramjets.

Figure 6.10 shows the probability density function obtained by measuring a quantity proportional to the density (by light diffusion) in a turbulent Bunsen burner flame. The profile of the signal measured as a function of time closely resembles that of Figure 6.7a. The signal is not zero between the peaks because of experimental noise. For this gas burner, the turbulent length scale is not easy to measure and probably varies within the flame (the length scale was estimated to be 0.3 cm at the point corresponding to the measurement shown). The kinetic energy of turbulence is easier to measure (by laser Doppler anemometry) and a value of $k^{1/2} = 0.3$ m/s was determined. For propane (used in this experiment) with an equivalence ratio close to unity, $S_L \cong 50$ cm/s and $e_L \cong 0.3$ mm (η, estimated using equation (6.13), is 1 mm).

Figure 6.11a gives the probability density of the temperature fluctuations, measured using very fine thermocouples, in the reactor shown schematically in

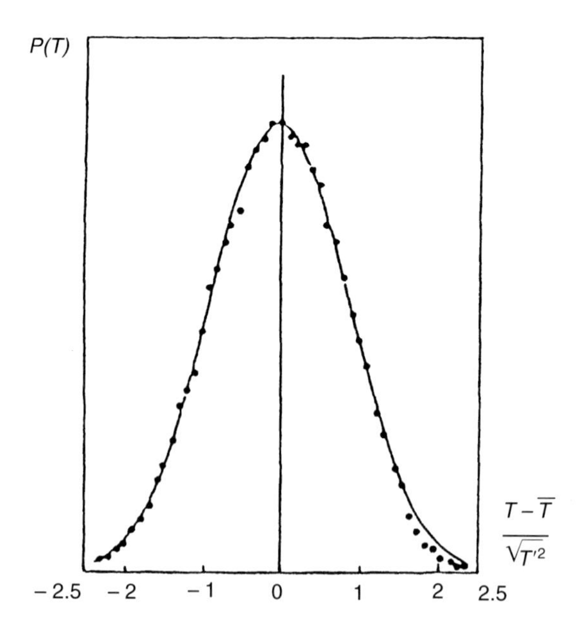

Figure 6.11a Probability density function of the temperature obtained in the burner shown in Figure 6.11b. Very fine thermocouples were used and the flame is of the "thickened" type.

Figure 6.11b. By using numerous injectors of small diameter, the scale of the turbulence produced in the flame is very small and, although not measured, must be of the order of the diameter of an injector (0.3 cm). The kinetic energy was found to be very high as a result of the high gas velocity; $k^{1/2} \approx 20$ m/s which implies a very small Kolmogorov scale of $\eta \approx 10$ μm (according to equations (6.10) and (6.11)).

Figure 6.12a shows the probability density of the temperature fluctuations recorded for the high-velocity turbulent flame stabilised in the apparatus

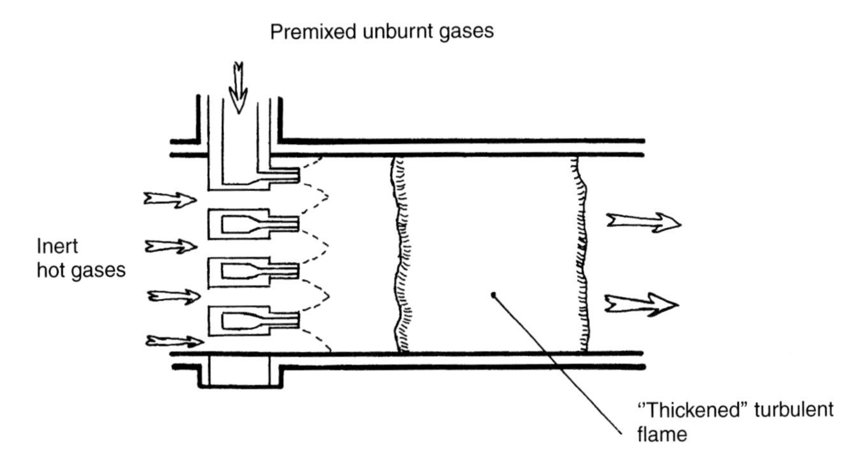

Figure 6.11b Schematic diagram of the jet-mixing burner used by Bellet et al. (1981), *C.R. Hebd. Séanc. Acad. Sci. Paris*, 293, series II, p. 259.

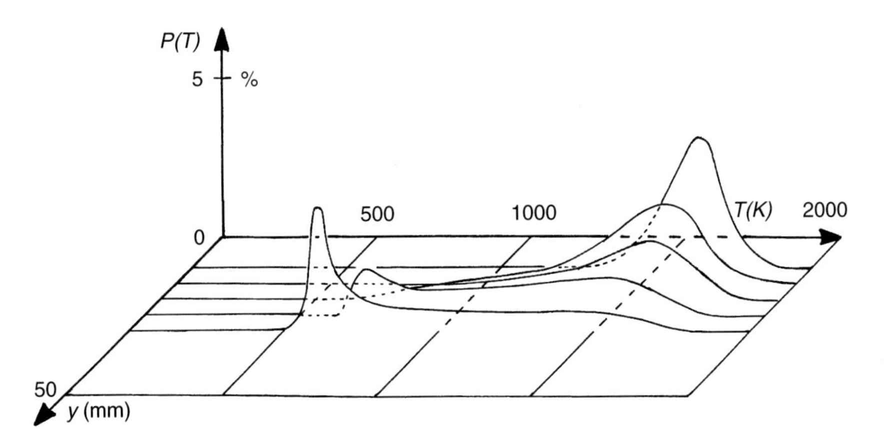

Figure 6.12a PDF of the temperature obtained in the combustion chamber shown in Figure 6.12b. Probability density function values are given over a range of positions (varying *y*, for *x* fixed).

shown in Figure 6.12b. Here the PDF profile should be typical of that for a "wrinkled-thickened" flame, but there is some doubt as to whether the resolution of the measuring technique is sufficient.

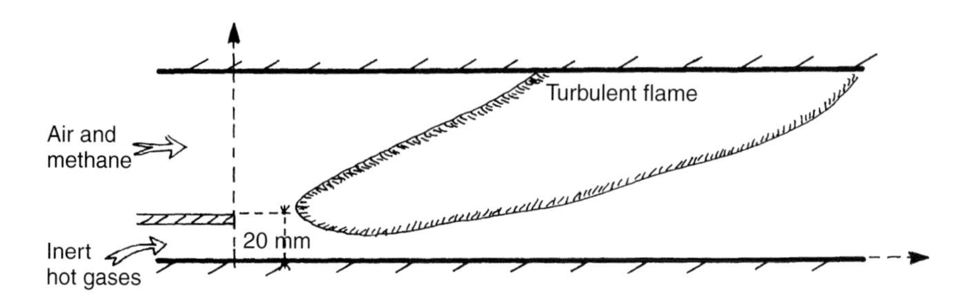

Figure 6.12b High-velocity combustion chamber with stabilisation by a pilot jet. The velocity of the premixed gases is 60 m/s, and of the pilot jet of hot gases 120 m/s; the temperature of the unburnt gases is 600 K, and of the pilot jet 1900 K (After Moreau, Boutier (1976, *16th Symposium (Int.) on Combustion*, The Combustion Institute, pp. 1747-1756).

6.3.3 Burning velocity and thickness of a premixed turbulent flame

Two very important characteristics of a premixed laminar flame are its normal burning velocity and its thickness. Is it possible to calculate an equivalent burning velocity and thickness for turbulent flames?

Firstly, it must be realised that these quantities can only be defined as average values for a turbulent flame. These averages are determined by considering the propagation of a large number of flames in various experiments performed under similar macroscopic conditions, and not simply the result of a single experiment measuring the propagation of a flame in a turbulent medium. For a turbulent flow, the only reproducible quantities are those obtained by averaging the values obtained for these quantities from a large number of separate and thus different events; and the same is true for turbulent flames. The application of this defining principle is simplified when the flame (or flow) under consideration has a fairly extensive spatial homogeneity, e.g. a flame propagating in one direction can often be assumed to be statistically homogeneous in all of the infinite number of planes perpendicular to the propagation direction. Under these conditions the average flame, as defined by the sum of all the small local portions of a single propagating flame, superimposed, can be considered as being a good approximation of the local average of a set of numerous flames. Hence, the statistical average for the set is replaced by a spatial average in a given plane (this approach again falls within the scope of the ergodic assumption).

Based on these principles, the first question to be answered concerns the intrinsic existence of a turbulent flame. For a known turbulent medium, whose characteristics are constant over time and across space, does an (average) turbulent flame in this medium have a constant (average) burning velocity and a constant (average) thickness, irrespective of how the flame was produced, a certain time after and at a certain distance from the point of ignition ?

Researchers have thus far been able to produce a definite answer to this question for all possible cases.

In the thickened-type flame regime, where the turbulence time τ_t is short enough to allow the temperature and concentration fluctuations to be small (see Fig. 6.8b), turbulence no longer plays an active role or, at most only affects the diffusion of mass and energy. Under these conditions, a coefficient of turbulent diffusion, d_t, can be used to replace the molecular coefficient d, with a simple transposition allowing the laminar flame theory to be used. Thus, for the ideal case where the turbulence and the coefficient of turbulent diffusion are both constant, there is indeed an intrinsic burning velocity for the turbulent flame, as well as an intrinsic thickness. Calculating these two parameters is easy, and simply involves replacing the molecular coefficient of diffusion with its turbulent equivalent in the laminar equations (as defined in chapter 5). In this way the average burning velocity and flame thickness for the turbulent flame, S_T and e_T, can be determined (where α and β are constants) from:

$$S_T = \alpha \left(\frac{d_t}{\tau_c}\right)^{1/2} \tag{6.19}$$

$$e_T = S_T \tau_c \tag{6.20}$$

Taking $Sc_t = 1$ and allowing for the fact that $d_t = \beta k^{1/2} l_t$ (see section 6.2.4) these equations can also be expressed as:

$$S_T = (\alpha\beta^{1/2})\, k^{1/2} \left(\frac{\tau_t}{\tau_c}\right)^{1/2} \tag{6.21}$$

$$e_T = (\alpha\beta^{1/2})\, l_t \left(\frac{\tau_t}{\tau_c}\right)^{1/2} \tag{6.22}$$

Unfortunately the treatment is more complex for the other flame types and, even for wrinkled flames, there is currently no evidence to support the existence of intrinsic burning velocities or thicknesses. For example, one theory may propose that the flame accelerates continually after its ignition; however it would be almost impossible to conduct an experiment designed to investigate this phenomenon with constant turbulence.

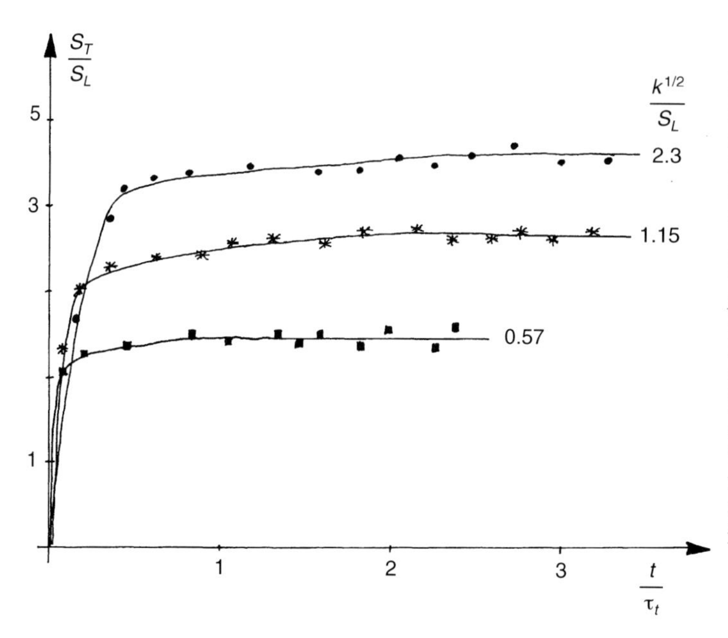

Figure 6.13a Establishment of propagation (at a constant velocity S_T) for a turbulent, wrinkled flame, based on the numerical simulations developed by Saïd and Borghi (1988) *22nd Symposium (Int.) on Combustion*. The Combustion Institute, 569-577. The time required for establishment is of the same order of magnitude as the integral turbulence time, τ_t, but is also dependent on $k^{1/2}/S_L$, as is the final value.

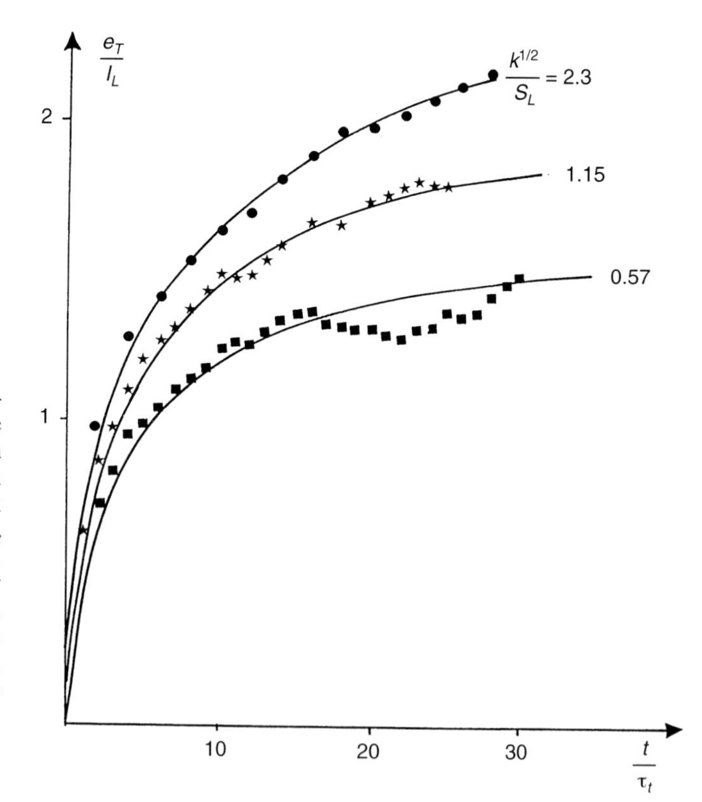

Figure 6.13b Establishment of the average flame thickness, e_T, at a constant value which depends on the length scale of the turbulence, l_t and on the ratio $k^{1/2}/S_L$. The time required for establishment is much longer than for the velocity (same numerical simulation as for Figure 6.13a).

Attempts have been made to solve this problem using numerical simulation. A stochastic numerical simulation has been recently performed (Borghi and Saïd[3]) which follows the random motion of fluid particles of unburnt and burnt gases and uses approximate models to represent their small-scale mixing and the movements of the laminar flamelet fronts. The result shows the establishment of an (average) turbulent flame, for which the average burning velocity and thickness do indeed tend towards finite values over time. These simulations have also revealed the following relationships between S_T or e_T and other turbulence characteristics (where C is a constant):

$$\frac{S_T}{S_L} = 1 + \frac{Ck^{1/2}}{S_L} \tag{6.23}$$

$$e_T = l_t f\left(\frac{k^{1/2}}{S_L}\right) \tag{6.24}$$

The simulation shows indeed that the flame thickness is a function of $k^{1/2}/S_L$ and does seem to increase considerably as $k^{1/2}/S_L$ tends to infinity. On the other hand, the time required to reach these intrinsic values is quite long, (cf. Fig. 6.13), and can become very long when $k^{1/2}/S_L$ is large. Other numerical simulations, containing fewer assumptions but with their own limitations, agree with these results. The existence of S_T and e_T as intrinsic values appears to be confirmed, (although only when $k^{1/2}/S_L$ is finite), and the time required to calculate these intrinsic values may limit their application in practical situations.

The simulations show that the ratio $k^{1/2}/S_L$ plays a very important role in wrinkling the flamelet fronts, which is also immediately apparent when this front is photographed using the same method as that used to obtain Figure 6.1b. Figure 6.14 indicates that the greater this ratio, the more the flamelet is wrinkled by the turbulence, without S_L allowing the wrinkles to be dissipated.

The proportional relationship between S_T and $k^{1/2}$ suggested by equation (6.23) appears to be valid. The available experimental results seem to support this claim, (even though there is an element of doubt associated with the method of measuring S_T and the results are highly dispersed), favouring a linear relationship so long as $k^{1/2}/S_L$ is not too large (i.e. so long as the flame remains wrinkled). Figure 6.15 plots a selection of these results.

3. Borghi R. and Saïd R. (1988) *22nd Symposium (Int.) on Combustion*, The Combustion Institute, 569-577.

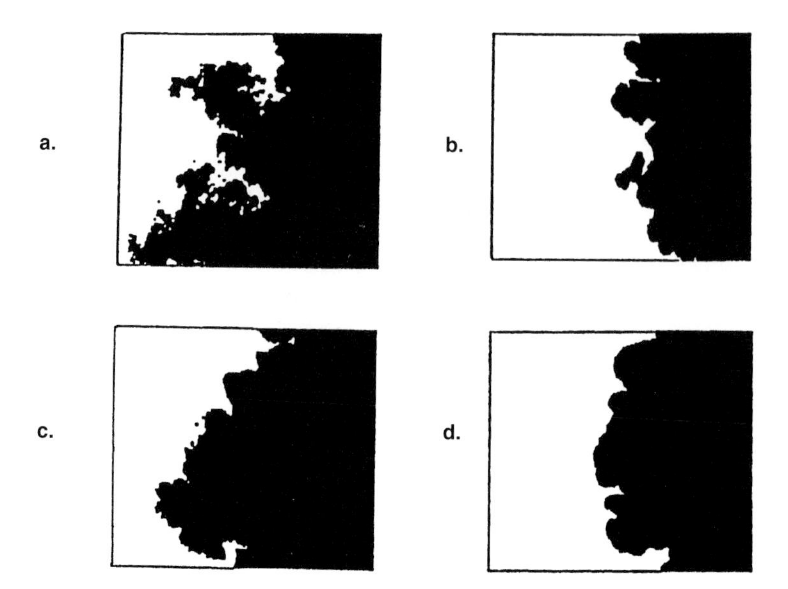

Figure 6.14 Sections of a premixed flame front propagating in a chamber of constant volume. Same photographic technique as for Figure 6.1b with burnt gases shown in black. The air/propane mixture has an equivalence ratio of 0.9 and the ratio $k^{1/2}/S_L$ has a value of 2.48, 1.55, 1.22, and 0.68 in Figures a), b), c) and d) respectively. Photographed by A. Pocheau (Marseille). See also: A. Floch, M. Trinité, F. Fisson, T. Kageyama, C.H. Kwon and A. Pocheau (1989) in: *Dynamics of Explosions and Reactive Systems: Flames, Progress in Astronautics and Aeronautics*, 131, 378-393, AIAA.

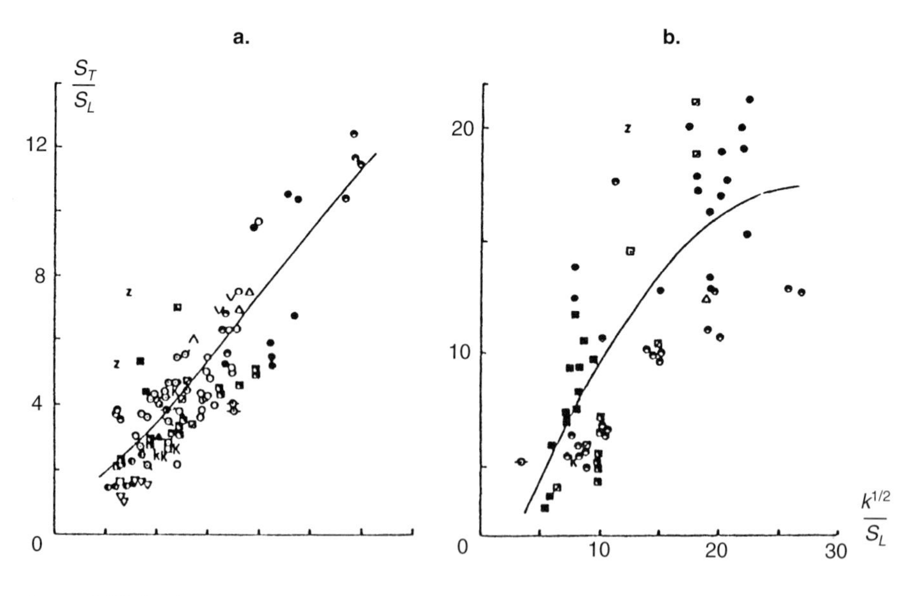

Figure 6.15 Various experimental results relating to S_T, obtained by D. Bradley, A.K. Lau, M. Lawes (1992) *Phil. Trans. R. Soc. Lond. A*, 338, 355-387. Note the almost linear relationship when $k^{1/2}/S_L$ is not too large, followed by subsequent tailing-off.

6.4 "MODELLING" PREMIXED TURBULENT FLAMES

Although the burning velocity and the thickness of the turbulent flame are rightly the first quantities to be considered, they are not the only quantities of interest, and indeed under certain conditions they are of little practical use. The simulation discussed in section 6.3 demonstrated that these two quantities are only meaningful after a certain time and at a certain distance from the ignition point, yet we are often concerned by what is happening in the flame before this point is reached. On the other hand, the practical conditions are often such that we cannot be certain that the turbulence is constant across the whole flame, and usually it is not even known. Moreover, increasing interest is being shown in the development of a mathematical model able to calculate the structure and characteristics of a flame and which is valid for more general conditions, without having to assume the existence of a turbulent burning velocity.

6.4.1 The context of the modelling problem

It is through the use of appropriate balance equations that the problem of modelling turbulent flames can be tackled, indeed this was the approach used for laminar flames. In this particular case, averaged equations are required similar to those used in chapter 3.

In section 3.5.2 of chapter 3, average balance equations were introduced for momentum, energy, and species for a turbulent reactive medium. The averaged balance equation for the reactive species can be written, based on (3.32):

$$\frac{\partial}{\partial t}\,\overline{\rho}\widetilde{Y}_i + \sum_1^3 \frac{\partial}{\partial x_\alpha}(\overline{\rho}\,\widetilde{v}_\alpha\widetilde{Y}_i) = \sum_1^3 \frac{\partial}{\partial x_\alpha}(-\overline{\rho}\,\widetilde{v'_\alpha Y'_i}) + \overline{\rho}\,\widetilde{w}_i \qquad (6.25)$$

The concept of modelling the turbulent diffusion term $-\widetilde{v'_\alpha Y'_i}$ has already been discussed and we will make further reference to it at the end of this chapter for the case of turbulent flames. We shall now consider the additional problem of modelling the average rate of reaction, $\widetilde{w}_i$.

The instantaneous rate of reaction, w_i, is a complex function of all the concentrations of all (or at least most of) the species present, as well as the temperature and pressure. It is only through understanding all the chemical processes involved that this function can be determined, as we saw in the previous chapters. Thus, the average rate of reaction can be written, by definition, in terms of a probability density function (PDF) which allows for the fluctuations in all the variables needed for ρw_i:

$$\bar{\rho}\,\widetilde{w}_i = \int\int\int \dots \int \rho w_i\,(Y_1, \dots Y_n, T, \rho)\,P(Y_1, \dots Y_n, T, \rho)\,\mathrm{d}Y_1 \dots \mathrm{d}Y_n \mathrm{d}T\,\mathrm{d}\rho \quad (6.26)$$

Since the function w_i is highly nonlinear and the fluctuations are not necessarily small (particularly for wrinkled flames where the PDF of T, for example, indicates two peaks), such fluctuations cannot be neglected and so the simple equation :

$$\bar{\rho}\,\widetilde{w}_i = \bar{\rho}\,w_i(\widetilde{Y}_1, \dots \widetilde{Y}_n, \widetilde{T}, \bar{\rho}) \quad (6.27)$$

is in general incorrect. $P(Y_1 \dots T, \rho)$ must therefore be known before the values of $\widetilde{w}_i$ can be calculated, which is a major complication. Moreover, for the limiting case of a wrinkled flame in which the chemical reactions occur infinitely rapidly, equation (6.26) becomes indeterminate since when w_i becomes infinite the PDF, P, becomes zero!

In fact, it is only towards the limiting case of thickened-type flames, at which τ_t is very small compared with τ_c, that the fluctuations in Y and T (and in ρ for subsonic flow) are small enough to allow equation (6.27) to be a good approximation, indeed it is highly inaccurate for all other cases.

Replacing the defining equation (6.26) with a simpler equation, which takes into account the fluctuations in a realistic but not overly complicated manner, thus generates a method of modelling the average rate of reaction. When a modelling method such as this is combined with a turbulence model able to calculate (in an approximate manner) the diffusion fluxes, the result is a model of turbulent combustion.

The use of such a mathematical model requires the integration of the associated partial differential equations, such as equation (6.25), taking into account the initial and boundary conditions. This is achieved nowadays by approximate numerical integration methods, including those for geometrically-complex flows.

Before giving some examples of this type of model, note that the first step in replacing equation (6.26) is to choose a means of representing the chemical processes, i.e. a kinetic reaction mechanism. The complexity of the model will be chiefly governed by the complexity of the reaction mechanism itself, although, fortunately, when the chemical reactions are very fast, the exact details of the mechanism are unimportant, as will be shown below.

6.4.2 The "Eddy Break-Up" model

Consider the flame range corresponding to wrinkled flames, and up to the extreme limit of this range where $\tau_c \ll \tau_t$, i.e. where all the chemical processes occurring do so very rapidly compared with the processes involved in turbulent flow and, in particular, where $\tau_c < \tau_k$.

Between 1970 and 1980, a model was proposed to describe this situation based on the physical fact that combustion in the flame can be divided into two consecutive stages: firstly the necessary mixing, down to a molecular scale, of the unburnt gases with a certain amount of burnt gases (since the unburnt mixture does not burn spontaneously!) followed by the chemical reactions which lead to the conversion of the unburnt gases. Thus, if the chemical reaction step occurs infinitely rapidly, then it must be the initial, turbulent mixing step which controls the whole process.

Moreover, as we have seen above (Figs. 6.8 and 6.11), the reacting medium contains only unburnt and completely burnt gases, i.e. at chemical equilibrium. By choosing a chemical species j present in the unburnt gas mixture (in a proportion Y_j), a variable, C, can be defined as:

$$C = \frac{Y_j^0 - Y_j}{Y_j^0 - Y_j^{\mathrm{eq}}} \tag{6.28}$$

which varies from 0 to 1 from the completely unburnt gases to completely burnt gases (since Y_j^{eq} represents the proportion of j at equilibrium). C has been termed the progress variable and is a measure of the extent of reaction (although it is defined differently from the extent of reaction discussed in chapter 1).

The proposed equation, obtained after some trial and error, expresses the average rate of reaction as a function of the Favre mean for C:

$$\widetilde{w}_i = -\,C_{\mathrm{EBU}}\,\frac{\widetilde{C}(1-\widetilde{C})}{\tau_\tau}\,(Y_i^0 - Y_i^{eq}) \tag{6.29}$$

where C_{EBU} is an empirical constant. This result can be explained, very approximately, by saying that the rate of combustion only depends on the probability of finding, at the same place, a fluid particle of unburnt gases and one of burnt gases since, once mixed, the chemical reactions are very rapid in all cases (and the probability of finding a burnt fluid particle is $\widetilde{C}$ and that of finding one which is unburnt is $(1-\widetilde{C})$). This equation is valid for all species i present in the unburnt or burnt gases (and in the latter case $Y_i^0 = 0$). For species which are reaction intermediates, existing neither in the unburnt or burnt gases (or in minute quantities), both their average rate of reaction and average mass fractions are zero.

The name "Eddy Break Up" was suggested by D.B. Spalding[4] in 1971 for this model, to emphasise the fact that the controlling process is turbulent mixing, through breaking-up of the eddies present in the fluid. The presence of the term $\widetilde{C}(1-\widetilde{C})$ in the equation can be more convincingly explained theoretically than above and thus a physical interpretation of C_{EBU} can be given, as was first shown by K.N.C. Bray[5]. However, the numerical value of C_{EBU} remains empirical.

In order to take into consideration the important role played by the parameter $k^{1/2}/S_L$ in flamelet wrinkling, more recent work has suggested that C_{EBU} is in fact a function of $k^{1/2}/S_L$, such that C_{EBU} tends towards a constant value as $k^{1/2}/S_L$ tends towards infinity, and that C_{EBU} increases as $k^{1/2}/S_L$ decreases. The equation proposed is:

$$C_{\text{EBU}} = C^0_{\text{EBU}}\left(1+\frac{4.4}{1+3.2\dfrac{k^{1/2}}{S_L}}\right) \tag{6.29 bis}$$

Equation (6.29) must again be associated with a turbulence model in order to calculate both τ_t and the turbulent diffusion flux.

Suppose, for example, that a coefficient for turbulent diffusion is defined, i.e. that:

$$\widetilde{v'_x Y'_i} = d_t\,\frac{\partial \widetilde{Y}_i}{\partial x} \tag{6.30}$$

for any species, and that τ_t and d_t have been determined in some way. Now consider the simple case of a planar turbulent flame propagating in a direction x, yet kept stationary in an infinite medium of given turbulence. In this case the model reduces finally to the integration of the equation for $\widetilde{C}$. This can be found directly from equation (6.25), and values for Y_j (where j is the chosen species), using equations (6.28), (6.29), and (6.30):

$$\frac{\mathrm{d}}{\mathrm{d}x}(\bar{\rho}\,\tilde{v}_x\widetilde{C}) = \frac{\mathrm{d}}{\mathrm{d}x}\left(\bar{\rho}\,d_t\,\frac{\mathrm{d}\widetilde{C}}{\mathrm{d}x}\right) + C_{\text{EBU}}\,\frac{\widetilde{C}(1-\widetilde{C})}{\tau_t} \tag{6.31}$$

The problem of characterising turbulent flame propagation is similar to that for the laminar flames discussed in chapter 5, except that the reaction term is different and hence the method of finding the solution is slightly different from that described in chapter 5. It appears that the solution to this equation was considered as far back as 1937 by Kolmogorov, Petrovskii and Pistunov,

4. Spalding D.B. (1971) *13th Symposium (Int.) on Combustion*. The Combustion Institute, p. 649.
5. Bray K.N.C. (1980) Turbulent Reacting Flows, *Topics in Applied Physics*, 44, 115-183. Ed. P.A. Libby, F.A. Williams, Springer Verlag.

(in relation to the spread of illnesses, and for constant $\overline{\rho}$). The solution to the equation (where α' is a different numerical coefficient from that in (6.19)) is:

$$S_T = \alpha' C_{\text{EBU}}^{1/2} \left(\frac{d_t}{\tau_t}\right)^{1/2} \quad (6.32)$$

$$e_T = S_T \tau_t \quad (6.33)$$

from which, using $d_t = \beta' k^{1/2} l_t$ (where β' is a constant which may perhaps be dependent on $k^{1/2}/S_L$), analogous expressions to those given by equations (6.23) and (6.24) are obtained:

$$S_T = \alpha' \beta'^{1/2} C_{\text{EBU}}^{1/2} k^{1/2} \quad (6.34)$$

$$e_T = \alpha' \frac{\beta'^{1/2}}{C_{\text{EBU}}^{1/2}} l_t \quad (6.35)$$

But where is the temperature in this theory? It is explicitly lost in equation (6.31) since it is not considered in the average rate of reaction. This is not as surprising as it may seem since we are assuming that the reactions take place very rapidly, and thus that the temperature of the medium has no influence whatsoever on their rates, since only the presence (or absence) of the reactants is important. This does not, however, prevent the calculation of the average flame temperature $\widetilde{T}$, which can be shown, using a few simple assumptions, to be directly related to $\widetilde{C}$.

The balance equation for the mean enthalpy $\widetilde{h}$ was derived in chapter 3; in steady-state and for one-dimensional flow, equation (3.18‴) becomes:

$$\frac{\mathrm{d}}{\mathrm{d}x}(\overline{\rho}\,\widetilde{v}_x \widetilde{h}) = \frac{\mathrm{d}}{\mathrm{d}x}\left(\overline{\rho}\, a_{\mathrm{T}} \frac{\mathrm{d}\widetilde{h}}{\mathrm{d}x}\right) \quad (6.36)$$

By considering the fact that the infinite medium undergoes no overall heat loss or gain, it follows that the solution to this equation is $\widetilde{h}$ = constant, because $\widetilde{h}(x = -\infty) = \widetilde{h}(x = +\infty)$ (recall that the continuity equation is $\mathrm{d}(\overline{\rho}\,\widetilde{v}_x)/\mathrm{d}x = 0$).

If we now consider the equations for $\widetilde{Y}_i$, in the same form as (6.25) with the rates of reaction (6.29) and the coefficient of diffusion (6.30) included, the result is the following simple equation for the case of a continuous, one-dimensional flow:

$$\frac{\mathrm{d}}{\mathrm{d}x}(\overline{\rho}\,\widetilde{v}_x \widetilde{Y}_i) = \frac{\mathrm{d}}{\mathrm{d}x}\left(\overline{\rho}\, d_t \frac{\mathrm{d}\widetilde{Y}_i}{\mathrm{d}x}\right) - C_{\text{EBU}}\,(Y_i^0 - Y_i^{\text{eq}})\,\widetilde{C}\left(\frac{1-\widetilde{C}}{\tau_t}\right) \quad (6.37)$$

By defining $Z = \widetilde{C} + (\widetilde{Y}_i - Y_i^{\text{eq}})/(Y_i^0 - Y_i^{\text{eq}})$, the resulting equation for Z is identical to that found for $\widetilde{h}$.

Thus, just like $\widetilde{h}$, Z = a constant is the solution to this equation (with in this case the constant equal to 1). The result is that for any species i (except for reaction intermediates, for which $\widetilde{Y}_i = 0$ as stated previously), $\widetilde{Y}_i$ is related linearly to $\widetilde{C}$:

$$\frac{\widetilde{Y}_i - Y_i^{\mathrm{eq}}}{Y_i^0 - Y_i^{\mathrm{eq}}} = 1 - \widetilde{C}$$

Thus, since by definition:

$$h = h(T, Y_1 \ldots Y_n) = \int_0^T \sum_1^n Y_i c_{pi}(T)\,\mathrm{d}T + \sum_1^n h_{i,0} Y_i$$

simply by making the assumption $c_p \cong$ constant $(= \overline{c_p})$ gives:

$$\widetilde{T} = T_0 + \left[\sum_1^n \frac{h_{i,0}}{\overline{c_p}} (Y_i^0 - Y_i^{\mathrm{eq}}) \right] \widetilde{C}$$

6.4.3 The case where the chemical reactions are not infinitely rapid

The problem becomes more complicated in cases where τ_c is not infinitely small compared with the other characteristic time scales.

The first complication relates to the kinetic reaction mechanism since, in reality, there is not one but several characteristic times for the chemistry, each of which must be compared with τ_k. Whilst certain reaction steps may be considered as being rapid, others cannot. Furthermore, more information must be known about the probability densities for the fluctuations in temperature and concentrations which in this case are no longer simply two peaks. Research into this subject continues and a full discussion of the latest models is beyond the scope of this work; however two specific points are of particular importance.

First, let us consider the two limiting cases of wrinkled and thickened flames. We have already discussed the two models available which can estimate their average rate of reaction. For wrinkled flames it is equation (6.29) which ignores the actual chemical reactions involved, whilst, for thickened flames, in the limiting case, the average rate of reaction can be calculated by applying the instantaneous rates of reaction to the average values (equation (6.27)). The results from these two equations can be plotted approximately for a given premixed flame as a function of the mean progress variable and the instantaneous rate can, for example, even be assumed to follow an Arrhenius relationship.

Figure 6.16a plots the two equations; the Eddy Break Up equation is a simple parabola whereas that derived from the Arrhenius relationship has a characteristic asymmetric form (see Figure 5.17a of chapter 5). The differences between the two cases can be clearly seen. For a wrinkled flame the average rate of reaction increases as $\widetilde{C}$ increases and combustion is easier, for a low extent of advancement of reaction, than for a thickened flame. On the other hand, the maximum reaction rate for the thickened flame is greater than for the wrinkled flame, due to the effect of an activation temperature, seen also with laminar flames.

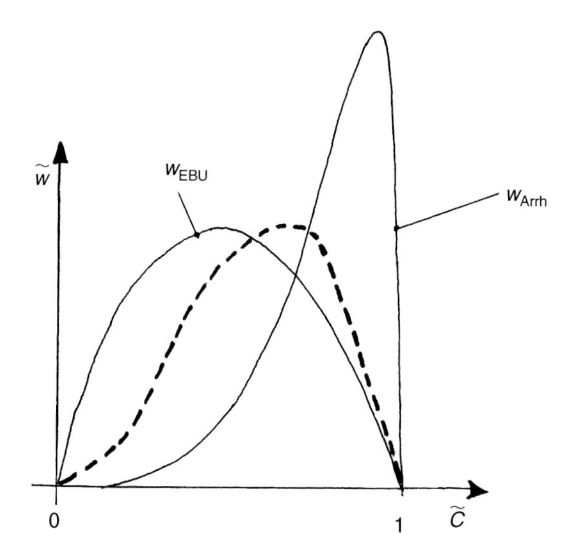

Figure 6.16a Variation of rate of reaction with the mean progress variable. Dashed line for typical experimental results.

In fact, the maximum value of each curve does not depend at all on the same quantities. Moreover, the total combustion time, which can be defined as the reciprocal of the area under the curve, is different for each curve.

If the average rate of reaction through the flame is defined as:

$$\overline{\widetilde{w}} = \frac{-1}{Y_i^0 - Y_i^{eq}} \int_0^1 \widetilde{w}_i(c)\,dc$$

then $\overline{\widetilde{w}}\tau_t$ is found to be a constant for a wrinkled flame whereas $\overline{\widetilde{w}}\tau_c$ is a constant for a thickened flame.

In Figure 6.16.b, which shows $\overline{\widetilde{w}}\tau_t$ as a function of the Damköhler number, $D_a = \tau_t/\tau_c$, the plots corresponding to these limiting cases can be clearly seen. For a given flame, as a function of the Damköhler number, the real plot $\overline{\widetilde{w}}\tau_t$ probably resembles closest the dashed curve. It is also likely that there is not simply one plot for all premixed flames, although the family of plots will all approximate the two limiting plots shown when D_a is very small or very large.

Figure 6.16b relates only to the average value of $\overline{\widetilde{w}}$ across the entire flame thickness ($0 \leqslant \widetilde{C} \leqslant 1$). The profile of a plot of $\widetilde{w}$ as a function of $\widetilde{C}$ for intermediate Damköhler numbers probably lies somewhere between the two limiting curves; the dashed curve in Figure 6.16a being one example. Here again it is likely that the plot is different for the various specific turbulent flames.

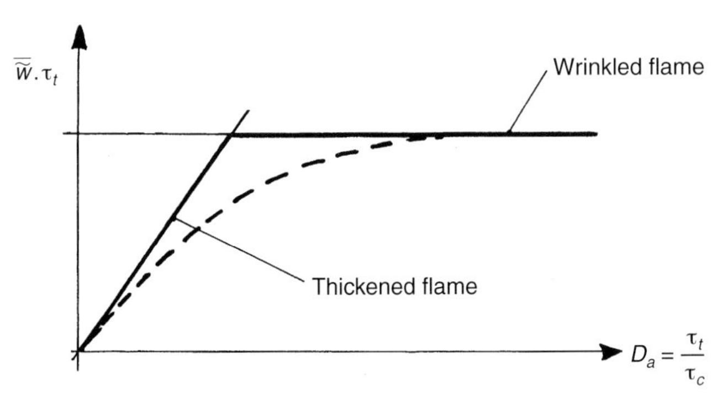

Figure 6.16b Approximate variation in average total rate of reaction in the flame as a function of Damkölher number.

A very approximate model for describing all flame types, whose use is justified only through its simplicity and by the previous remarks, can be formed by taking w_{Arrh} on the left side (where $w_{\text{Arrh}} < w_{\text{EBU}}$) and w_{EBU} on the right side (where $w_{\text{EBU}} < w_{\text{Arrh}}$).

Research is currently underway to develop a model able to predict the average rate of reaction which is valid for the various flame types shown in Figure 6.5. A discussion of this particular research work is beyond the scope of this book.

6.5 DETAILED EXAMPLE: A TURBULENT GAS BURNER

6.5.1 Velocity field

A simple example of a premixed burner which can operate in turbulent conditions is that of a gas Bunsen burner. To understand fully how the flame develops, information must be ascertained relating to the velocity field and the turbulence. This information can be determined experimentally using a Doppler laser anemometer.

Figure 6.17 shows the typical results obtained when the velocities are measured. The velocity vectors plotted were measured in a diametrical plane and the position of the zone corresponding to the average flame is also indicated (between the lines corresponding to mean progress variables of 0.1 and 0.9). The lengths of the velocity vectors at the burner mouth, in the flame region and even beyond, are virtually constant, whilst the expansion of the hot gases causes the flow to deviate away from the flame. The turbulence generated by the premixed gases is modified in the flame; a phenomenon which will be discussed in more detail in section 6.8. Furthermore, results concerning the kinetic energy and length scale of the turbulence will be considered, two parameters which are essential for calculating the characteristic time τ_t.

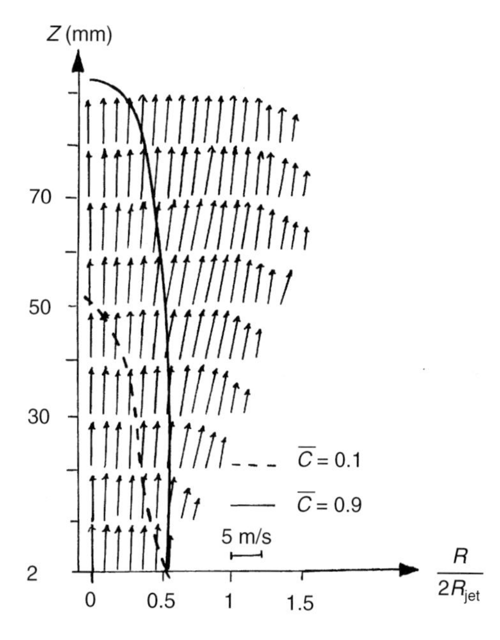

Figure 6.17 Average velocity field measured in the turbulent flame of a Bunsen burner (air—CH_4). V_{jet} = 7.2 m/s; R_{jet} = 5.5 mm. (After J.P. Dumont, D. Durox and R. Borghi (1993) *Comb. Sci. Tech.*, 89, 219-251).

6.5.2 The average rate of reaction field

The field corresponding to the mean progress variables $\overline{C}$ or $\widetilde{C}$, can be measured in a flame by analysing statistically several hundred photographs similar to the photograph shown in Figure 6.1b. In fact $\overline{C}(x)$ simply describes the probability of finding burnt gases in a small volume at point x. Figures 6.18a and b show the transversal profiles of $\overline{C}$ and $\widetilde{C}$; the plots are slightly asymmetric because of a slight error in the experiment.

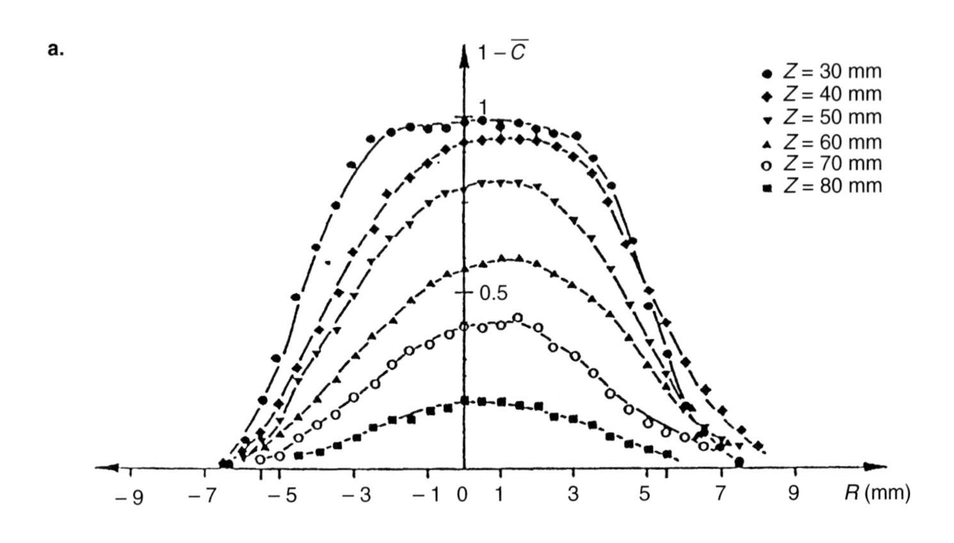

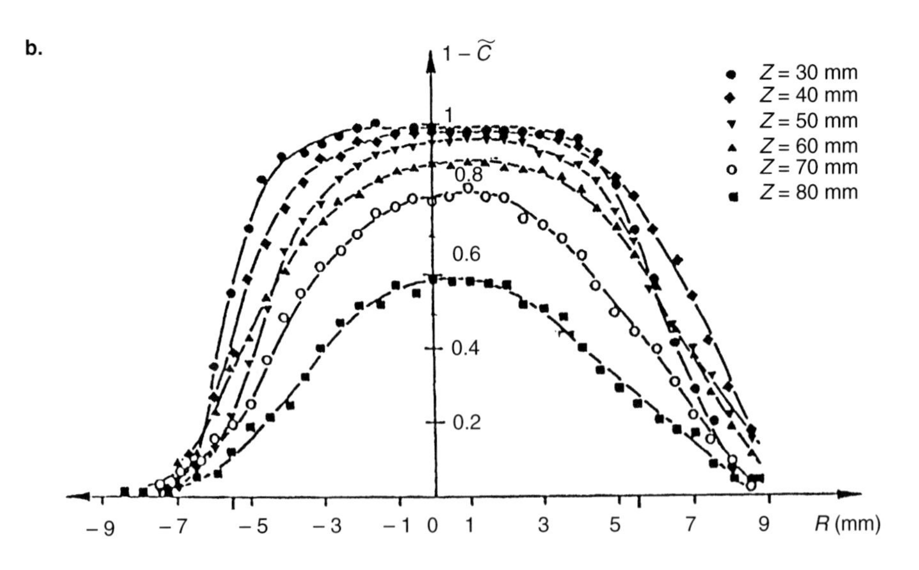

Figure 6.18 Profiles of the mean reaction progress for the flame shown in Figure 6.17, at various heights Z above the burner. There is slight symmetry error, and the difference is apparent between the conventional average of Reynolds $\overline{C}$ and that of Favre $\widetilde{C}$ (same reference as for Figure 6.17).

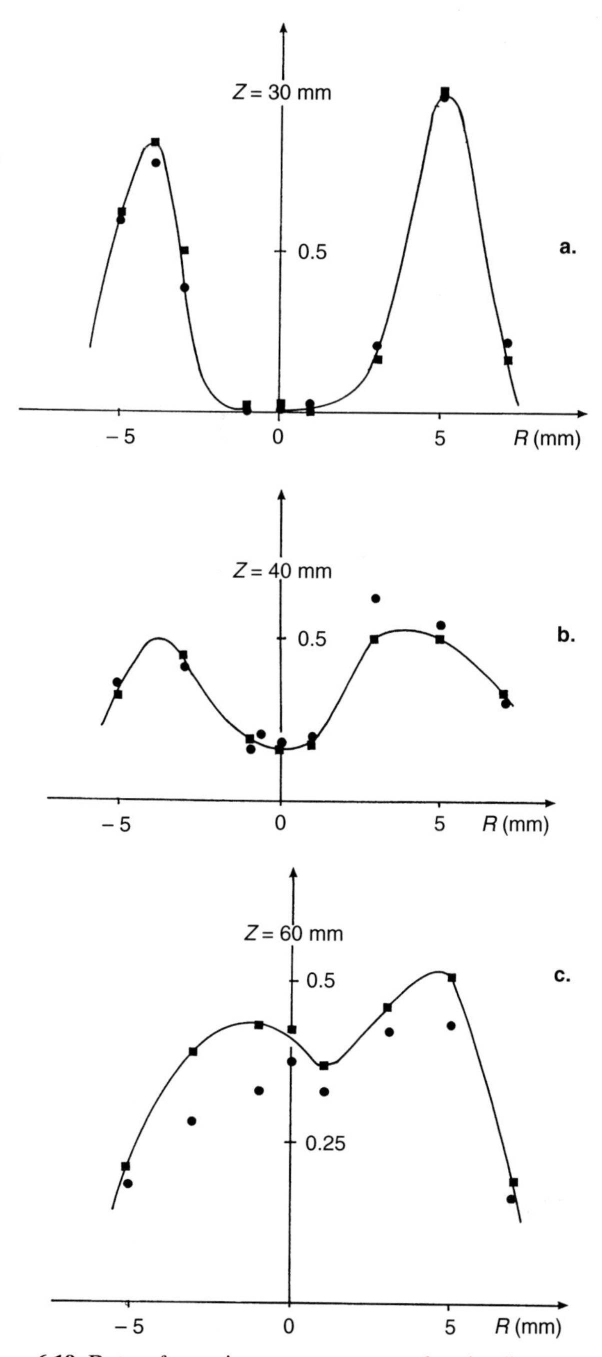

Figure 6.19 Rate of reaction measurements for the flame shown in Figure 6.17. The squares (joined by the solid line) are measurements made from the flame length. The circles are the values deduced from measurements of k, l_t, and of the average extent of advancement of reaction, using the Eddy Break-Up model, and adjusted such that the two methods coincide at Z = 30 mm and R = 5 mm (same reference as for Figure 6.17).

Based on the measurements of $\widetilde{C}$, k and l_t, the average rate of reaction can be deduced from the Eddy Break-Up equation (or at least accurate to one multiplying factor) by simply calculating $(1 - \widetilde{C})\,\widetilde{C}\tau_t$, with $\tau_t = l_t/k^{1/2}$.

Moreover, the instantaneous photographs of the flame front (Fig. 6.1b) also allow the rate of reaction to be calculated directly for a wrinkled flame. Indeed, the rate of reaction per unit volume is $\rho_0 S_L \overline{\Sigma}$, where ρ_0 is the density of the unburnt gases, S_L the laminar burning velocity of the flame front (with respect to the unburnt gases) and $\overline{\Sigma}$ the average flame surface per unit volume. This formula assumes that locally the flame is laminar and has not been modified by the turbulence. Using the recorded turbulence measurements, the value of η, the Kolmogorov length scale, can be estimated at different regions in the flame: with the result that η is always greater than 0.3 mm (in this case, the laminar flame thickness for a methane-air mixture at normal temperature and pressure) other than immediately over the burner rim where the turbulence scales are smaller and the kinetic energy greater.

The comparison of the results obtained from the two methods (one involving the measurement of $\widetilde{c}$, $k^{1/2}$ and l_t, and the other the direct measurement of $\overline{\Sigma}$ (from analysing photographs) is quite interesting. Figures 6.19a, b and c show that, to within the accuracy of one multiplying constant, the two methods give very similar results across the entire flame zone (and from 30 mm to 60 mm above the burner). This agreement can be interpreted as an indication that the Eddy Break-Up model is quite realistic, at least for the case of a wrinkled flame.

This agreement is all the more remarkable in view of the result indicated by Figure 6.17 which showed that the thickness e_T of the average turbulent flame is far from being constant with distance from the burner. It is thus impossible, under these conditions, to be able to define an average turbulent flame established obliquely in the flow, and the concept of an average flame thickness, e_T, and velocity, S_T, have no intrinsic meaning. A situation therefore exists where such concepts are of no practical interest, although the equations for the average rate of reaction still hold and are shown to be realistic.

6.6 THE STRUCTURE OF TURBULENT DIFFUSION FLAMES

We shall now consider turbulent flames fed by separate supplies of fuel and oxidant, which is a common design feature in industrial burners. The most straightforward example of this type of flame is the "jet flame", which is a high velocity jet of gas burning in a parallel annular flow of air moving at a lower velocity. Jet flames resemble the cigarette lighter example described in chapter 3, except that the jet velocity is much higher, and creates turbulence.

In practice, complex arrangements of large numbers of jets are used. Evidently thc flows produced are also more complicated, sometimes involving well-organised vortices designed to recirculate the gases and stabilise the flame. The flow in many burners is devised such that the streamlines spiral around the same axis as the jet of fuel; in what are known as "swirl" burners. Velocity gradients create the turbulence in all cases, and its intensity and length scale vary depending on the design of the equipment.

As a basis for the discussion of the structure of turbulent diffusion flames, we will consider the most simple flame. This flame, studied by Burke and Schumann, was introduced in chapter 3 and is the flame which establishes in a uniform flow between a current of fuel and a current of oxidant. The turbulence is assumed to have been created upstream of the flame, as a result of velocity gradient-induced shear stresses in the fluid near the walls or in the wake of a grid, and is assumed to be maintained along the full length of the flame.

6.6.1 Description of the structure of turbulent diffusion flames

In discussing Burke and Schuman's laminar diffusion flame, (in chapter 3), we showed that the flame zone consists of a reaction zone separating two diffusion-convection zones. By applying simple theory (again in chapter 3) it was demonstrated that if the combustion reaction is assumed to be a single reaction occurring infinitely rapidly then the reaction zone will be infinitely thin. When reverse reactions are taken into account, even if they also occur infinitely rapidly, then the reaction zone will be of finite thickness and will occupy part of the surrounding convection-diffusion zones. In all cases, the overall thickness of the diffusion flame is the sum of the two convection-diffusion zones, which depend on the (average) coefficient of diffusion, the velocity of the gases, and also on its position, since the flame thickness increases away from the point where the two flows meet.

When the chemical reactions are not very rapid (compared with the physical phenomena of convection and diffusion) the flame may be extinguished either locally or entirely. The local extinction of a jet flame occurs near the outlet of the fuel supply tube, lifting the flame off the mouth. In the ignition zone of a lifted flame a "triple flame" is formed. This small premixed flame, which is rich on one side and lean on the other, stabilises the diffusion flame's reaction zone. The characteristics of this type of flame are described in greater detail in chapter 5.

Consider Burke and Schumann's laminar diffusion flame, and what happens if the turbulence in the flow is increased in some way. As the turbulent eddies have grown, it is easy to see how they could move around and wrinkle the laminar flame zone. So long as the turbulence is not too great, the structure of the turbulent diffusion flame (as shown by the instantaneous diagram in Figure 6.20a) will resemble that of a wrinkled flame. The turbulent diffusion flame thus consists of a long "flamelet", wrinkled along its entire length, which is alternately stretched and compressed by the turbulent motion. As for

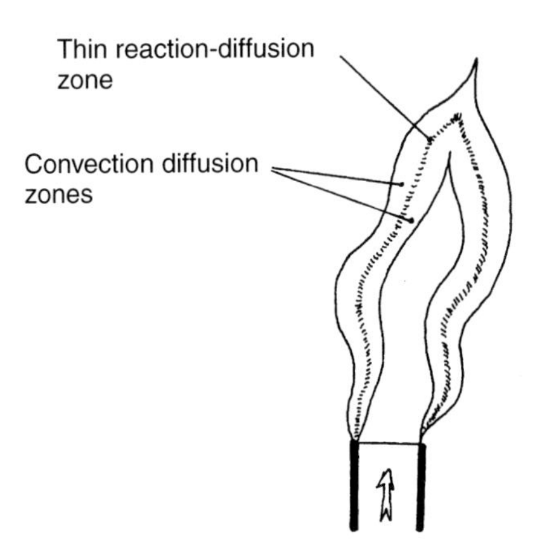

Figure 6.20a Slightly turbulent diffusion flame.

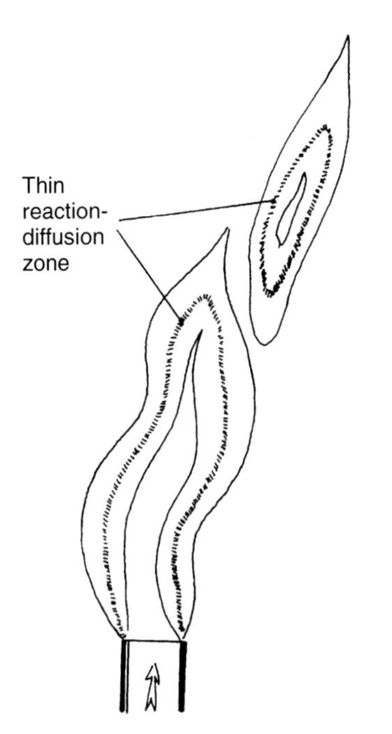

Figure 6.20b Large-scale interaction of "flamelets".

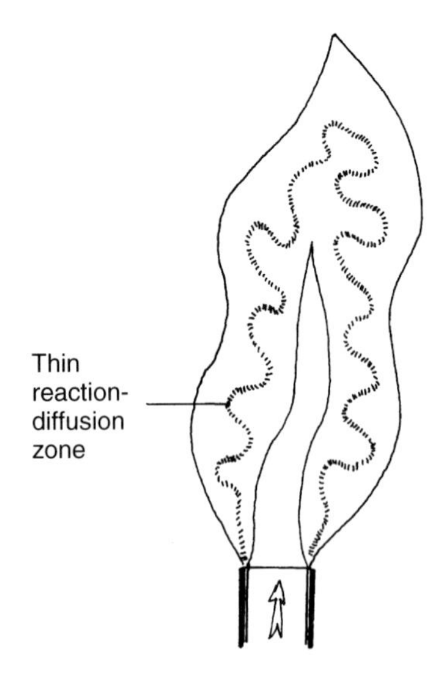

Figure 6.20c Small-scale interaction of "flamelets".

a premixed flame, for an individual flamelet to be apparent, the turbulence scales must all be greater than the thickness of the flamelet itself. But, unlike premixed laminar flames, the thickness of a diffusion flame is not an intrinsic characteristic, instead it depends on the distance from the burner, and increases with distance away from the fuel gas supply. In addition, the flamelet in a turbulent flame is not a simple laminar diffusion flame but can be stretched and compressed by the turbulence. Thus, it is difficult to determine the thickness with any precision. But for the region not far from the burner tube, and on condition that the Reynolds number for the turbulence is not too large, then the length scales for the turbulence remain large, at least compared with the thickness of the flamelet.

When the turbulence increases further, two phenomena may occur:

- The turbulence may stretch the flamelet in such a way that it extinguishes locally, at the point where the stretch rate reaches a critical value. Indeed, if a portion of a flamelet were isolated in a turbulent flow and during its motion tracked, it would be found to be alternately stretched and compressed by the local velocity gradient. In the former case, it would resemble a diffusion flame between two opposing jets, one of fuel and the other of air, as shown schematically in Figure 6.21. The stream lines in such a flame would diverge from their initial direction. Diffusion flames between two opposing jets (termed "counterflow flames") blow out if the stretch rate (i.e. the normal gradient $\partial v/\partial x$ of the burning velocity) is too great. If the chemical processes are characterised by a single characteristic time, τ_c, the extinction can be shown to occur when the number $\tau_c\ (\partial v/\partial x)$ is larger than a certain critical value. If the stretching, and hence the characteristic velocity gradient, of the flamelets is considered to be due to small-scale turbulence, then it is proportional to τ_K^{-1}, where τ_K is the

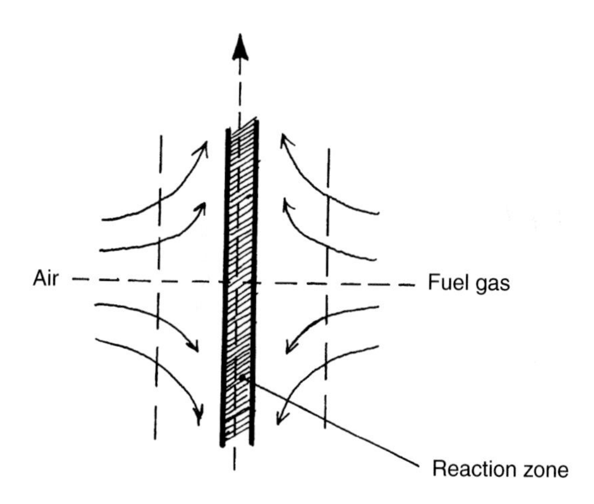

Figure 6.21 Diffusion flamelet, stretched in a local counterflow flame.

Kolmogorov time scale defined by (6.18), and the flamelet can be extinguished at any point where τ_c/τ_K exceeds a critical value. In this case the turbulent flame will no longer be formed by just one wrinkled flamelet but by several flamelets separated by extinction zones. These extinguished flame zones are, however, temporary and may reignite later as others are blown out, and so on.

- The second case is when the turbulence is too strong, but which also initiates interactions between two flamelet portions. The large length scales involved allow interaction between portions which are quite far apart, and lead to the formation of flamelet loops, as shown in Figure 6.20b. In contrast, the small turbulence scales, similar in size to the flamelets thicknesses, will lead to local thickening of these flamelets without actually resulting in the reaction zone closing in on itself, as indicated in Figure 6.20c. These local thickening events occur when the small turbulence scales, of size η, are similar to the thickness of the diffusion flamelet at the point in question. For a given integral scale, l_t, the Kolmogorov scale η reduces in size as the turbulent kinetic energy increases, since it is proportional to $Re_T^{-3/4}$. This local thickening phenomenon occurs with increasingly greater frequency as the Reynolds number Re_T increases.

Large-scale interactions will evidently modify the flame structure, since the flamelet will no longer be a single surface. However, small-scale interactions will induce more profound modifications since the structure of the flamelet itself is changed. When these interactions become too numerous, the flame can no longer be considered as being composed of one or more wrinkled flamelets, since the latter are themselves altered by the turbulence. As already stated, it is not easy to obtain the values of k and l_t at which too many interactions occur, although the resulting structural modification is only important for the chemical processes taking place if it extends as far as the central reaction zone. Since the degree of diffusion flamelet interaction is determined by the Reynolds number for the turbulence, Re_T, then flamelet presence is probably restricted to flows for which the value of Re_T is not too large compared to a value at which a significant fraction of the flamelet thickness is occupied by the reaction zone. When the combustion reactions can be likened to a global reaction with a very high activation temperature, the reaction zone only occupies a very small fraction of the flamelet and the turbulent flame can then be considered to be composed of flamelets up to very high Reynolds numbers. However, in practice, the thickness of the reaction zone is finite, which limits the validity of this type of structure for the turbulent flame. The limiting value of Re_T is not well known and is in fact the subject of some controversy.

6.6.2 The various types of turbulent diffusion flames

A diagram featuring the two characteristic numbers $k^{1/2}/S_L$ and l_t/e_L can be used to summarise the preceding discussion. However, for diffusion flames, S_L and e_L have absolutely no physical significance, representing simply the groups $(d/\tau_c)^{1/2}$ and $(d\tau_c)^{1/2}$ respectively (assuming here that ν, the coefficient of viscosity in the formula for the Reynolds number, Re_T, is very close to d, the coefficient of diffusion). Hence the local structure of a turbulent diffusion flame can be determined as a function of the values of these two parameters at this point.

Figure 6.22 again plots $Re_T = 1$, and the only area of interest is the truly turbulent region where $Re_T > 1$. Also plotted is $Re_T = Re^*_T$, which is the limit of the flamelet region. The line τ_c/τ_K = a constant, which relates to the local extinction of the flamelet, is also shown.

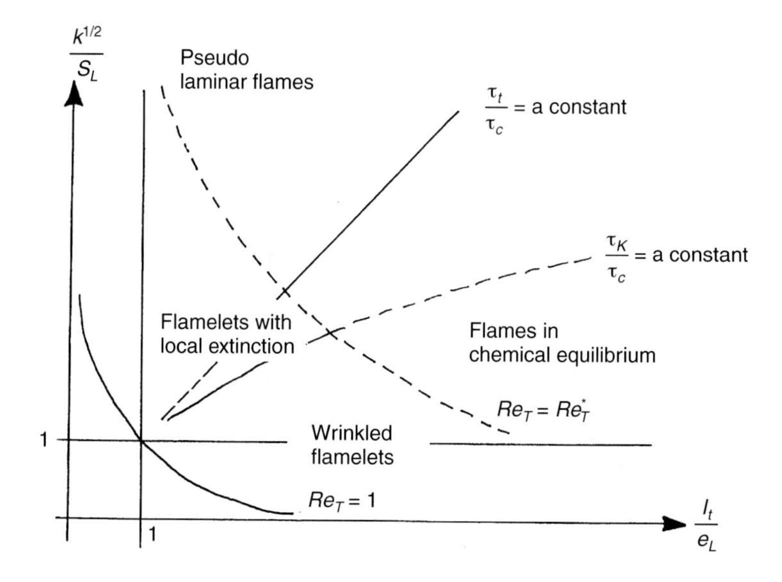

Figure 6.22 One classification of the various types of turbulent diffusion flame.

In the flamelet region, the flame has the instantaneous shape shown in Figure 6.20b and, in the region where localised flamelet extinction occurs, it is as shown in Figure 6.23a. These flamelets are interrupted at various points by areas where stretching has led to their extinction. Since the stretching is not continuous, a flamelet can reignite at a point where it was extinguished an instant before. The most likely reason for this reignition is the propagation of a premixed flamelet in the pocket of mixed gases created in the extinguished area. This premixed flamelet will propagate from each remnant of diffusion flamelet on both sides of the extinguished gas pocket, and each premixed flamelet will comprise two parts. It is thus based on a "triple flame" structure that each extinguished zone will reignite, as indicated in Figure 6.23b.

Figure 6.23a Instantaneous shape of a flame with extinguished flamelets at certain points.

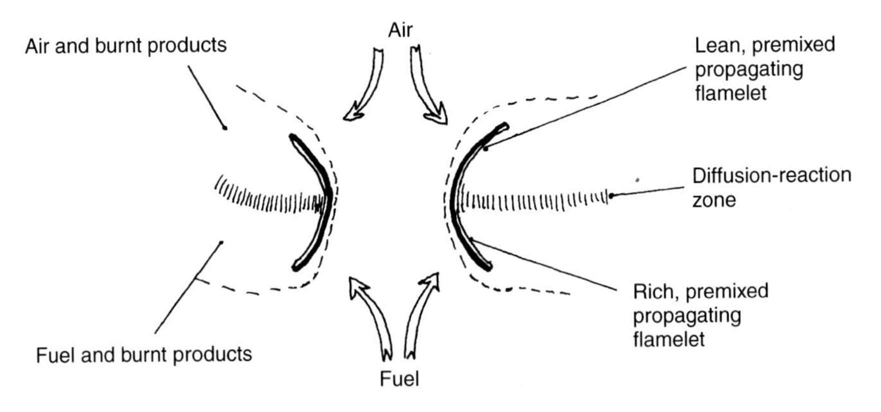

Figure 6.23b Structure of "triple" flamelets, following local extinction through stretching by "counter-current" flow.

In the region where $Re_T > Re_T^*$, it is no longer possible to differentiate between the flamelets in the strictest sense of the term, i.e. in that they are not internally modified as a result of the turbulence. However if only the reaction zones of the flamelets are drawn, disregarding the surrounding diffusion-convection zones, then the flame structure will once more resemble that of

Figure 6.20b, for $\tau_K < \tau_c$, with extinction, or that of Figure 6.23a, for $\tau_K > \tau_c$, with no extinction. Calculating the processes involved along the reaction zones, and particularly the local rates of reaction, is nonetheless significantly more difficult because the simplified picture of a counterflow flame can no longer be applied (Fig. 6.21).

Consider now the limit where $\tau_c \gg \tau_t$. With τ_t so small, any turbulent fluctuations in concentration and temperature in the flame will also be very small, since they will have had the time to dissipate before any significant chemical reaction could occur (although this does not imply that the velocity fluctuations are small; since they are continually renewed). Furthermore, if Re_T is large, then turbulence ensures that the diffusion phenomena are very efficient, since the effective turbulent diffusion coefficient is, by definition, proportional to $Re_T\nu$. This type of turbulent flame can be termed a pseudo laminar flame. The turbulent fluctuations are considerable, but on a very small scale ($l_t = k^{1/2}\tau_t$ can be small even if $k\tau_t$ is large) and produce a diffusion flame which hardly fluctuates, and which has thick diffusion-convection zones surrounding a reaction zone, as is the case for a normal laminar flame.

Since S_L and e_L no longer have any physical meaning in this case, it is also possible to establish the preceding classification by using more classical dimensionless numbers, e.g. the Damköhler number $Da = \tau_t/\tau_c$ $(= l_t/k^{1/2}\tau_c)$ and the Reynolds number for turbulence. The diagram shown in Figure 6.24 is obtained in this way.

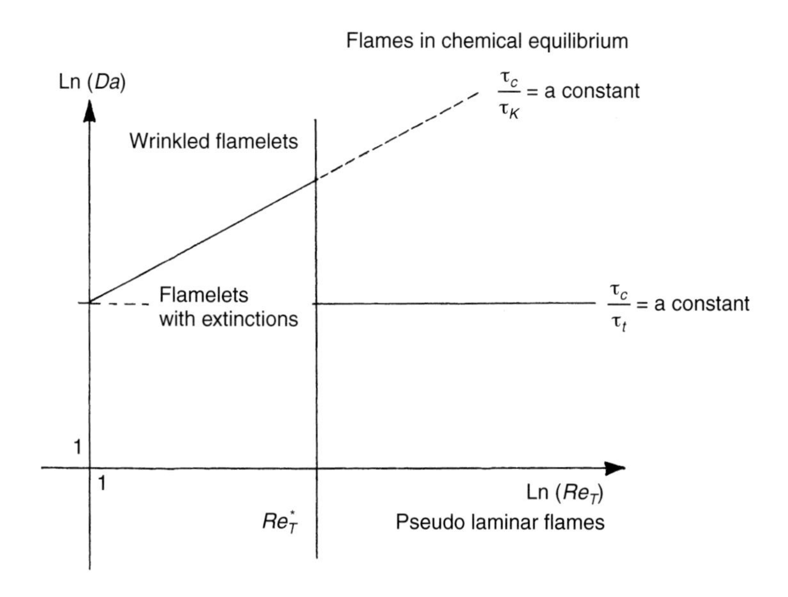

Figure 6.24 The various types of turbulent flame, classified in the plane (Da, Re_T).

6.6.3 Experimental evidence

Wrinkled diffusion flames can be represented in a highly visual way. For example, Figure 6.25 shows the changes in temperature across a flame, as a function of time. The lines of constant temperature are plotted in the (t, x) plane (where t is the time and x a coordinate across the flame). The two high temperature zones are apparent, within the two diametrically-opposed reaction zones. The two figures show these profiles as cross-sections at different heights above the burner. The profiles broaden and narrow, in an almost cyclic manner, reflecting the compression and stretching of the flamelet. These profiles were plotted based on measurements of the radiation emitted as recorded with a very fine refracting fibre traversing the flame. The flame has a low velocity (approximately one metre per second) and the burner has a mouth diameter of about one centimetre, hence quite low local Reynolds numbers are involved. The fact that the flamelet structures are very periodic is related to this fact, since natural convection plays a very important role in the behaviour of the flow, in terms of its stability, and its transition to turbulence.

In flames with higher velocities, and thus with higher Reynolds numbers, it is more difficult to visualise the processes involved. Nevertheless, instantaneous transversal profiles have been produced by measuring the temperature and concentration of the OH radical in flamelets (wrinkled or not). The experiment was based on the principle of Rayleigh diffusion (of the gas molecules themselves) and the laser-induced fluorescence of the OH radicals (in their ground state). Figure 6.26a gives some examples of instantaneous fluorescence signal profiles (roughly proportional to the number of moles of OH) obtained at a given position above the burner.

Each profile corresponds to one laser pulse, although these pulses are too far apart in time to allow to follow the changes over time. The presence of OH can be seen in the two diametrically-opposed flamelets. A certain number of double-peaked profiles can also be observed (some of which are indicated by arrows), which are evidence of interactions between two closely-spaced portions of a flamelet. There is no large-scale interaction between flamelets here although this does occur further away from the burner.

Figure 6.26b shows a sample of the results measured, for the same flame, of instantaneous temperature. Temperature profiles fairly similar to those for OH, although wider, were obtained and the maximum values of these profiles could be measured for each pulse. Figure 6.26b indicates, for three distances x/D above the burner (related in this way to burner diameter) the probability density function for the maximum temperature derived from each instantaneous flamelet profile. For unstretched flamelets, the maximum temperature should always be the same, which is the case for Burke and Schuman's laminar flame. For stretched flamelets, it should decrease with stretching until a temperature is reached below which there is no longer a stable, stretched flame.

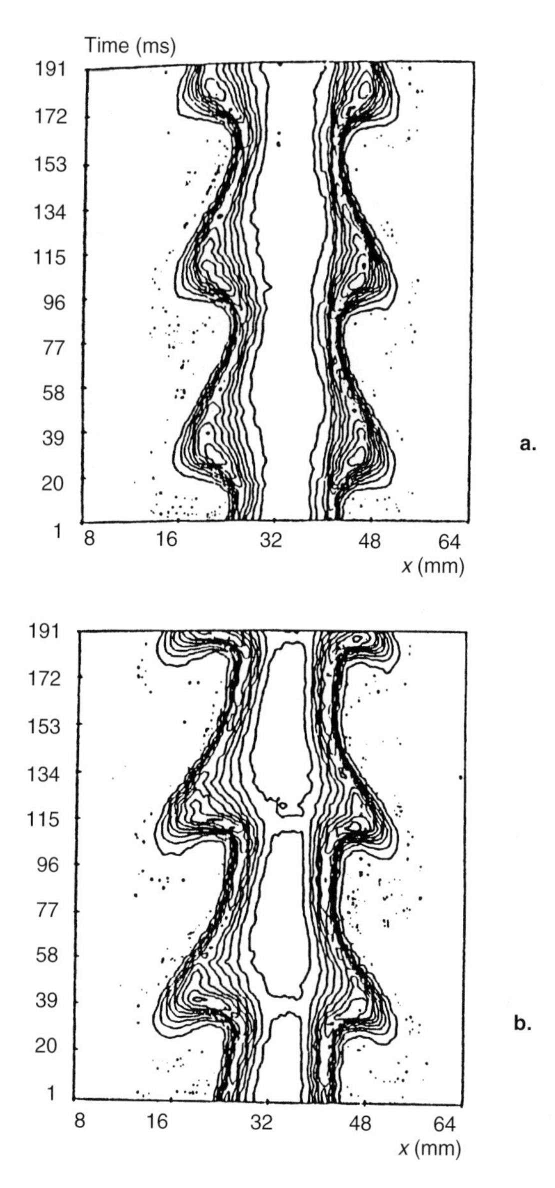

Figure 6.25 Temperature pulsations in a low Reynolds number diffusion flame. Radial profiles have been measured at a fixed height above the burner (different for a. and b.) each one corresponding to successive times. From these measurements one can obtain a surface in the three-dimensional space (T, t, x) and the isolines (at constant temperature) of this surface are plotted on the (t, x) plane. The temporal pulsations can be observed, symmetrical about the flame axis, successively stretching and compressing the flame. (After V. Vilimpoc and L.P. Goss (1988) *22nd Symposium (Int.) on Combustion*, The Combustion Institute, 1907-1914).

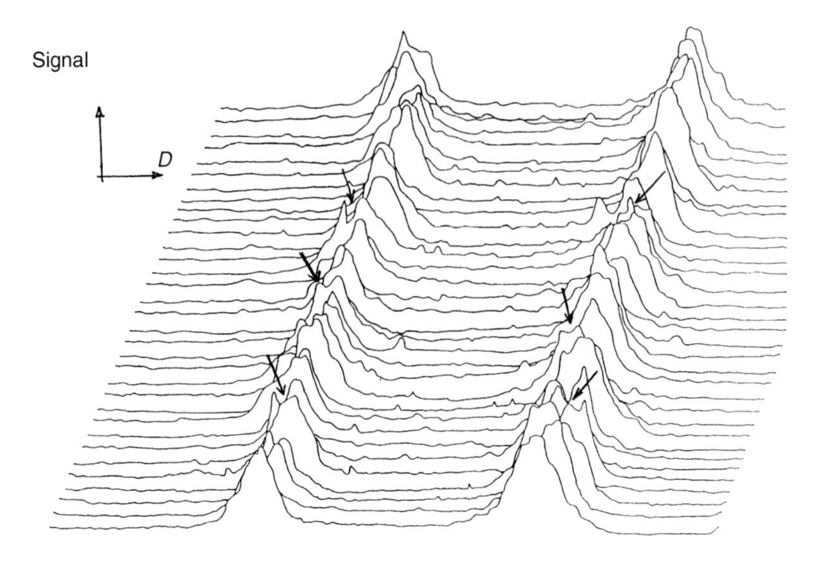

Figure 6.26a Profiles along a diameter of instantaneous OH fluorescence in a turbulent H_2—air flame. Each profile corresponds to one instantaneous measurement, and diametrically-opposed flamelets are apparent. The disturbance of the peaks in the measured signal (with two closely-spaced maxima) is caused by flamelet interactions. Unpublished experiment performed by D. Stepowski (1986) Faculté des Sciences de Rouen.

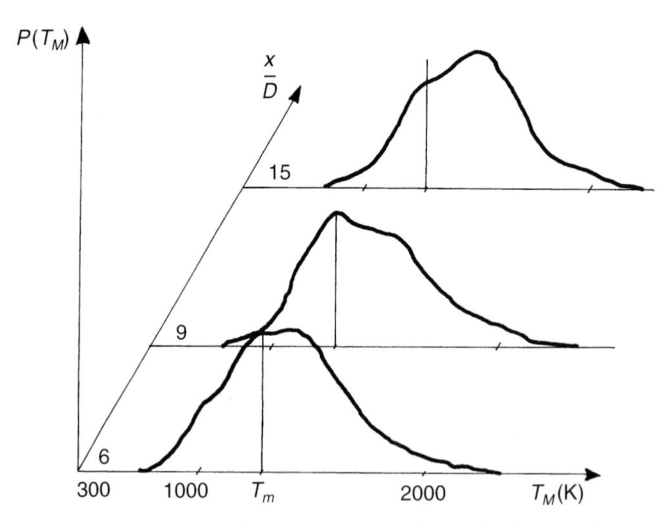

Figure 6.26b Probability density function of the maximum temperature T_M profile along a diameter, at three different heights in the flame, x/D. The value T_m corresponds to the minimum possible calculated value of the maximum temperature for a stretched, stationary flamelet, such as that shown in Figure 6.21. If the stretching is any greater, the flame blows itself out (and if any weaker the maximum temperature is greater). (After D. Stepowski, K. Labbaci and R. Borghi (1986) *21st Symposium (Int.) on Combustion*, The Combustion Institute, 1561-1568).

The figure demonstrates the wide dispersion in maximum flamelet temperatures, and even shows that temperatures below the minimum have a reasonable likelihood of existing. This suggests that extinction events occur in this particular flame, although the result of these events is not reflected in the measurement of temperatures as cold as that of the gases prior to combustion. The gas medium is thus complex, since gases of various compositions mix together within it: some of the chemical reactions involved may continue to completion, be halted or recommence; these events being reflected by extinctions or reignitions of the flame.

One interesting result which can be deduced from the measurements of the temperature and species present in turbulent diffusion flames concerns the joint probability density function of the temperature and local equivalence ratio, which is the mixture fraction (Z). This variable can be defined simply as the fraction of the fluid, at a given point and time, supplied by the fuel feed tube. This variable characterises the turbulent mixture (whilst neglecting the fact that the characteristics of the fluid may or may not have been changed because of a chemical reaction), and is similar to the variables Z_O and Z_P discussed in section 3.4 of chapter 3. If the fuel is pure, and allowance is made for the composition of the surrounding air, then a stoichiometric mixture fraction (Z_s) can be defined, which corresponds to the fuel/oxygen ratio required for the overall balanced stoichiometric reaction. Hence Z/Z_s is a variable which varies in the same way as the local equivalence ratio of the mixture.

When the chemical reactions can be assumed to occur infinitely rapidly (and with certain other assumptions, see sections 5.9 and 5.10 of chapter 5 and section 6.7.2 later in this chapter), it can be shown that a direct relationship exists between the local, instantaneous temperature and the mixture fraction (also local and instantaneous) in the flame. The probability density function $P(T, Z)$, if lines of equal value are plotted in the plane $T(Z)$, must therefore be reduced to the line which represents this relationship (where $T = T_{ad}(Z)$). When the flame consists of flamelets such as those in Figure 6.21, this implies that there is again a relationship between T and Z, where $T = T_F(Z, a)$, but now with the additional parameter a, the stretch rate of the flame.

Simultaneous measurements of T and Z in a flame allow lines to be constructed of equal value of $P(T, Z)$ in the T, Z plane. More qualitatively, a point can be considered in this same plane for each simultaneous measurement of T and Z, and the shape of $P(T, Z)$ can be determined from the surface density of the points. This has been done in Figures 6.27a, b, and c for a turbulent jet flame of methane (diluted with nitrogen) in air. The measurements were taken at a certain point in the flame (for $Z/D = 30$ and $R/D = 1.1$). Z can be found through measuring the nitrogen concentration–since nitrogen is unreactive (or very nearly so)–using Raman diffusion of N_2. The three Figures a, b, c correspond to the same flame but for three different combustion jet velocities, increasing progressively from a to c. The continuous lines on the

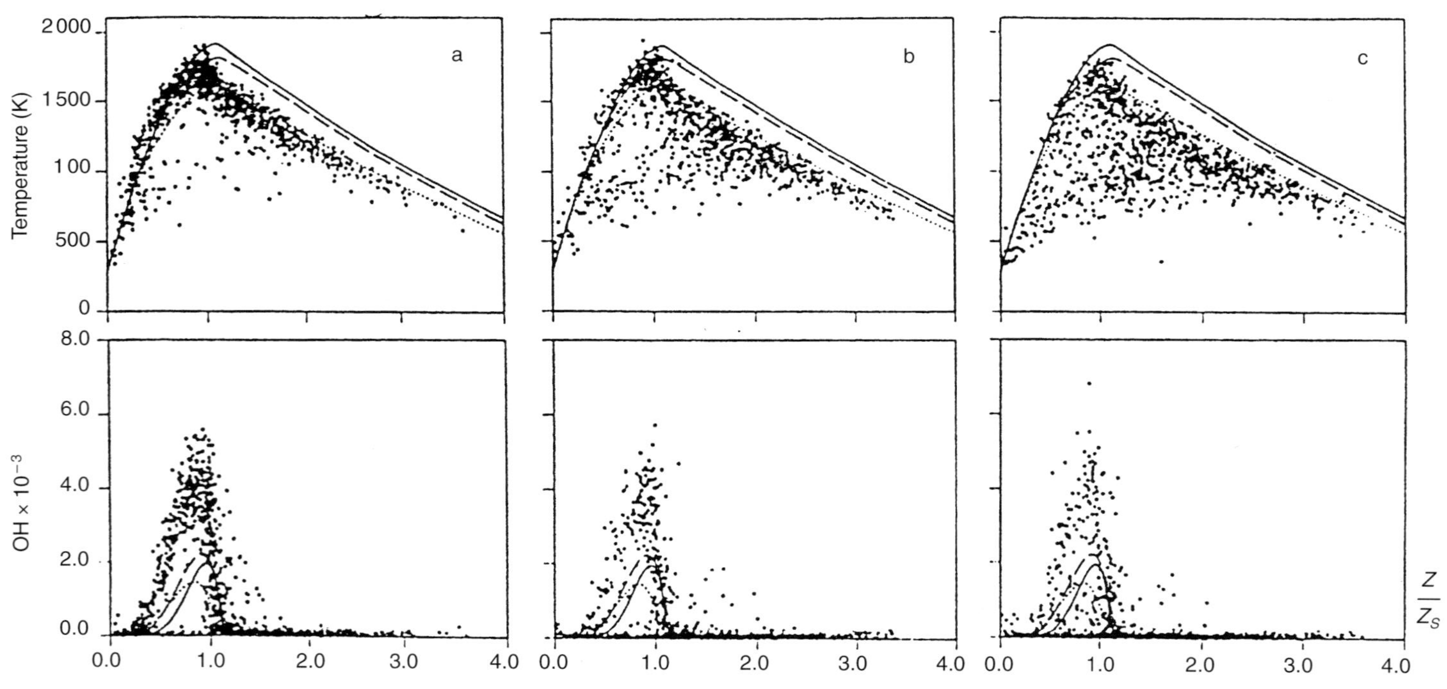

Figure 6.27 A series of simultaneously-recorded measurements of T, OH and Z in a turbulent diffusion flame between a jet of a methane-nitrogen mixture and a current of air. Each point corresponds with a signal collected as the result of a laser pulse striking a small flame element. All three cases correspond to the same point in the flame, but for increasing supply velocity of the combustible jet. The measurements increasingly deviate from the condition corresponding to chemical equilibrium (shown by the solid curves). The dashed curves represent verified relationships for stationary stretched flamelets in a counterflow flame. The closest curve to chemical equilibrium corresponds to a stretching rate of 50 s^{-1}, the other of 250 s^{-1} (if the stretching is any greater the flame will extinguish itself). (After R.S. Barlow, R.W. Dibble and S.H. Stårner, R.W. Bilger (1990). *23rd Symposium (Int.) on Combustion*, The Combustion Institute, pp. 583-589).

figures give the adiabatic temperature $T_{ad}(Z)$ relationships and the dashed lines the $T_F(Z, a)$ relationships for different stretching rates.

These figures clearly show that the scatter of the points becomes greater at lower temperature with increasing jet velocity; this is an indication of the presence of increasing numbers of unburnt fluid particles. The flames are clearly not at chemical equilibrium since, if they were, all the measured points would be situated on the line corresponding to the relationship $T_{ad}(Z)$. For the flame with the lowest velocity, except for a few points below 1000 K, it may be stated that the flame consists entirely of steady-state flamelets stretched like the counterflow flames, whereas for the other two, the fluid particles seem to be less and less like those predicted if such flamelets were present.

Figure 6.27 also shows simultaneous measurements of Z and the mass fraction of the OH radical. From the figure, the dispersion caused by turbulence can again be seen; and, likewise, deviations from the limiting case of infinitely rapid chemical reactions, i.e. at chemical equilibrium, and from the case where counterflow flamelets occur (dotted lines). It can also be noted that, for flames corresponding to higher jet velocities, the number of points with zero OH concentration increases, which verifies that the associated fluid particles contain unreactive, although not necessarily cold, gases.

6.7 MODELLING NON-PREMIXED, TURBULENT FLAMES

6.7.1 What should be calculated and how?

As for premixed, turbulent flames and for all turbulent flow, it must be stressed again that all calculations relate only to average quantities in a statistical sense. Indeed, it is impossible to calculate accurately the non-stationary changes in a flame since they will depend on specific limiting conditions which themselves cannot be determined. However, average quantities, the standard deviation of these quantities, or any other statistically-defined characteristic can be calculated.

Three types of quantity are of practical interest from amongst the mean characteristics: flame length measured from the burner mouth, its stabilisation range as a function of controllable parameters (gas velocity, global equivalence ratio of the mixture, etc.), and the total amount of the various products which can be released by the flame (or the maximum temperature which can be attained at its core). Described in layman's terms, these concepts are too vague to be calculated, and must be defined with more precision (if a little arbitrarily at times) in order to be able to determine their values. Two of these quantities will be dealt with here, namely flame length and the maximum temperature in the flame. The temperature of interest is the statistical average temperature. In general, the value considered in the calculations is the mean, weighted by the density (the Favre mean as defined in section 6.2 of this chapter), although the measurement techniques do not offer direct access to this mean (different techniques yield differently-defined averages). The value of this average temperature along the burner's axis of symmetry allows the flame length to be defined: either as being the distance from the mouth of the burner to the maximum temperature position, or as the distance between the mouth and a certain temperature value chosen for its relevance to the combustion chamber being studied.

In addition to convection, turbulence, turbulent diffusion, and chemical reactions, two physical processes which often play a key role in non premixed flames are heat transfer by radiation and natural convection due to buoyancy. Non premixed flames generally produce soot (i.e. those which burn carbon in some form) which, when heated, radiates heat in quite an efficient manner. The walls of the reactor vessel are also good radiators when heated to a high temperature (i.e. when they are not cooled). These two radiation sources can make significant contributions to the energy balance of the burning gaseous medium. Moreover, if the velocities of the jets of fuel or oxidant are not very high, or if quite extensive zones exist in the flame where the gas velocity is not much greater than about one metre per second, then gravity (in the form of a buoyancy effect) will play an important role in determining the gas flows, as a consequence of temperature differences.

These two processes will not be considered in this chapter since they are not essential for a qualitative understanding of turbulent flames, indeed, they only play a minor role in certain specific practical examples. The hydrogen-air jet flame is one such case and will often be used as an example.

The approximate method which we shall use to calculate the lengths and temperatures of non-premixed turbulent flames is the same as that used for laminar flames (see chapters 3 and 5) and for premixed turbulent flames (see this chapter). The method involves the use of balance partial differential equations to describe models of turbulent diffusion and to determine the average rate of reaction. The numerical calculations frequently used in this method to obtain physical values may appear at times to be somewhat cumbersome. However, the value of the method is its range of application, since it is suited to many types of burner of varying geometry, in different combustion chambers or reactors, and for a variety of fuel-oxidant combinations. The calculations for these various cases are complicated and can only be solved by the use of a numerical computer program. Furthermore, this method can be generalised to take into account buoyancy effects, radiation, and a more detailed calculation of the chemical processes involved.

This section will only consider cases in which the chemical reactions leading to the release of heat occur very rapidly in comparison with the rates governing the other processes. It is, in fact, for this reason that aspects relating to the stabilisation region of the burners will not be discussed. As for premixed turbulent flames, the inclusion of chemical reactions of finite rate brings up complications which are beyond the scope of this book.

6.7.2 Calculating non-premixed turbulent flames, assuming rapid chemical reactions

If $\widetilde{Y}_i$ is used to denote the average mass fraction of a species i in the flame, then this quantity can be inserted in the general balance equation (6.25), originally formulated for the case of a premixed flow. However, once again, this equation is only of value if the rates of reaction, $\widetilde{w}_i$, are known.

In his initial studies in 1962, H.L. Toor considered the simple (some would say overly simplistic) case of a single, irreversible chemical reaction involving fuel and oxidant, written symbolically as:

$$K + \nu Ox \rightarrow P$$

Thus $w_{Ox} = -\, kY_{Ox}Y_K = \nu_s w_K$, where k is the rate constant and ν_s the stoichiometric ratio ($\nu_s = \nu M_{Ox}/M_K$) for the reaction. The assumption that the reaction occurs infinitely rapidly, does not automatically infer that the rate of reactions w_i are infinite but rather that, in reality, $k \rightarrow \infty$. It then follows that the product $Y_{Ox}Y_K \rightarrow 0$ in order for the rate w_{Ox} to remain finite, and the greater the value of k, the smaller must be the Y_{Ox} **or** Y_K values in the flame zone.

Consider the variable $Z^* = Y_{Ox} - \nu_s Y_K$, and assume that the species Ox and K have the same coefficient of diffusion $d_{Ox} = d_K = d$, Z^*, with the Y_{Ox} and Y_K variables here representing instantaneous mass fractions in the flame rather than average values. Thus Y_{Ox} and Y_K obey the usual form of the instantaneous balance equation (as in section 3.2 of chapter 3, equation (3.17) with Fick's Law of Diffusion):

$$\frac{\partial(\rho Y_i)}{\partial t} + \sum_1^3 \frac{\partial}{\partial x_\alpha}(\rho v_\alpha Y_i) = \sum_1^3 \frac{\partial}{\partial x_\alpha}\left(\rho d \frac{\partial Y_i}{\partial x_\alpha}\right) + \rho w_i \tag{6.38}$$

Also, Z^* satisfies the equation with no reaction term:

$$\frac{\partial(\rho Z^*)}{\partial t} + \sum_1^3 \frac{\partial}{\partial x_\alpha}(\rho v_\alpha Z^*) = \sum_1^3 \frac{\partial}{\partial x_\alpha}\left(\rho d \frac{\partial Z^*}{\partial x_\alpha}\right) \tag{6.38 bis}$$

Z^* can therefore be considered to be the mass fraction of an inert species subjected only to the processes of convection and diffusion, in the turbulent (non-isothermal) flow within the flame. Z^* can be thought of as being a "tracer", i.e. an inert species injected into the flow at such a low concentration that it does not alter the flow. Since this tracer is chemically inert, it is subjected to, and thus controlled only by turbulent physical phenomena. If it is possible to understand and model the turbulence, it should also therefore be possible to determine Z^* accurately. Because Z^* is an instantaneous variable in the turbulent flow, it can best be described through $P(Z^*, \rho)$, the probability density of the random variables associated with Z^* and ρ, at each point in the flow.

The variable Z^* is particularly interesting when the reaction occurs infinitely rapidly. Earlier analysis showed that in this specific case the product $Y_{Ox}Y_K$ is zero (or nearly), i.e. either $Y_{Ox} = 0$ and $Y_K \neq 0$, or $Y_{Ox} \neq 0$ and $Y_K = 0$, or $Y_{Ox} = Y_K = 0$. These conditions imply that when $Z^* < 0$, $Z^* = -\nu_s Y_K$ and $Y_{Ox} = 0$, and when $Z^* > 0$, $Z^* = Y_{Ox}$ and $Y_K = 0$. Thus, if we know Z^*, a non reactive, virtual variable, then we can determine the reactive variables Y_{Ox} and Y_K.

Hence, if $P(Z^*, \rho)$ is known it is possible to calculate $\overline{Y_{Ox}}$, $\overline{Y_K}$ as well as $\widetilde{Y}_{Ox}$ and $\widetilde{Y}_K$, without using either the averaged equations involving $\widetilde{Y}_i$ or $\widetilde{w}_i$. Indeed, the preceding considerations yield simply that:

$$\overline{Y_{Ox}} = \int_{Z^*>0} \int_{\rho} Z^* P(Z^*, \rho)\, dZ^*\, d\rho$$

$$\overline{Y_K} = -\frac{1}{\nu_s} \int_{Z^*<0} \int_{\rho} Z^* P(Z^*, \rho)\, dZ^*\, d\rho \qquad (6.39)$$

or:

$$\widetilde{Y}_{Ox} = \int_{Z^*>0} Z^* \widetilde{P}(Z^*)\, dZ^*$$

$$\widetilde{Y}_K = -\frac{1}{\nu_s} \int_{Z^*<0} Z^* \widetilde{P}(Z^*)\, dZ^* \qquad (6.39\ \text{bis})$$

with:

$$\widetilde{P}(Z^*) = \frac{1}{\overline{\rho}} \int_{\rho} \widetilde{\rho P}(Z^*, \rho)\, d\rho$$

$$\overline{\rho} = \int_{Z^*} \int_{\rho} \rho P(Z^*, \rho)\, dZ^*\, d\rho \qquad (6.40)$$

The solution to the problem thus reduces to that of calculating $\widetilde{P}(Z^*)$. Later we will discuss how to determine $\widetilde{P}(Z^*)$ approximately (as indeed all quantities related to turbulence must be approximate), but first let us consider how this method can be generalised, as a result of some additional assumptions, to the case of multiple, reversible chemical reactions as originally demonstrated by R.W. Bilger[6].

6. Bilger R.W. (1976) Turbulent Diffusion Flames, *Progress in Energy and Combustion Sciences.* Vol. 1, p. 87.

This proof retains the assumption that all the species have coefficients of diffusion which are approximately equal, and which are also equal to the thermal diffusivity, such that all Lewis numbers are equal to unity. The enthalpy then satisfies an instantaneous balance equation very similar to equation (6.38), which can be written using equation (3.18) from chapter 3:

$$\frac{\partial}{\partial t}(\rho h_t) + \sum_1^3 \frac{\partial}{\partial x_\alpha}(\rho v_\alpha h_t) =$$

$$\sum_1^3 \frac{\partial}{\partial x_\alpha}\left(\rho d\,\frac{\partial h}{\partial x_\alpha} + \sum_{\beta=1}^3 \tau_{\alpha\beta}\, v_\beta\right) + \frac{\partial p}{\partial t} + \sum_1^3 F_\beta v_\beta \quad (6.41)$$

By neglecting long-distance forces, F_β, and considering a flow of sufficiently low velocity such that $h_t \approx h$ and $\tau_{\alpha\beta}{\cdot}v_\beta$ and $\partial p/\partial t$ can be neglected, an equation can be derived in h which is identical in form to that for Z^* (see 6.38 bis).

Furthermore, several variables can be defined, such as Z^*, which are unchanged by chemical reaction, namely those representing the numbers of atoms, and hence the masses of each chemical element present in the flame. For example, consider a hydrogen-air flame and assume that the only species present in non-negligible amounts are O_2, H_2, H_2O, N_2, OH, H, O, H_2O_2, NO and NO_2, then:

$$Z_N = Y_{N_2} + (M_N/M_{NO})\,Y_{NO} + (M_N/M_{NO_2})\,Y_{NO_2}$$

$$Z_H = 2\,(M_H/M_{H_2O})\,Y_{H_2O} + Y_{H_2} + (M_H/M_{OH})\,Y_{OH} + Y_H + 2\,(M_H/M_{H_2O_2})\,Y_{H_2O_2}$$

$$Z_O = Y_{O_2} + (M_O/M_{H_2O})\,Y_{H_2O} + (M_O/M_{OH})\,Y_{OH} + (M_O/M_{NO})Y_{NO}$$

$$+ 2\,(M_O/M_{NO_2})Y_{NO_2} + 2\,(M_O/M_{H_2O_2})Y_{H_2O_2} + Y_O$$

As a consequence of the conservation of elements in all chemical reactions, and based on the assumption of identical coefficients of diffusion, then clearly Z_N, Z_O, and Z_H all satisfy an equation identical to (6.38 bis).

Let us now consider in more detail the case of the jet flame produced by two flows, one rich in fuel and the other in oxidant (for example, a jet of pure hydrogen in air). Each flow has a well-defined composition and temperature and the subscripts j and e will be used to denote characteristics relative to the fuel-rich jet and the flow of oxidant respectively. Note that all the Z_k parameters defined previously, as well as h, can be related linearly to a variable Z which obeys equation (6.38 bis) and is such that the flame entry conditions are $Z_j = 1$ and $Z_e = 0$. Clearly Z_k, related to Z by:

$$\frac{Z_k - Z_{ke}}{Z_{kj} - Z_{ke}} = Z \quad (6.42)$$

also satisfies (6.38 bis), and similarly for h, related to Z by:

$$\frac{h - h_e}{h_j - h_e} = Z \quad (6.42\ \text{bis})$$

So long as limiting conditions other than those at the gas entry point, (e, j), are not incompatible with (6.42) and (6.42 bis), these two formulae calculate all Z_k and h values as a function of Z only, a simple variable which describes the turbulent mixture. This is true for the jet flame in the infinite medium we are considering, since the other limiting conditions are $\partial Y_i/\partial n_{\text{limit}} = 0$, $\partial h/\partial n_{\text{limit}} = 0$ (normal derivatives at the boundaries of the region in which the calculations are made) which results in $\partial Z_k/\partial n_{\text{limit}} = 0$, $\partial h/\partial n_{\text{limit}} = 0$, $\partial Z/\partial n_{\text{limit}} = 0$, which is compatible with (6.42) and (6.42bis). However, this is not true if, for example, the jet flame develops close to a wall which is impermeable to the transfer of species but not to heat transfer. In this case, the expression $\partial Y_i/\partial n_{\text{wall}} = 0$ is true at this wall, but $\partial h/\partial n_{\text{wall}} \neq 0$ and these two expressions are not compatible with either equation (6.42) or (6.42 bis).

For the case currently being considered of a jet flame in an infinite medium, the important result is that Z_k and h may be calculated as a function of Z alone, as defined above. This specific, virtual, variable is the mixture fraction defined in section 6.6.3.

By including the assumption that the chemical reaction occurs infinitely rapidly (at all points), this implies, for the multi-component mixture, that chemical equilibrium is reached at all points and at all times. We know that in this particular case all the chemical species can be calculated as a function of the temperature and the elementary composition of the mixture, hence $Y_i = Y_i^{\text{eq}}(Z_k, \forall k, T)$, where the function Y_i^{eq} is known. Since it has just been shown that all the Z_k are (linear) functions of Z, as well as h $(Y_i, \forall i, T)$, it follows that all Y_i, as well as T may be expressed as a function of Z alone. Let these functions be denoted $Y_i^{\text{eq}}(Z)$ and $T^{\text{eq}}(Z)$.

It is now easy to calculate $\widetilde{Y}_i$ and $\widetilde{T}$ as previously, using the probability density for the fluctuations in Z (rather than the averaged balance equations):

$$\widetilde{Y}_i = \int_Z Y_i^{\text{eq}}(Z)\,\widetilde{P}(Z)\text{d}Z \tag{6.43}$$

$$\widetilde{T} = \int_Z T^{\text{eq}}(Z)\,\widetilde{P}(Z)\text{d}Z \tag{6.43 bis}$$

Equations (6.39 bis) are only simply specific cases of equation (6.43). Indeed, Z^* in equation (6.39 bis) can also be linked to the Z here, through $(Z^* - Z_e^*)/(Z_j^* - Z_e^*) = Z$. Moreover, the segments of the linear relationships linking Y_{Ox} and Y_{K} to Z are nothing more than approximations of the general relationships involving $Y_i^{\text{eq}}(Z)$. Finally, the temperature has not been calculated when assuming a single reaction system since the composition of the reaction products is independent of temperature. Had it been calculated, however, then the following assumption would be required: that the Lewis numbers were all equal to unity, that the flow velocity was quite low, and that no non-adiabatic wall disturbed the flame; in fact, these were the assumptions made for the general case.

The relationships $Y_i^{eq}(Z)$ and $T^{eq}(Z)$ are represented schematically in Figure 6.28, with their corresponding approximations to a single, irreversible reaction. Some experimental results for plots of this type have already been given (see Figure 6.27). Experimentally, Z can be measured by defining it as being linearly related to the total mass fraction of nitrogen in the flame (ignoring any NO and NO_2). Unfortunately, calculation of the plots involving $Y_i^{eq}(Z)$ and $T^{eq}(Z)$ relies on the assumption that the Lewis numbers are equal to unity. It would be useful to avoid this approximation since these numbers may differ greatly from one; which would introduce interesting phenomena. However, no other way of doing this is currently known.

Although the method described here (often known as the "conserved scalar method", since Z is a scalar quantity which is conserved in chemical reactions) can take into consideration all the details of the turbulence, it is quite difficult

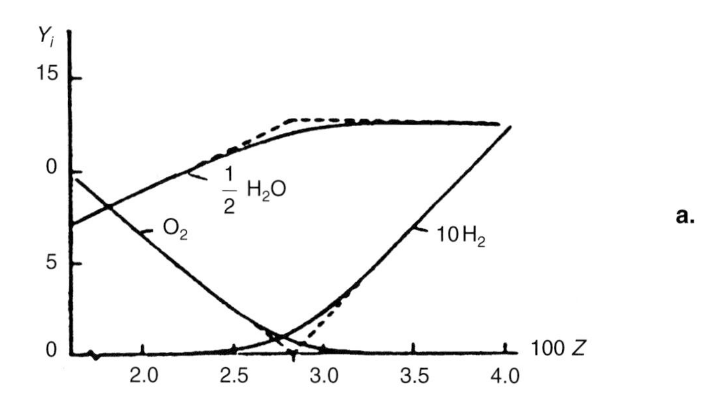

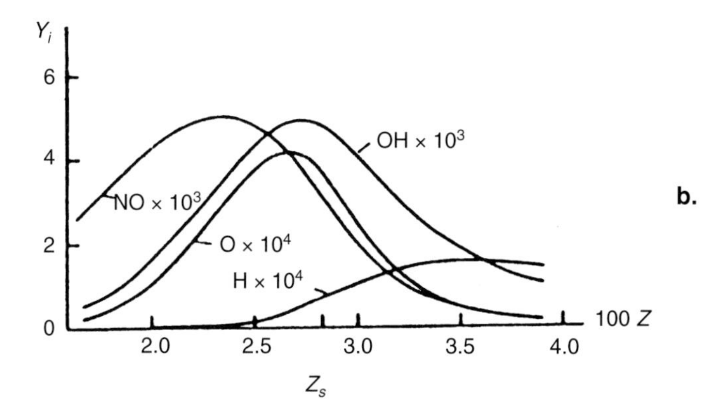

Figure 6.28 Relationships between the mass fractions Y_i of various species and the mixture fraction Z, due to chemical equilibrium, for an H_2—air flame. The value of Z corresponding to the stoichiometric mixture is $Z_s = 0.028$. (After R.W. Bilger (1976) *Progress in Energy and Combustion Sciences*, 1, p. 87).

to apply due to the complexity involved in calculating $\widetilde{P}(Z)$. A more heuristic method is usually preferred, based on the use of the equation proposed by Magnussen, which is similar to the "Eddy Break-Up" model for premixed turbulent flames.

In 1975, B. Magnussen[7] proposed a model for expressing the $\widetilde{w}_i$ of turbulent diffusion flames in cases where the reactions can be assumed to occur infinitely rapidly. The model is based on the principle that the average rate of reaction must be controlled by the rate of turbulent mixing, i.e. that $\widetilde{w}_i$ should be proportional to τ_t^{-1}. Moreover, to ensure that $\widetilde{w}_i$ is zero when either the average mass fraction of the fuel $\widetilde{Y}_K$, or of the oxidant $\widetilde{Y}_{Ox}$, is zero (a trivial, but necessary condition), Magnussen proposed that:

$$\widetilde{w}_i = -C_{Mg}\,\frac{Y_i^0 - Y_i^{eq}}{\tau_t}\,\mathrm{Min}\,(\widetilde{Y}_K, \nu_s\widetilde{Y}_{Ox}) \tag{6.44}$$

where ν_s is the specific stoichiometric ratio of the equivalent global reaction. Thus the average balance reaction (6.25) is used with (6.44), to determine $\widetilde{Y}_K$ and $\widetilde{Y}_{Ox}$; in combination with the turbulence equations to determine d_t and τ_t; and together with the average flow and average energy equations for velocity and temperature fields.

The physical justification of the $\mathrm{Min}(\widetilde{Y}_K, \nu_s\,\widetilde{Y}_{Ox})$ factor in equation (6.44) is weak. However, this formula yields results which are qualitatively realistic provided that a suitable value for the empirical constant C_{Mg} had been chosen; this constant is a numerical value similar to the C_{EBU} constant described in section 6.4, and ensures that the calculated results are correct within one order of magnitude.

6.7.3 Calculating an inert species in a non-premixed turbulent flame

We have seen that the calculation of a turbulent flame requires calculating the probability density $P(Z)$ of a "virtual inert species" Z diffusing in the turbulence. This is a fundamental problem when analysing turbulence, and one which is encountered in other applications. However, it had not been studied in detail prior to the discovery of its usefulness in analysing turbulent flames and, more generally, in reactive turbulent flows.

Research has been conducted since the start of the seventies in an attempt to calculate $P(Z)$ directly in a flow of constant ρ or $\widetilde{P}(Z)$ in a flame. The most satisfactory methods propose a new balance equation for $\widetilde{P}(Z)$, using the variables Z, x_α and t (position and time). Other, approximate methods suggest

7. Magnussen B. (1975) *14th Symposium (Int.) on Combustion*, The Combustion Institute.

calculating $\widetilde{P}(Z)$ using a standard form which can be determined as a function of the average value $\widetilde{Z}$ and of the variance $\widetilde{Z'^2}$. Thus, $\widetilde{Z}$ and $\widetilde{Z'^2}$ can be calculated (as functions of x_α and t) using a balance equation, from which $\widetilde{P}(Z; x_\alpha, t)$ is deduced.

What profile of $\widetilde{P}(Z)$ might be expected in a jet flame? Obviously the answer to this question depends on which part of the flame is being considered. In the fuel jet, before the fuel can mix with the surrounding air, $\widetilde{P}(Z)$ is characterised by one Dirac peak since Z is exactly equal to unity. In the wake immediately downstream the fuel supply tube, $\widetilde{P}(Z)$ probably has two Dirac peaks, one at $Z = 1$ and the other at $Z = 0$, with different surface areas, but whose sum is one. When a fluid particle from the fuel jet passes the observation region, then $Z = 1$, and when an air fluid particle from the surrounding air passes then $Z = 0$. Thus, since the case of interest is the position very close to the meeting point of the two fluid particles, then practically only these two possibilities can exist. A long way downstream of the jet flame, $\widetilde{P}(Z)$ will return to a single Dirac peak at $Z = 0$ since the molecules provided from the fuel supply tube will, in any case, be mixed with an excess of molecules from the surrounding air. In the flame itself, away from the mouth, there will be a "bell" shape distribution around the average value. It is also very probable that $\widetilde{P}(Z)$ will have an increasingly Gaussian profile with increasing distance from the jet supply point, with a progressively diminishing "thickness" (i.e. variance $\widetilde{Z'^2}$) (shown schematically in Figure 6.29).

The concept of calculating $P(Z; x, t)$ using a balance equation was suggested by E. O'Brien and C. Dopazo[8] in 1974. The model balance equation they used was:

$$\frac{\partial}{\partial t}\rho P(Z) + \sum_1^3 \frac{\partial}{\partial x_\alpha}(\overline{v}_\alpha P(Z)) =$$

$$\sum_1^3 \frac{\partial}{\partial x_j}\left(\rho D_p \frac{\partial P(Z)}{\partial x_j}\right) + \frac{\partial}{\partial Z}\left(\rho \frac{C_D}{\tau_t}(\overline{Z} - Z)P(Z)\right) \quad (6.45)$$

which includes non-stationary and convection terms, in addition to a term for diffusion through physical space x, and a term representing the decrease of the fluctuations in Z around its average value $\overline{Z}$. This latter effect causes the PDF to become progressively narrower around $Z = \overline{Z}$ with increasing distance downstream of the mouth. This is shown in Figure 6.29 which represents mixing of fluid particles down to the molecular scale.

Whilst this equation was initially derived for the case where ρ is a constant, it can be generalised by introducing $\overline{\rho}$, $\widetilde{P}(Z)$ and $\widetilde{v}_\alpha$.

8. O'Brien E. E. (1980) Turbulent Reacting Flows, in: *Topics in applied physics*. Ed. P.A. Libby and F.A. Williams. Springer Verlag.

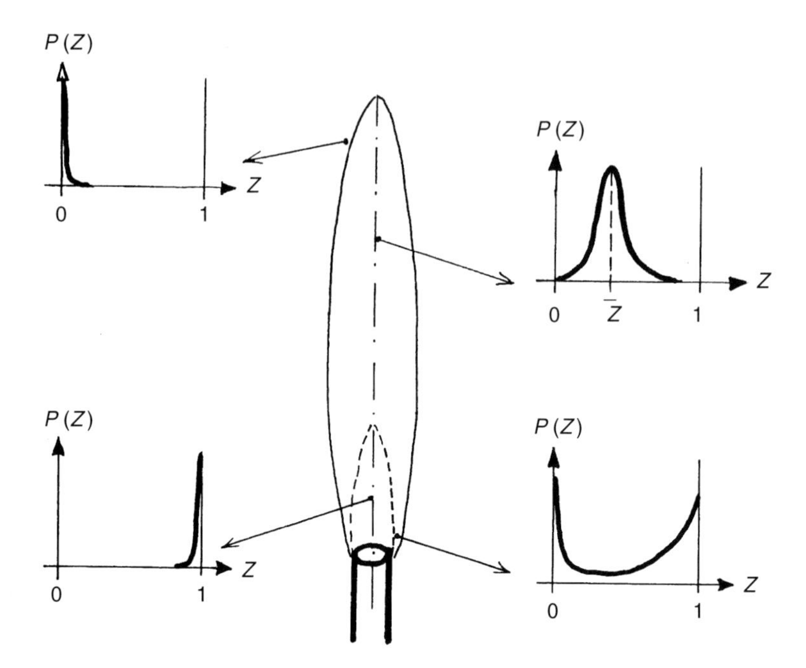

Figure 6.29 Form of the *PDF* profiles, $P(Z)$, at various positions in a turbulent jet flame.

The critical feature in all these models is the last term, and particularly the involvement of the integral time scale for the velocity fluctuations, $\tau_t = l_t/k^{1/2}$, where $1/\tau_t$ is a turbulence-induced frequency of mixing.

Some calculations of $\widetilde{P}(Z)$ have been performed using equations of this type for non-reactive flows. The comparisons with experiment are reasonably satisfactory, and are at least as good as the classical $k - \varepsilon$ type models, since these models were used to calculate k and l_t. Models other than that described by equation (6.45) exist; particularly the coalescence-redispersion model which incorporates an integral form of the last term of (6.45). This model is based upon a physical model which simulates the turbulent mixing between several fluid particles caused by collisions, during which two fluid particles coalesce and then separate again into identical particles. Under these conditions, the last term (which represents this mixing on the scale of a fluid particle) can be written as:

$$\text{mixing} = \frac{C'_D}{\tau_t}\left[2\int\int P(c')P(c)\delta\left(Z - \frac{c+c'}{2}\right)\mathrm{d}c\,\mathrm{d}c' - P(Z)\right] \qquad (6.46)$$

(where δ is the Dirac distribution) if it is simply assumed that two fluid particles, for which Z is respectively c and c', yield after coalescence and separation two fluid particles for which Z is equal to $(c + c')/2$. Indeed, the rate of appearance of particles of composition Z is due to the sum of all collisions such that $(c + c')/2 = Z$, and the rate of disappearance of these particles is due to all the collisions of a particle of composition Z with any other particle. Furthermore, these rates are also proportional to $1/\tau_t$, which is an average collisions frequency. C'_D is again a constant which is adjusted to correlate the equation with experimental results.

Figure 6.30 shows examples of the $\widetilde{P}$ profiles calculated for an axisymmetrical jet flame with rotational symmetry along the flame axis. Since Z must be between 0 and 1 (inclusive) and because $\widetilde{Z'^2}$ only becomes very small for quite large x/D values, where Z is very small, the Gaussian form is not particularly suitable. The results shown here use the model given by equation (6.46), although the model (6.45) gives quite similar results. In fact, the results are

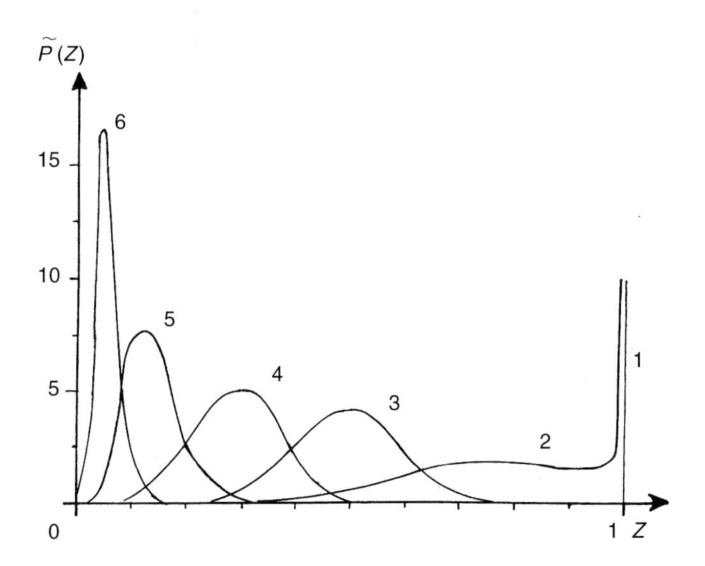

Figure 6.30 Calculation, through the use of a balance equation, of the probability density function (PDF) of the mixture fraction in a diffusion jet flame. The profiles 1, 2, 3, 4, 5, and 6 correspond to points in the flame axis which are increasingly further away from the fuel supply jet (After J. Janicka, W. Kolbe and W. Kollmann (1978) *Proc. of the 1978 Heat Transfer and Fluid Mechanics Institute*. Ed. C.T. Crowe and W.L. Grosshandler, Stanford Univ. Press).

quite close to the measured values, bearing in mind that the latter are affected by certain experimental errors. The value of the constant C'_D which provides a suitable fit with experimental results is, as an average for most cases, approximately 0.9.

Research into this subject continues, despite this undeniable success, because in principle $\widetilde{P}$ should be found to be Gaussian when the fluctuations are very small and because certain faults nonetheless exist in certain cases, particularly in relation to the coefficient D_p of equation (6.45). In addition, the influence of variations in ρ and the differences between $\widetilde{P}$ and P merit further studies.

6.7.4 Comparison between calculation and experiment

In cases in which the assumption that the chemical reactions occur infinitely rapidly is valid, then the results of calculations using the methods described by equations 6.7.2 and 6.7.3 are reasonably satisfactory.

For example, for the case of a turbulent jet flame of hydrogen at high velocity ($v_j \simeq 50$ m/s, $v_e \simeq 9$ m/s), then the use of a $k - \varepsilon$ type model for turbulence and representing $P(Z)$ by an approximate function reproduces the experimental trends perfectly.

Z being the mixture fraction (defined in section 6.6.3), one has drawn in the (r, x) plane of figure 6.31a lines of equal experimental $\widetilde{Z}$ values and on each one marked the point where r is equal to zero, i.e. on the axis of the flame, for instance here I for $\widetilde{Z} = 0.6$. Then one considers the point where, for the same x-value, $\widetilde{Z}$ is twice smaller, therefore here J for $\widetilde{Z} = 0.3$. Of course, there is no obvious reason for choosing $\widetilde{Z}/2$ rather than $\widetilde{Z}/3$ or $\widetilde{Z}/4$, etc. The different IJ values, called R_Z, are functions of mixing conditions along axis and radius. In a similar way, but with the fuel mass fraction, Y_K, one obtains R_F, whose value depends on the combustion chemistry; R_F is the radius for the line where $\widetilde{Z} = Z_S$ (defined by the stoichiometric proportion). Thus it is possible to know something about the flame geometry: i.e. to find the x-value for which $R_F = 0$, with R_Z showing the width of the mixing zone. It appears in Figure 6.31b that the total flame length is slightly underestimated, which is probably connected with an overestimation of R_Z in the zone with the highest temperatures. However, the comparison between experiment and calculation is, in general, good.

For the same flame, Figure 6.32 gives the profiles of $\widetilde{Z}$ and $\widetilde{v}/v_j$ along the axis. The same flame length as that shown in 6.31b is found at a value of x/D where $\widetilde{Z} = Z_S$. The "potential core" of the jet, where $\widetilde{Z}$ and $\widetilde{v}/v_j$ are very close to unity, is predicted by the calculations to extend for about 5 or 6 jet diameters, although measurements have not been obtained in this zone. For $\widetilde{Z}$, there are certain differences between calculation and experiment in the highest temperature zone but, on the whole, the agreement is good (although to a certain extent this might be expected due to the use of logarithmic scales). It should also be noted that the experimental measurements are, by nature, very difficult to obtain and hence may ultimately be shown to be inaccurate.

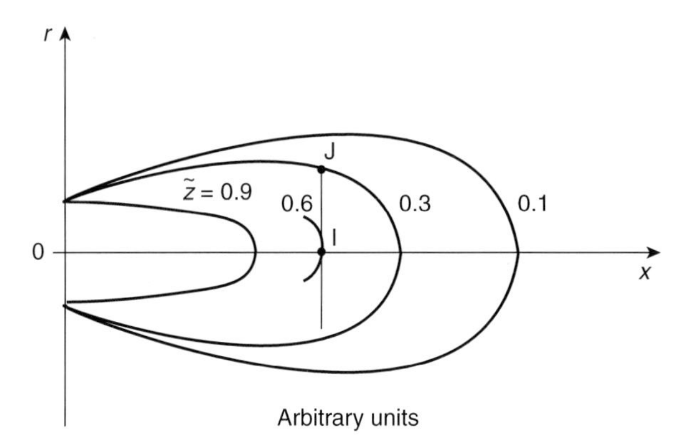

Figure 6.31a Lines of equal $\widetilde{Z}$ values in the r/x plane, used to define R_Z. IJ is the R_Z-value for a point such I on the axis of the flame. Using the fuel mass fractions, analogous lines can be drawn with equal Y_K values. The iso-line for $Z = 0$ intercepts the x-axis for an infinite value of x, but not, of course, the iso-line for $Y_K = 0$.

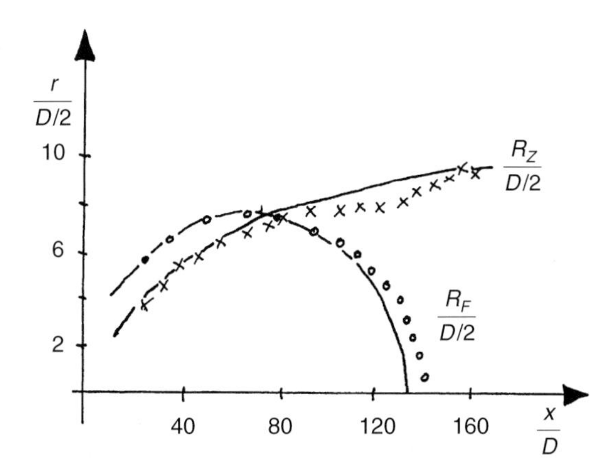

Figure 6.31b Comparison between calculation and experiment for an H_2–air jet flame. Flame radius and mixing zone (After R.W. Bilger and R.E. Beck (1975) *15th Symposium (Int.) on Combustion*. The Combustion Institute, p. 541).

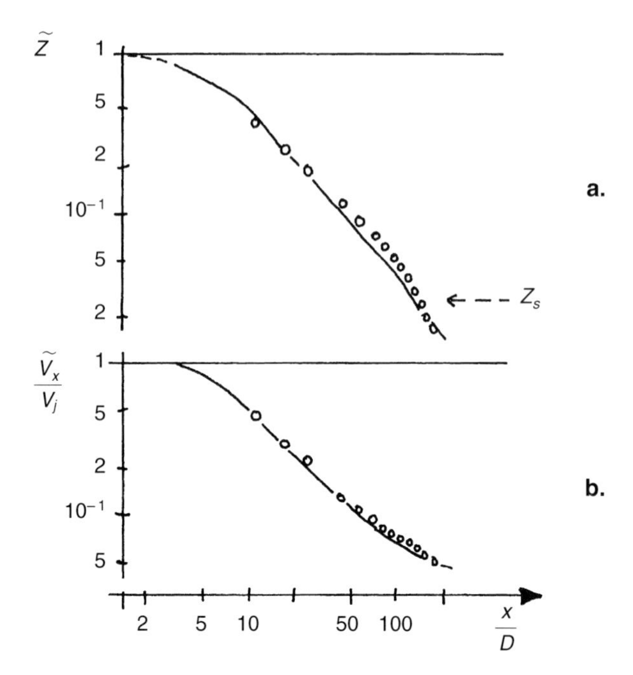

Figure 6.32 Comparison between calculation and experiment for an H_2–air jet flame. Profiles of "mixture fraction" and average velocity along the flame axis (on logarithmic scales). Same flame and reference as for Figure 6.31b.

Figures 6.33a and b show the mass fraction profiles of the major chemical species in addition to the temperature in a jet flame (note: again it is the average values for these quantities which are plotted). The agreement appears to be satisfactory, especially considering the measurement difficulties involved.

What then is the reasoning behind using methods such as those which have just been described, which take into account fluctuations? Would not the agreement between calculation and experiment be just as good if the fluctuations were neglected and the turbulent flame treated as a pseudo laminar flame? In reply to these questions, note firstly that, with reference to Figures 6.33, the existence of H_2 and O_2 at the same point, (when it is known that chemical reactions occur very rapidly), can only be explained by the presence of fluctuations (in phase opposition) in these two species. Thus, by neglecting fluctuations, we are led to "believe" that the reactions do not occur very rapidly, which clearly is not true. Moreover, this effect is accompanied by a drop in mean temperature, since unburnt reactants exist at this point due to an

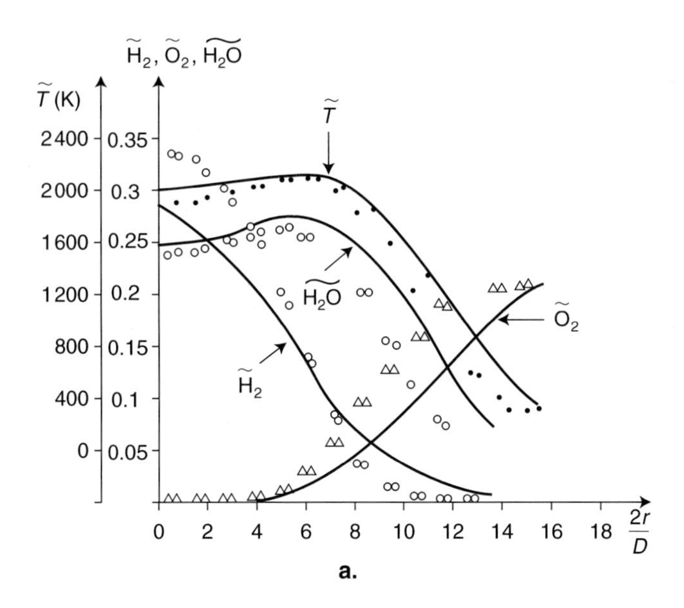

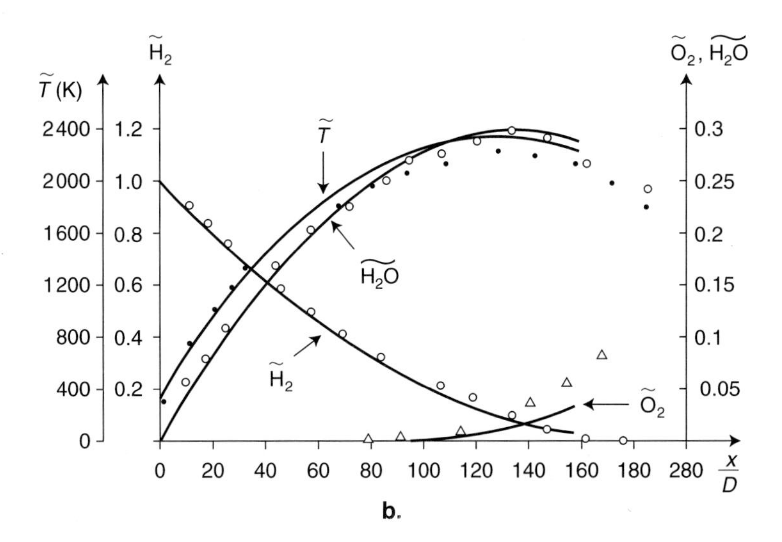

Figure 6.33 Comparisons between calculation and experiment for an H_2—air jet flame. Profiles of temperature and chemical species.
a. Radial profiles at $x/D = 85$.
b. Profiles along the flame axis. Same flame and same reference as for Figure 6.31b.

absence of small-scale mixing. This is clearly seen from Figure 6.34, where the temperature has been arbitrarily recalculated as if fluctuations were absent, and the resulting maximum temperature found to be 20% greater! This error is the direct result of neglecting these fluctuations.

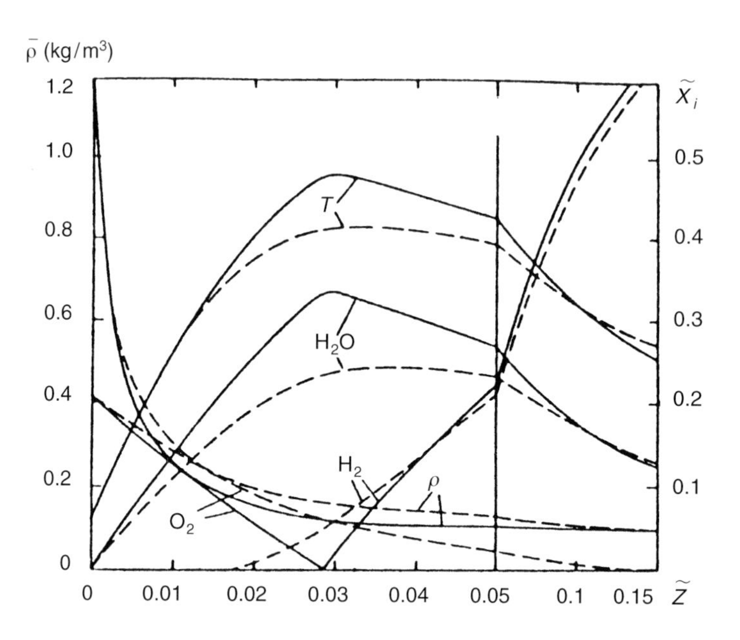

Figure 6.34 Influence, according to theoretical calculations, of turbulent fluctuations on average flame structure. The theoretical results have been plotted, for a given flame cross-section, as a function of average mixture fraction (with a jump in the scale at 0.05). Full line: calculation which neglects all fluctuations; broken line: calculation assuming a given degree of fluctuation in Z: $(\widetilde{Z'^2})/\widetilde{Z} = 0.05$ (After R.W. Bilger (1976) *Progress in Energy and Combustion Sciences,* 1, p. 87).

6.8 THE INFLUENCE OF COMBUSTION ON TURBULENCE

We have stated above that one condition for turbulent flames is that the flow velocity of the gases are sufficiently rapid to generate turbulence, and in general become turbulent before the gases enter the flame zone. However, this does not mean that turbulent fluctuations in the velocity (and pressure) are not modified by combustion. On the contrary, there is every reason to believe that the release of heat will cause the gases to expand greatly, in general leading, in turn, to an increase in their velocity. Furthermore, this change in velocity may even occur upstream of the flame, since pressure waves may propagate against the flow in a subsonic flow regime.

The influence of heat release on the mean flow is fully taken into consideration by the balance equations introduced in section 3.5.2 of chapter 3. The question which must now be asked is how can the influence of combustion on the turbulent velocity fluctuations be calculated? One solution is to return to a turbulence model (such as the $k - \varepsilon$ model), suitably adapted to this specific case. This section will explain how combustion can modify turbulence, based on this model. Only the principles required to study the relevant phenomena will be discussed, without entering into details about them or their modelling, since such a treatment would be outside the scope of this book.

The first problem relates to defining the kinetic energy of turbulence in the flame. For a flow of constant density, the kinetic energy of turbulence k is defined as $\sum_{\alpha=1}^{3} \overline{v'_\alpha v'_\alpha}/2$, with Reynolds' averages and fluctuations. The same definition can be retained for turbulent flames, where fluctuations in the density are found, although we have seen the advantage of introducing averages weighted for ρ (Favre averages). But, it is more common to define the kinetic energy of turbulence in terms of these averages, i.e. such that :

$$\widetilde{k} = \frac{1}{2} \sum_{1}^{3} \widetilde{v'_\alpha v'_\alpha} = \frac{1}{2\overline{\rho}} \sum_{1}^{3} \overline{\rho (v_\alpha - \widetilde{v}_\alpha)^2} \tag{6.47}$$

Clearly $\widetilde{k}$ differs from k (when ρ is not constant) since $\widetilde{v}_\alpha$ is different from $\overline{v}_\alpha$. It is not difficult to derive the balance equation for $\widetilde{k}$ from the conservation and Navier-Stokes equations for v_β, making the scalar product by v_β. Classical averaging, and some algebraic manipulation, yield an equation for $\sum_{1}^{3} \widetilde{v_\beta v_\beta}/2$. Moreover, if the balance equation for $\widetilde{v}_\beta$ (Reynolds equation) is scalar multiplied by $\widetilde{v}_\beta$ then, after some algebraic manipulation, the balance equation for $\sum_{1}^{3} \widetilde{v}_\beta \widetilde{v}_\beta/2$ is obtained. Simply subtracting these two equations gives:

$$\widetilde{k} = \sum_{1}^{3} \frac{\widetilde{v'_\beta v'_\beta}}{2} = \sum_{1}^{3} \frac{\widetilde{v_\beta v_\beta}}{2} - \sum_{1}^{3} \frac{\widetilde{v}_\beta \widetilde{v}_\beta}{2}$$

The following equation is obtained if the molecular viscosity flux terms are neglected:

$$\frac{\partial}{\partial t}(\overline{\rho}\widetilde{k}) + \sum_{1}^{3} \frac{\partial}{\partial x_\alpha}(\overline{\rho}\,\widetilde{v}_\alpha \widetilde{k}) = \sum_{1}^{3} \frac{\partial}{\partial x_\alpha}\left(-\overline{\rho} \sum_{1}^{3} \frac{\widetilde{v'_\alpha v'_\beta v'_\beta}}{2} + \overline{p' v'_\alpha}\right)$$

$$- \overline{\rho} \sum_{\alpha,\beta=1}^{3} \widetilde{v'_\alpha v'_\beta} \frac{\partial \widetilde{v}_\alpha}{\partial x_\beta} - \sum_{1}^{3}\left(\overline{v'_\alpha}\frac{\partial \overline{p}}{\partial x_\alpha} - \overline{p' \frac{\partial v'_\alpha}{\partial x_\alpha}}\right) - \overline{\rho}\,\widetilde{\varepsilon} \tag{6.48}$$

The last term, $\overline{\rho}\,\widetilde{\varepsilon}$, has a more complex definition than its equivalent when ρ is constant (given by equation (6.14)). In fact, it may be said that $\overline{\rho}\,\widetilde{\varepsilon}$ represents the terms remaining from equation (6.48).

The first term on the right-hand side of equation (6.48) represents the turbulent diffusion of $\widetilde{k}$, and the second term is a production term for $\widetilde{k}$ due to the average velocity gradients. The problem of the production of $\widetilde{k}$ through combustion is complex and not yet completely understood, however, there appear to be four main factors which can influence $\widetilde{k}$ in a flow in which combustion is occurring:

1. Combustion can produce, through the expansion of gas, significant changes in the average velocity gradients and thus lead to additional turbulence.

2. The temperature rise leads to an increase in the viscosity of the gaseous medium and thus to a reduction in the Reynolds number for the turbulence which, if it drops sufficiently, can dampen the turbulence.

3. On the other hand, density fluctuations in a combusting medium can play a specific role. Indeed, through using averages weighted by ρ, the term $\sum_{1}^{3} \overline{v'_\alpha}\,\partial\overline{p}/\partial x_\alpha$, appears clearly in equation (6.48) where v'_α is the fluctuation around $\widetilde{v}_\alpha$, and hence $\overline{v'_\alpha} \neq 0$ only if ρ fluctuates. From a physical perspective, this term represents the work done by the force which constitutes the average pressure gradient for the fluid particles of fluctuating density and, when $\partial\overline{p}/\partial x_\alpha$ is due to the force of gravity, the result is the work done by buoyancy. This term may be positive or negative depending on the respective signs of $\overline{v'_\alpha}$ and $\partial\overline{p}/\partial x_\alpha$ producing or destroying the turbulence.

4. Finally, a coupling of the fluctuations of pressure, velocity, and density is also included in the equation for $\widetilde{k}$ through a term $\sum_{1}^{3} \overline{p'(\partial v'_\alpha/\partial x_\alpha)}$ analogous to the term above. This term is non-zero only if ρ fluctuates since, if ρ is constant, then $\partial v'_\alpha/\partial x_\alpha = 0$. It is, however, very difficult to obtain an order of magnitude for this term since the pressure fluctuations at a given point depend on the entire flow field. Moreover, the pressure in a turbulent flow is almost impossible to measure.

Several experimental studies have been conducted to measure the kinetic energy of turbulence in flames. The laser Doppler anemometry technique can be used which measures the velocity of very fine (refractory) particles dispersed in the flow upstream of the flame. Depending on the measurement analysis technique either $k = \sum_{1}^{3} \overline{v'_\alpha v'_\alpha}/2$, and some of its components ($\overline{v'^2_x}/$ or $\overline{v'^2_y}/2\ldots$) can be obtained, or $\widetilde{k}$ and its components. In fact, the relative changes in k and $\widetilde{k}$ are not greatly different, at least qualitatively, although their values at a given point in the flow may differ greatly.

Figure 6.35a gives the results obtained along the axis of the turbulent gas burner shown in Figure 6.17. The kinetic energy of turbulence is not truly changed by the presence of the flame; its value remains more or less equal, or even less than that in the upstream flow. In fact, as can be seen in Figure 6.35b, the profile of the average velocity is quite flat within the flame and the velocity gradients which do exist are located outside the flame in the hot gases. Furthermore, the gradients of the average pressure $\partial\overline{p}/\partial x_\alpha$ are zero or close to zero since the flame is unconfined. Thus two important terms relating to the production of $\widetilde{k}$ in equation (6.48) are small. However, although $\overline{k}$ is not greatly changed in the flame, Figure 6.35a shows that the distribution between $\overline{v'^2_Z}$ and $\overline{v'^2_R}$, changes quite markedly, the flame making the turbulence less isotropic.

Figure 6.36 illustrates another case, where a premixed turbulent flame flows at high velocity in a tube of uniform cross-sectional area (already shown in Figure 6.12b). Figure 6.36a gives the longitudinal profiles of $\overline{v'^2_x}$ which more or less follow a streamline while Figure 6.36b shows the average velocity profiles.

In this case, a sharp increase in turbulence is observed in the flame zone (the turbulence can also be shown to be less isotropic) and correlated with this is a steep transversal velocity gradient which is due to the confinement and expansion of the gases. There is also a pressure gradient $\partial\overline{p}/\partial x$ associated with this experiment although it would seem that it is the transversal gradient $\partial\overline{v_x}/\partial y$ which is mainly responsible for the increase in the kinetic energy of turbulence.

Lastly, Figures 6.37a and 6.37b show the case of a turbulent premixed flame at low velocity stabilised by a wire. Although the velocity gradients are slight they are still present (around 20 s^{-1} compared with 1000 s^{-1} in the previous case). It seems that the increase in k observed in Figure 6.37b is due, in part, to the pressure gradient $\partial\overline{p}/\partial x$. Moreover, this increase was removed when the lateral walls were removed, which substantially cancelled out $\partial\overline{p}/\partial x$.

When the turbulent flame being studied is premixed and wrinkled, consisting only of pockets of unburnt and burnt gas, some authors recommend defining three types of fluctuation: turbulence in the unburnt gases, turbulence in the burnt gases, and an additional term related to the fluctuating passage of the flame front which is not "true" turbulence, and is simply associated with the average velocities and the kinetic energies in the unburnt and burnt gas areas. Whilst this approach has a certain validity, at the present time it complicates rather than simplifies the picture and is only valid for premixed, wrinkled flames.

In addition to $\widetilde{k}$, the kinetic energy, another characteristic of turbulence which is just as important is $\widetilde{\varepsilon}$, or the integral length scale l_t. In reality, little is

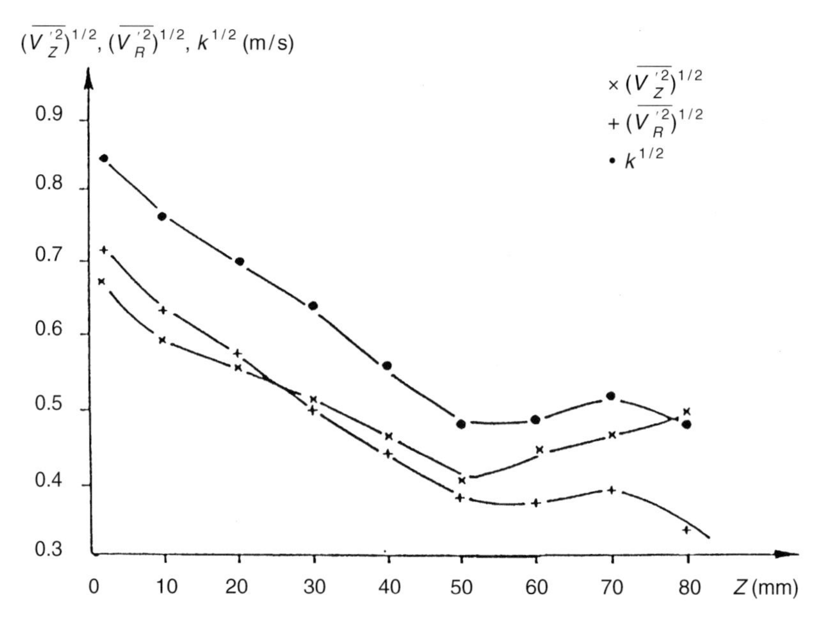

Figure 6.35a Measured velocity fluctuation along the axis of a turbulent Bunsen burner flame (that shown in Figure 6.17). *Z* is the axial coordinate and *R* the radial coordinate. The turbulence decreases in the unburnt gases along *Z*, then increases slightly on crossing the flame (After J.P. Dumont, D. Durox and R. Borghi (1993) *Comb. Sci. Tech.*, 89, 219-251).

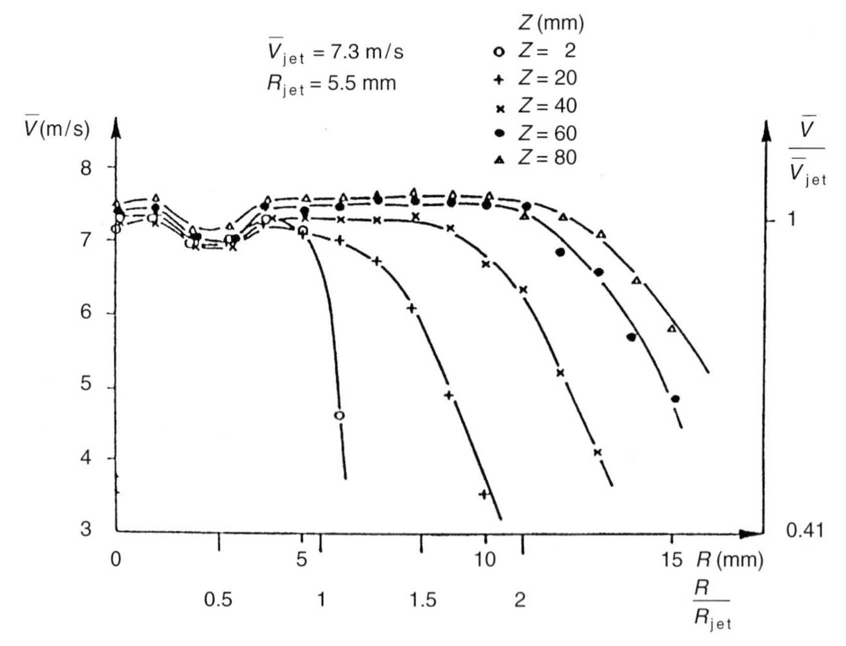

Figure 6.35b Average axial velocity profiles in a turbulent Bunsen burner flame. There is a slight fault in the uniformity of the velocity profile at around R/R_{jet} = 0.5. The flow of the gas coming from the flame, at almost constant velocity, widens progressively as *Z* increases as a result of the increase in the amount of burnt gas. Same reference as for Figure 6.35a.

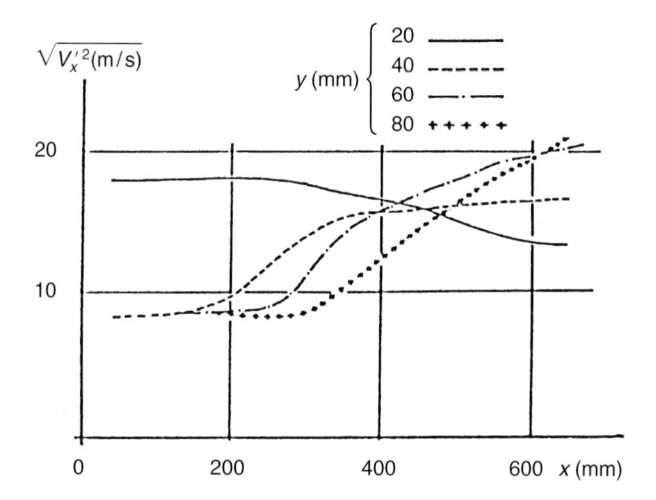

Figure 6.36a Measured axial velocity fluctuations on crossing the turbulent flame shown in Figure 6.12. The measurements were taken as a function of the axial coordinate x, for a given value of y. Depending on the value of y, the presence of the flame increases the fluctuations at different values of x since it develops obliquely (see diagram in Figure 6.12b). For y = 20 mm, the fluctuations of the velocity decrease since the flame does not cut the line y = 20 mm (After P. Moreau and A. Boutier (1977) *16th Symposium (Int.) on Combustion*, The Combustion Institute, 1747–1756).

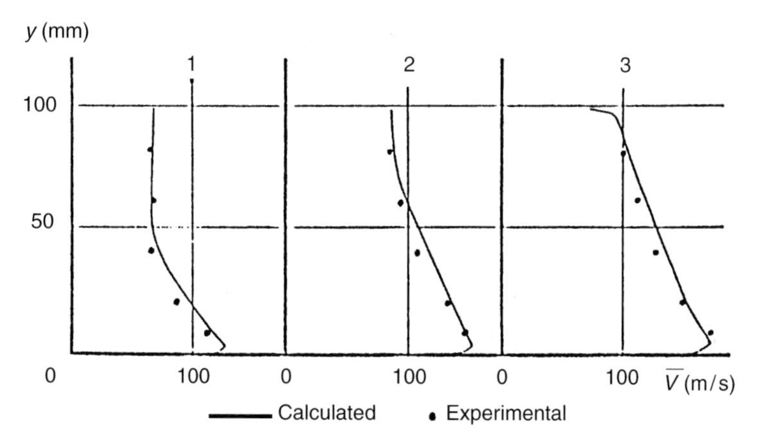

Figure 6.36b Transversal profiles of the average axial velocity in the flame shown in Figure 6.12. The curves 1, 2, and 3 correspond to x = 12.2 cm, x = 32.2 cm, and x = 42.2 cm respectively. These profiles indicate a velocity gradient which is not damped as x increases, thus explaining the production of turbulence seen in Figure 6.36a. This velocity gradient is the result of an axial pressure gradient which accelerates the hot (lighter) gases more than the unburnt (heavier) gases. Same reference as for Figure 6.36a.

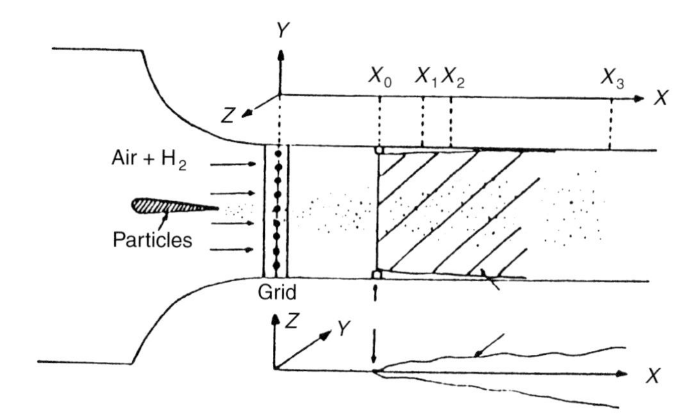

Figure 6.37a Turbulent premixed flame stabilised behind a wire in grid-generated turbulence. The flame, viewed in a direction parallel to the wire, is "V"-shaped (After R. Borghi and D. Escudié (1984) *Combustion and Flame*, 56, 149-164).

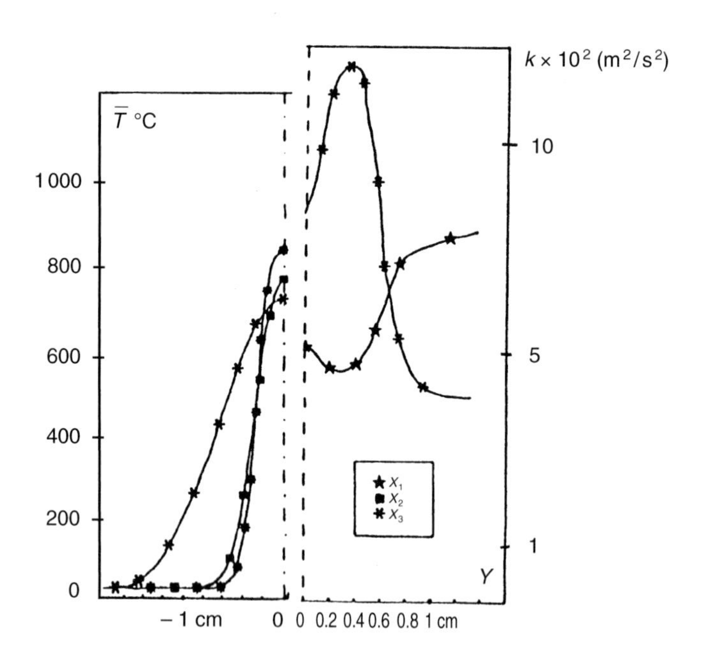

Figure 6.37b Transversal profiles of average temperature and kinetic energy of turbulence k, measured in the "V-shaped" flame at the x-axis values indicated in Figure 6.37a. The equivalence ratio of the mixture is low, such that the maximum temperature is 850°C. k decreases with x outside the flame, and then decreases before rising again greatly inside the flame zone. Same reference as for Figure 6.37a.

known about l_t or $\tilde{\varepsilon}$, because measurement is more difficult than that of $\tilde{k}$. In principle, correlated measurements at two points are necessary (or at two different times at the same point according to certain hypotheses).

One rare example of such measurements is given in Figure 6.38 and relates to the turbulent Bunsen flame shown in Figure 6.17. l_t increases in the burnt gases, with this increase appearing to follow the expansion of the gases resulting from combustion.

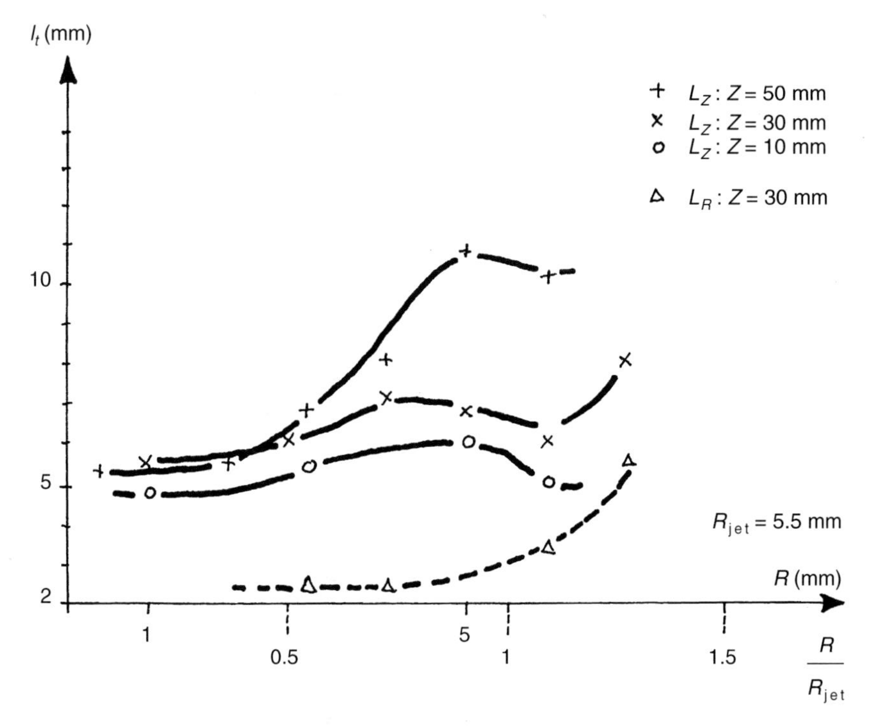

Figure 6.38 Measurements of the integral turbulent length scale, l_t, for the turbulent Bunsen burner flame shown in Figure 6.17. L_Z is the longitudinal correlation scale and L_R the transversal scale. In the flame region itself, an increase in the scales can be seen. Same reference as for Figure 6.35a.

WORKED EXAMPLES

■ 1) Calculating the decrease in turbulence behind a grid

Behind a grid placed in a continuous, one-dimensional flow, the average velocity V remains constant along the x-axis and the kinetic energy of turbulence k decreases as a function of x. To calculate $k(x)$ the $k - \varepsilon$ model can be applied. In this case, the equations reduce to:

$$V\frac{dk}{dx} = -\varepsilon; \quad V\frac{d\varepsilon}{dx} = -C_2\frac{\varepsilon^2}{k}$$

(see the general equations in section 6.2.4). Solve this system of equations, using Re_T, to find $k(x)$, $\varepsilon(x)$, $Re_T(x)$, and $\tau_t = k/\varepsilon$.

■ 2) Calculating the length of diffusion flames

a) For a laminar flame resulting from a circular jet of fuel in air at rest, equation (5.24′) described in chapter 5 can be used to find L_f/r_j as a function of V_j for propane. For a turbulent flame, experiments have yielded the following equation (Günther's formula):

$$L_f/r_j = 12((\nu_s Y_{K,j}/0.23) + 1)(\rho_j/\rho_{flame})^{1/2}$$

where ρ_{flame} is the density corresponding to the maximum flame temperature. Compare L_f/r_j for the two cases (d_j can be taken as 3×10^{-5}).

b) For a turbulent flame, use Günther's formula to estimate the flame lengths for methane, hydrogen, and a jet containing 50/50 H_2 and N_2.

CHAPTER 7

DETONATION AND SUPERSONIC COMBUSTION

7.1 SHOCK AND DETONATION: PHENOMENA AND DEFINITIONS

Chapter 5 discussed the propagation of a combustion wave through the coupling of chemical reactions with the exchange of heat and matter between unburnt and burnt gases. This case, known as deflagration, was dealt with in chapter 5 and involves a pressure drop from the unburnt to the burnt gases. However, as noted in chapter 5, combustion can propagate via another mechanism involving rapid increases in both pressure and temperature; known as detonation. In practice, a detonation consists of a shock wave followed by a combustion wave analogous to that for deflagration. Since the combustion wave produces the shock wave, the latter precedes the former as they both move towards the unburnt gases. However, the shock wave does induce combustion as it passes, and is so effective at doing so that ultimately the two waves join. They "feed" off each other, and once the phenomenon is established the relatively small distance which separates them remains constant.

In general, a "shock wave" is defined as a pressure rise, irrespective of origin, which has virtually zero length (and thus exists for a very short time). A "pulse" is the term given to a small-scale displacement of low amplitude; and it is said to be "transversal" when in the same plane as the wave (or, more generally, in the plane tangential to the surface of the wave at the point under consideration) and "longitudinal" when in the direction of propagation. Fluid media (gases or liquids) only allow the transmission of longitudinal pulses. Once produced in an element of volume M_0, a longitudinal pulse can propagate to the next M by pushing neighbouring volume M_1 which pushes M_2, etc. It can be shown that a one-dimensional longitudinal pulse, such as that produced by a sound wave, propagates at a speed :

$$a = \left(\frac{1}{\chi\rho}\right)^{1/2} \tag{7.1}$$

where ρ is the density, related to the specific volume v by $v = 1/\rho$, and:

$$\chi = -\left(\frac{1}{v}\frac{\partial v}{\partial p}\right)_{\text{isentropic}}$$

is the adiabatic, isentropic compressibility coefficient (adiabatic and reversible). For an ideal gas:

$$p = \frac{\rho RT}{M} = \frac{RT}{Mv}$$

and:

$$a = v\left(-\frac{\partial p}{\partial v}\right)^{1/2} = \left(\gamma\frac{RT}{M}\right)^{1/2} \tag{7.2}$$

where γ is the ratio c_p/c_v of the heat capacities at constant pressure and volume.

The reasoning developed by Becker in 1922, explains, at least qualitatively, how a pulse can produce a planar shock wave. Consider a piston accelerated

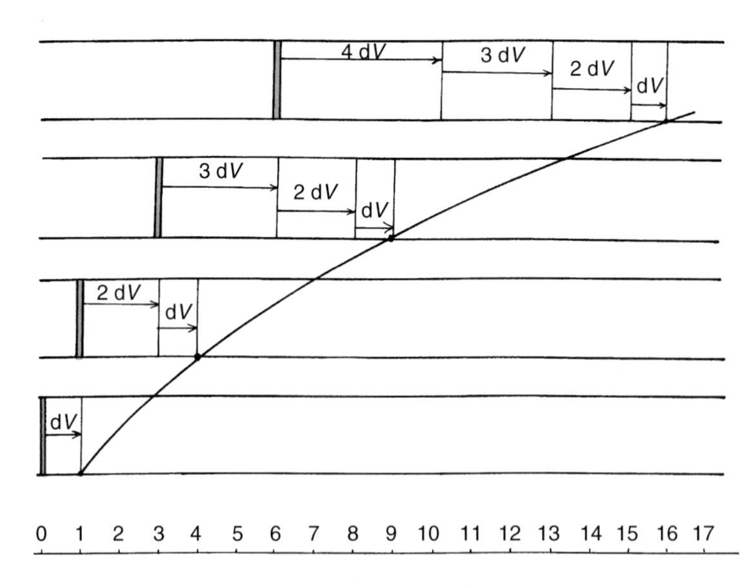

Figure 7.1 Profile near a piston in a cylindrical tube showing the change in velocity of a gas, pushed by the piston, from 0 to V in increments of dV.

along an infinitely long cylindrical tube, containing a gas at rest, and from zero up to a constant value V. Now imagine that this velocity V is attained by a series of increments dV, each occurring over equal time periods (Fig. 7.1). The first increment produces an initial pulse resulting in a slight adiabatic compression, which propagates at a speed a and slightly heats the medium. The section of compressed gas between the piston and the wave front propagates at a velocity dV, whilst ahead of the front the gas is at rest and not compressed. Since the speed increases when the temperature increases, the second increment produces a second compression which propagates faster than the first, at a speed $a + \mathrm{d}a = a'$, and can therefore catch it up. The same reasoning applies to all subsequent pulses. The first section produced (that which has advanced furthest) propagates at a velocity dV, the second at a velocity 2 dV, and so on. The compression fronts, originally

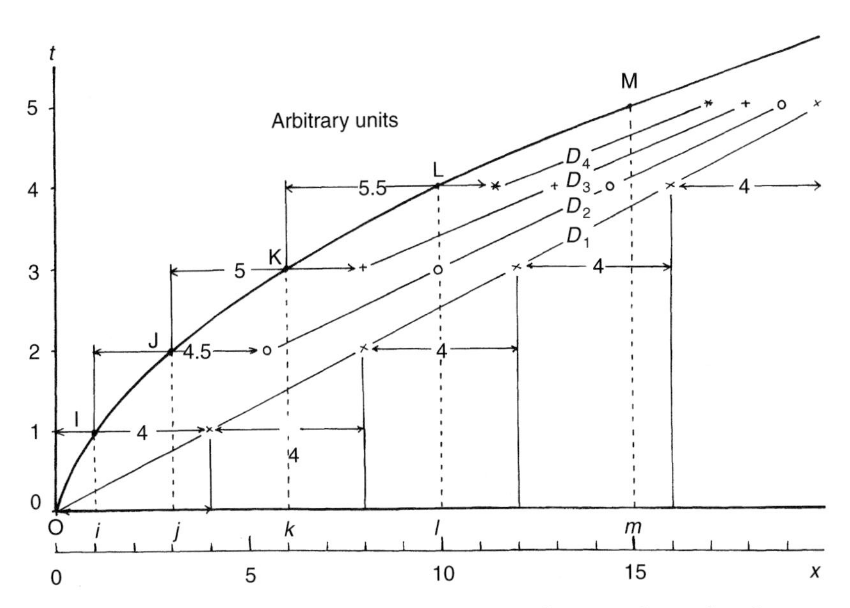

Figure 7.2 Idealised representation of shock wave formation by coalescence of compression waves (arbitrary units). The velocity at I is equal to dV, at J to 2 dV, etc., thus $ij = 2\,\mathrm{O}i$, etc.

×: first compression wave front at t = 1 then 2, etc. Its velocity, a, is taken to be equal to 4.

○: second compression wave front at t = 2 then 3, etc. Its velocity, greater than a, is taken to be equal to 4.5.

+: third compression wave front at t = 3 then 4, etc. Its velocity, greater still than a, is taken to be equal to 5.

*: fourth compression wave front at t = 4 then 5, etc. Its velocity, greater still than a, is taken to be equal to 5.5.

The straight lines obtained, D_1, D_2, D_3, D_4... come closer together with increasing t; as a result, the successive wave fronts ultimately coalesce.

separated, ultimately coalesce into a single front (Fig. 7.2) with neither wave able to advance ahead of the other. Indeed the fastest, which are those closest to the piston where the temperature is highest, slow down as they move towards lower temperature regions and ultimately synchronise at the propagation velocity of the single wave front, which is greater than the speed of sound. Consequently, the resulting shock wave propagates supersonically.

To a first approximation, the single front ultimately produced by this model should be a surface across which not only the temperature but all the intensive variables change discontinuously. Since the piston is perpendicular to the walls of the tube, this front is also planar and perpendicular to the walls (at least if friction at the wall is disregarded). In reality the front cannot have zero thickness, since this would be impossible in a medium comprising individual molecules. The front is discontinuous from a microphysical perspective but continuous from a macrophysical perspective. In fact, the front is very thin and its thickness has been estimated to be between 10^{-6} and 10^{-8} m, which is the same order of magnitude as the mean free path of gas molecules.

Evidently, the flow velocity of the fluid (dV at the beginning and V at the end) is not the speed of sound for the medium under consideration. In fact, they are not even the same kind of quantities, even if they do have the same units; the gas velocity is the velocity at which the mass of fluid displaces, whereas the speed of sound relates to the transferral of the pulse, hence the displacement of the pressure wave. This is why the term "speed" has been used in the above to highlight the difference between "velocity" and "speed".

Between the front and the piston there is a column of gas. A section of this column near to the piston is constantly being accelerated. Once the piston attains its final velocity V, **if** the piston is to continue to advance at this velocity then work must be done against the pressure difference produced.

In a detonation, the shock is fed by the combustion zone immediately behind it. For a stable detonation, the distance separating the shock and the combustion is such that the shock is just fed by the combustion and no more.

A deflagration can develop into a detonation, but the exact conditions under which this occurs are not fully understood. The initial conditions are certainly involved, as discussed in chapter 5 in connection with laminar flames and deflagrations. Possibly the transition from deflagration to detonation occurs so rapidly that the deflagration cannot be detected. Also, a deflagration may encounter conditions of turbulence or of confinement (hence of geometry), resulting in temperature, pressure and density conditions which lead to detonation after a certain time.

7.2 ONE-DIMENSIONAL OR "PLANAR" SHOCK WAVES IN GASES

Before looking at how detonations propagate, we will first consider in this section the highly idealised propagation of a pure, one-dimensional, shock wave unaffected by any chemical reaction. In this case the enthalpy of the system only changes as a result of the physical properties (of temperature only for an ideal gas) and hence the representative point remains on the same Hugoniot curve (introduced in chapter 5) before and after the shock. On the contrary, when a reaction is involved, as for the case of deflagration (chapter 5) and detonation (later in this chapter), there is a transition from the Hugoniot curve for the reactants to that for the products.

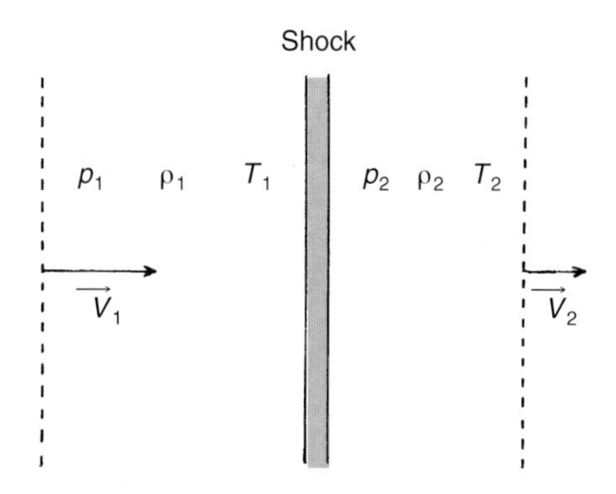

Figure 7.3 Planar shock wave. V_1, p_1, ρ_1, and T_1 are the flow velocity, pressure, density, and temperature before the shock and V_2, p_2, ρ_2, T_2 the same parameters after the shock.

The conservation equations already described in chapter 5 are applied here to a one-dimensional shock. The wave is referred to a coordinate system which moves with the shock, and normal to the flow (see Fig. 7.3) effectively reducing the shock's velocity to zero in this system. The gases enter the shock with a velocity V_1 and move away from it with a velocity V_2. With subscripts 1 and 2 used to refer to the system before and after the shock, equations (5.7), (5.8), (5.8′) and (5.11) from chapter 5 become:

$$p_2 - p_1 = \left(\frac{1}{\rho_1} - \frac{1}{\rho_2}\right)(\rho_1 V_1)^2 \tag{7.3}$$

$$V_1 = \frac{1}{\rho_1}\left(\frac{p_2 - p_1}{\dfrac{1}{\rho_1} - \dfrac{1}{\rho_2}}\right)^{1/2} \tag{7.4}$$

$$V_2 = \frac{1}{\rho_2}\left(\frac{p_2 - p_1}{\frac{1}{\rho_1} - \frac{1}{\rho_2}}\right)^{1/2} \tag{7.5}$$

$$2(h_2 - h_1) = (p_2 - p_1)\left(\frac{1}{\rho_1} + \frac{1}{\rho_2}\right) \tag{7.6}$$

ρ_1 is the density of the system before the shock,
p_1 the pressure before the shock,
h_1 the specific enthalpy (i.e. enthalpy per unit mass) before the shock, and
ρ_2, p_2, h_2, these same quantities after the shock.

Equation (7.6) plots the Hugoniot curve. The pressure, temperature and density, in state 1 **or** in state 2 are related by the gas state equation.

The difference $V_g = V_1 - V_2$ is given by the expression:

$$V_g = V_1 - V_2 = \left(\frac{1}{\rho_1} - \frac{1}{\rho_2}\right)\left(\frac{p_2 - p_1}{\frac{1}{\rho_1} - \frac{1}{\rho_2}}\right)^{1/2} \tag{7.7}$$

If the gas is stationary (in relation to a fixed reference system) before the shock arrives (Fig. 7.3), then the shock speed is:

$$c = -V_1 = -\frac{1}{\rho_1}(\text{tg}\,\theta)^{1/2} \tag{7.8}$$

where (tg θ) is the gradient, with the sign changed, of the straight line through the points J_1 and J_2 on the dynamic adiabatic (often referred to as the Hugoniot curve), H, of Figure 7.4, with J_1 denoting the system before the shock and J_2 afterwards. To avoid having a negative speed c (given by (7.8)) in the following equations, c is replaced by the positive quantity $-c$, which will be called C (since the direction of propagation (Fig. 7.3) is obvious). The following can be written:

$$C = -c = \frac{1}{\rho_1}\left(\frac{p_2 - p_1}{\frac{1}{\rho_1} - \frac{1}{\rho_2}}\right)^{1/2} \tag{7.9}$$

Moreover, according to expression (7.2), the speed of sound in the gas at point J_2 after the shock is:

$$a_2 = \frac{1}{\rho_2}\left(-\frac{\partial p}{\partial\left(\frac{1}{\rho}\right)}\right)^{1/2} \quad \text{in } J_2 = \frac{1}{\rho_2}(\text{tg}\,\tau)^{1/2} \tag{7.10}$$

where (tg τ) is the gradient of the tangent at the point J_2, with its sign changed (Fig. 7.4).

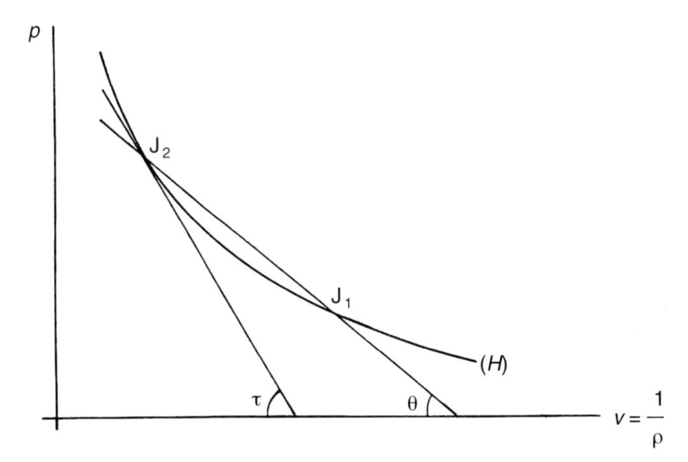

Figure 7.4 Shock-induced transition, with no reaction, from point J_1 to point J_2 on the dynamic adiabatic of the system in the pressure/(specific volume (= 1/density) plane.

For a more in-depth description of shock waves, refer to Shapiro[1].

7.3 ONE-DIMENSIONAL (OR PLANAR) DETONATIONS IN A GAS: THE CHAPMAN-JOUGUET CONDITION

In a similar way to how we addressed pure shocks, we shall now consider the highly idealised case of a **one-dimensional**, or "planar", detonation (which will subsequently be modified to correct the overly-simplified aspects of this model) which **propagates**, leaving aside for the moment the problem of its **possible stabilisation**.

Whilst it is true that the shock precedes combustion, it is difficult to identify the path followed (Fig. 7.5) by the representative point in the system from the initial point I_1 on the dynamic adiabatic, H_0, for the reactants to the final point reached on the adiabatic for the products, H, above H_0. Consider the Rayleigh line passing through the point I_1 and intersecting H at J and K. Imagine that, depending on the slope, the representative point in the system moves as a result of the shock from point I_1 to a point I_2 on H_0 and then, as a result of the

1. Shapiro A.H. (1983) *The Dynamics and Thermodynamics of Compressible Fluid Flow*, 2 vol., Wiley.

chemical reaction, from point I_2 on H_0 to point J or to point K on H. In this case, the conservation expressions (7.3) and (7.6) become:

$$p_f - p_1 = \left(\frac{1}{\rho_1} - \frac{1}{\rho_f}\right)(\rho_1 V_1)^2 \tag{7.11}$$

$$2(h_f - h_1) = (p_f - p_1)\left(\frac{1}{\rho_1} + \frac{1}{\rho_f}\right) \tag{7.12}$$

Equation (7.11), which is the analogue of equation (5.7) in chapter 5, is therefore the equation for the Rayleigh line relating to detonation. In equation (7.12), $(h_f - h_1)$ is positive, which was not the case for $(h - h_0)$ in equation (5.11) of chapter 5, for deflagration. As before, the subscripts 1 and 2 are used to indicate the conditions before and after the shock respectively, whilst the subscript f is now introduced to represent the final system after both shock and reaction.

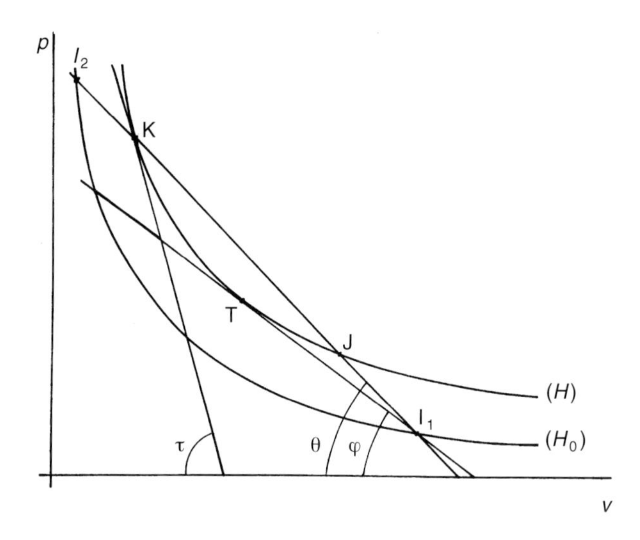

Figure 7.5 Detonation-induced transition (due to shock and reaction) from the point I_1 on the dynamic adiabatic, H_0, of the reactants to a point J, K, or T on that of the products (v = specific volume = 1/density).

These expressions hold for the values of p, ρ and h at both points J and K, so long as the difference in specific enthalpies, $(h_f - h_1)$, includes the contribution from the reaction.

The position of points J and K obviously depends on the position of the Rayleigh line I_1JK, hence on its slope (with sign changed), $(\rho_1 V_1)^2$. Consider the limiting case where J and K are superimposed, the line I_1JK having just merged with the tangent I_1T to the dynamic adiabatic, H, of the products. If point T is the operating point then the angles θ, ϕ and τ are all equal. Moreover, if the detonation is stable, the shock wave and the combustion wave must both propagate at the same speed. Since C denotes the speed of the shock and D that of the detonation, the equality $C = D$ must be introduced into expression (7.9). Finally, for this hypothesis, the expressions (7.7), (7.8), (7.9) and (7.10), with the subscript f replacing 2, yield:

$$D = V_g + a_f \quad \text{or} \quad D - V_1 = a_f - V_f \tag{7.13}$$

Applying the assumption used above when writing equation (7.8), i.e. that the unburnt gases are at rest before the shock, or in other words that $D = V_1$, then, with Ma_f denoting the Mach number at T:

$$Ma_f = \frac{V_f}{a_f} = 1 \tag{7.14}$$

This is the Chapman-Jouguet expression (shortened to CJ) and point T is the Chapman-Jouguet point. Under these conditions the flow velocity V_f is sonic.

For the same initial system, represented by the point I_1, then at all points, such as point K, the slope (with its sign changed) of the dynamic adiabatic is greater than that of the Rayleigh line I_1JK; tg τ is greater than tg θ. It follows, still based on equations (7.8), (7.9) and (7.10), that:

$$\frac{D - V_g}{a_f} = \frac{V_f}{a_f} < 1 \tag{7.15}$$

The flow velocity V_f is subsonic with respect to the coordinate system moving with the shock; although it can be shown that the detonation velocity is supersonic under these conditions. This is the case for what are known as strong detonations. On the other hand, the velocity V_f is supersonic for any point such as J, in which case weak detonations are produced. In both cases, the flow velocity drops across the detonation, but remains supersonic in the latter case whereas for the former, as shown above, it reduces from supersonic to subsonic. For propagating, non-stabilised detonations, both strong and weak detonations are transient states which both evolve towards the Chapman-Jouguet point T. Note in passing that below point C in Figure 5.8, in the deflagration region, and unlike the situation above point B, the slopes (with sign changes) of all the Rayleigh lines passing through I are always less than that of the tangent originating from point I.

In all the above cases, we have assumed that, in detonation, the chemical reaction (from reactants to products) proceeds to completion instantaneously. In reality, the transition must be progressive, and a better representation considers (see Fig. 7.6), in addition to the final Hugoniot of the products, a family of Hugoniots corresponding to increasing extents of reaction, ranging from the initial Hugoniot for the reactants to the final Hugoniot of the products.

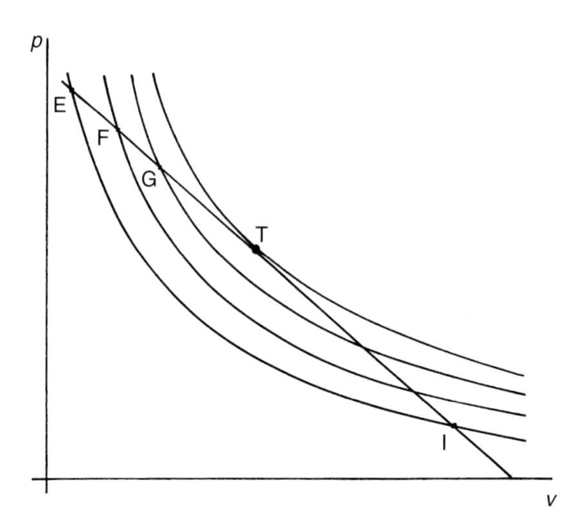

Figure 7.6 Shock-induced transition from point I to point E on the dynamic adiabatic of the reactants, then to F, G...T on those of the products for increasing extents of advancement of reaction. The extent of advancement of reaction is unity at point T (if the reaction is complete) (v = specific volume = 1/density).

The path followed is therefore as given below. The initial transition, corresponding to the pure shock, is from point I to a point E on the Hugoniot of the reactants before moving across the family of Hugoniots, corresponding to the products, at F, G, etc. and ultimately to T, for increasing extents of advancement of the reaction. All these points are aligned on the same Rayleigh line so long as the speed, C, from expression (7.9) (which is therefore the speed of detonation, D) has the same value from one end to the other. Under these conditions, the transition from point I to point E represents the passage of the shock, and that of point E to point T that of the reaction zone. The former involves an abrupt rise in pressure, temperature and density, whilst the latter causes a drop in pressure and density with the temperature continuing to increase slowly in the combustion zone. This model suggests, but does not prove, that for propagating, non-stabilised detonations, the Chapman-Jouguet point is reached "from the strong detonation side", moving from E towards T, which eliminates weak detonations. In fact, for non-stabilised detonations, it has been possible to show the existence, albeit transient, of strong detonations, but not of weak detonations.

After the detonation has passed, the medium naturally returns to rest. The pressure, temperature and density all decrease, whereas the specific volume increases. The resulting expansion propagates into the burnt gases in the opposite direction to the direction in which the detonation propagates.

In spite of the approximations made, measurements of the speed D, the pressure p and the temperature T after detonation yield values which are quite close to those predicted by the Chapman-Jouguet condition, as can be seen from the examples given in Table 7.1 below.

TABLE 7.1 APPROXIMATE VALUES OF THE SPEED, EXPERIMENTAL AND CALCULATED, AND MEASURED VALUES OF THE TEMPERATURE AND PRESSURE, ASSOCIATED WITH THE DETONATION OF A SELECTION OF GAS MIXTURES (initial pressure = 1 bar, initial temperature = 300 K). (After Van Tiggelen et al.[2]. See also Barrère and Fabri[3], Fickett and Davis[4], Nettleton[5]).

Mixture	Speed ($m \cdot s^{-1}$)		Temperature (K)	Pressure (bar)
	Experimental	Calculated		
$H_2 + 0.5O_2$ (1)	2825	2850		
$H_2 + O_2$	2315	2300	3700	18
$H_2 + 2O_2$	1920	1925		
H_2 + air (2)	1940	1940	2950	16
$CO + 0.5O_2$ (1)	1760		3500	19
$CH_4 + 2O_2$ (1)	2320	2390	3700	
C_2H_2 pur (3)	1870			
$C_2H_2 + 3O_2$	2330	2340		
$C_3H_8 + 5O_2$ (1)	2360	2360		

(1) Stoichiometric mixture.
(2) Deduced here by interpolation between $H_2 + 0.5\,O_2 + 1.5\,N_2$ and $H_2 + 0.5\,O_2 + 2.5\,N_2$.
(3) Explosive decomposition of pure C_2H_2 (see chapter 1).

The values given do not take into consideration the slight dependency of the speeds on the diameter of the tube in which the detonation propagates and which can undoubtedly be accounted for, at least partly, by heat losses across the walls of the tube.

---

2. Van Tiggelen et al. (1968) *Oxydations et combustions*, vol. II, 838-840 and following, Éditions Technip, Paris.
3. Barrère M., Fabri J. (1971) *Ondes de choc avec combustion* in: *Chocs et ondes de choc* (Jaumotte, ed.) Masson, Paris.
4. Fickett W., Davis W.C. (1979) *Detonation*, University of California Press.
5. Nettleton M. A., *Gaseous detonations*, Chapman and Hall.

7.4 ONE-DIMENSIONAL SHOCKS IN A GAS: PROPERTIES AS A FUNCTION OF MACH NUMBER

If the medium can be considered as being an ideal gas for which the molar mass and heat capacities c_p and c_v are constant, equations (7.4), (7.5) and (7.6) can be solved analytically to give the ratios p_2/p_1 of the pressures, T_2/T_1 of the temperatures, ρ_2/ρ_1 of the densities, and the Mach number, Ma_2, given by (V_2/a_2), in medium 2 after the shock as a function of the Mach number, Ma_1, given by (V_1/a_1), in medium 1 before the shock. With γ still denoting the ratio c_p/c_v, the solutions are given by:

$$\frac{p_2}{p_1} = \frac{2\gamma Ma_1^2 - (\gamma - 1)}{\gamma + 1} \tag{7.16}$$

$$\frac{T_2}{T_1} = \frac{2\gamma(\gamma - 1)Ma_1^4 - (\gamma^2 - 6\gamma + 1)Ma_1^2 - 2(\gamma - 1)}{(\gamma + 1)^2 Ma_1^2} \tag{7.17}$$

$$\frac{\rho_1}{\rho_2} = \frac{(\gamma - 1)Ma_1^2 + 2}{(\gamma + 1)Ma_1^2} \tag{7.18}$$

$$Ma_2 = \left[\frac{(\gamma - 1)Ma_1^2 + 2}{2\gamma Ma_1^2 - (\gamma - 1)}\right]^{1/2} \tag{7.19}$$

These expressions show that p_2/p_1, T_2/T_1 et ρ_2/ρ_1 all increase as Ma_1 increases, as shown by Figure 7.7 for $\gamma = 1.405$. Note that these ratios are only relevant if Ma_1 is greater than one. Indeed, it can be shown that if Ma_1 is less than 1 then the final specific entropy s_2 is less than the initial value s_1, which contradicts the Second Law of Thermodynamics (since the system is virtually isolated). At the limiting case, when Ma_1 tends towards 1, p_2/p_1, T_2/T_1, ρ_1/ρ_2 and Ma_2 tend towards 1. Consequently, a shock can only exist in supersonic motion with respect to the upstream medium (medium 1), or in other words it can only either propagate at supersonic speed or be stabilised in a supersonic flow.

Planar shock waves are classical phenomena in aerodynamics and in supersonic gas dynamics. In addition to planar shock waves perpendicular to the flow velocity of the gases, angled and curved shock waves (in relation to the flow) are also observed.

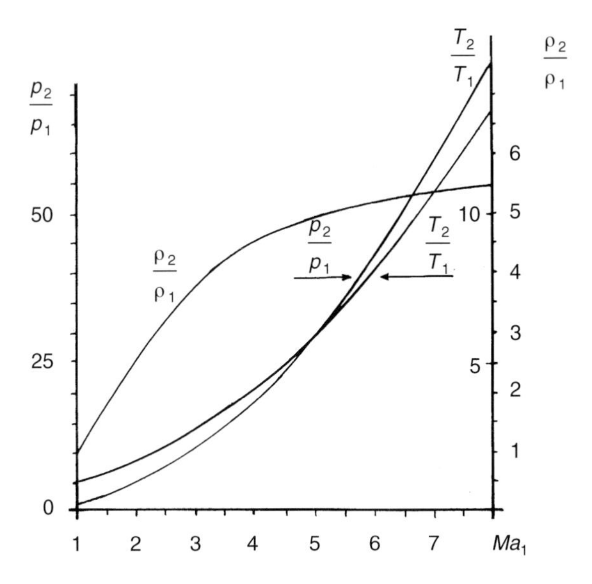

Figure 7.7 Calculated temperature, pressure and density ratios (with subscripts 1 and 2 denoting quantities before and after the shock respectively) as a function of Mach number, Ma_1, before the shock.

7.5 ONE-DIMENSIONAL DETONATIONS IN A GAS: PROPERTIES AS A FUNCTION OF MACH NUMBER

By continuing to apply the assumption made above that the medium can be considered to be an ideal gas whose molar mass and specific heat capacities, c_p and c_v, are constant, then detonation can be treated in a similar manner to the approach adopted above. The difference compared to a pure shock is that an exothermic chemical reaction occurs and that, consequently, the heat released q by the reaction must be taken into consideration in the equations. As before, the subscript 1 denotes the pre-shock medium, hence before detonation, and f the final state after the combustion wave has passed, i.e. after detonation. The equations derived in the previous section, which ignored the effects of mass and heat transfer on the pressure, temperature and specific volume ratios (after and before the shock), can still be considered as being valid for detonation in the zone between the shock wave and the combustion wave. Thus, for the detonation:

$$\frac{p_1}{\rho_1 T_1} = \frac{p_f}{\rho_f T_f} = \frac{R}{M} = r = c_p - c_v (\text{J.K}^{-1}.\text{kg}^{-1}) \tag{7.20}$$

$$h_f - h_1 = c_p(T_f - T_1) - q \tag{7.21}$$

T_f is not the same as T_1 and $(h_f - h_1)$ is not an enthalpy of reaction at a given temperature and pressure; instead it is the change in enthalpy produced by two factors; firstly that T_f is not the same as T_1 and secondly that the chemical composition of the system has changed. Since this term is positive (see section 7.3 above) then, on the basis of equation (7.21), the system must have gained energy from the surrounding medium, when considered as a whole, while the combustion reaction, if it had taken place at a given temperature, would have released energy. Combining equations (7.20) and (7.21) gives:

$$h_f - h_1 = \left(\frac{p_f}{\rho_f} - \frac{p_1}{\rho_1}\right)\frac{\gamma}{\gamma - 1} - q \tag{7.22}$$

where γ still denotes the ratio c_p/c_v. The following dimensionless quantities can be introduced:

$$Y = \frac{p_f}{p_1} \qquad X = \frac{\rho_1}{\rho_f} \qquad \text{and} \qquad B = \frac{2(\gamma - 1)q\rho_1}{p_1}$$

where B is known and X and Y are unknowns. The Hugoniot equation (7.12) becomes:

$$2\gamma(X \times Y - 1) - B = (\gamma - 1)(Y - 1)(X + 1) \tag{7.23}$$

giving:

$$\frac{\mathrm{d}Y}{\mathrm{d}X} = -\frac{(\gamma + 1)Y + (\gamma - 1)}{(\gamma + 1)X - (\gamma - 1)} \tag{7.24}$$

The Chapman-Jouguet condition can be checked by comparing this slope with that of the Rayleigh line JK (Fig. 7.5), i.e. $-(Y - 1)/(1 - X)$, which gives:

$$Y = \frac{X}{(\gamma + 1)X - \gamma} \tag{7.25}$$

thus by introducing the Mach number for medium 1:

$$Ma_1 = \frac{V_1}{a_1} = \frac{D}{a_1} = D\left(\frac{\rho_1}{\gamma p_1}\right)^{1/2}$$

and the expressions (7.9) with $C = D$ and (7.4) with $D = V_1$:

$$Ma_1^2 = \frac{(Y - 1)}{\gamma(1 - X)} \tag{7.26}$$

Eliminating Y by combining equations (7.25) and (7.26) yields:

$$X = \frac{\frac{1}{Ma_1^2} + \gamma}{\gamma + 1} \tag{7.27}$$

and in the same way:

$$Y = \left(\frac{1}{Ma_1^2} + \gamma\right)\left(\frac{Ma_1^2}{\gamma + 1}\right)$$

Finally, by substituting these values of X and Y into the Hugoniot equation (7.23), the following quadratic equation is obtained in Ma_1^2:

$$Ma_1^4 - \left(\frac{B(\gamma+1)}{\gamma} + 2\right) Ma_1^2 + 1 = 0 \tag{7.28}$$

which has two roots, the product of which equals one. As we saw above, the only valid solution is that which has a value greater than unity. The approximate value of this solution is:

$$Ma_1^2 = \frac{B(\gamma+1)}{\gamma} = \frac{2q(\gamma^2 - 1)\rho_1}{\gamma p_1} \tag{7.29}$$

giving:

$$D = [2q(\gamma^2 - 1)]^{1/2} \tag{7.30}$$

which is of course zero for $q = 0$.

Similarly:

$$\frac{T_f}{T_1} = \frac{p_f \rho_1}{p_1 \rho_f} = XY = Ma_1^2 \frac{\left(\frac{1}{Ma_1^2} + \gamma\right)^2}{(\gamma+1)^2} \tag{7.31}$$

giving approximately:

$$\frac{\gamma^2 Ma_1^2}{(\gamma+1)^2}$$

Despite the approximations made, the values determined in this way, e.g. for the ratio T_f/T_1, do not differ greatly from those measured, such as those given in table 7.1 above.

7.6 LIMITS OF DETONABILITY

The approaches used to define flammability limits can be repeated to determine limits of detonability. Uncontrolled detonations can cause serious damage as a result of the high local pressures produced, which is why it is important to know the limits of detonability precisely. The values can be determined either by producing a shock wave experimentally or by analysing the transition from deflagration to detonation. However, as already stated in the first section of this chapter, the problem is complex, and these limits depend to a large extent on the specific conditions which initiate detonation and, to a lesser extent, on the geometry (one- or multi-dimensional) of the propagation. Measurements are therefore difficult to make and many, particularly from older experiments, lack precision and must therefore be applied with care, especially when analysing situations where safety is at risk.

For comparison, a few approximate values of the flammability limits and limits of detonability (in percentage by volume of fuel) are given in Table 7.2 below. The initial pressure is approximately 1 bar and the temperature close to 300 K. The tube diameter is large enough (10 mm or more) to ensure that any possible slowing down due to the walls is negligible; the indicated values for detonation are those determined in a confined medium.

TABLE 7.2

Mixture	Lower limit		Upper limit	
	Deflagration	Detonation	Detonation	Deflagration
H_2—air	4	18	59	75
H_2—O_2	4	15	90	95
C_2H_2—O_2	2.8	3	89	93
C_3H_8—O_2	2.4	2.5	43	55

The limits of detonability are clearly narrower than those for flammability.

7.7 APPEARANCE OF COMPLEX THREE-DIMENSIONAL STRUCTURES

As we have just seen, the speeds measured are fairly similar to those predicted by the Chapman-Jouguet condition (7.13). However, the planar, one-dimensional front described in the previous section is a considerable approximation to the fronts encountered in practice. Furthermore the front is unstable, particularly near the limiting conditions for propagation, and can distort and twist such that its speed varies, sometimes in a cyclic manner, and the laminar flow may become turbulent. Small-scale sinusoidal distortions of the combustion front can be further amplified by a mechanism analogous to that described for deflagration (chapter 5, section 5.2), and so on.

There is strong evidence to support the assertion that detonations have a cellular structure. A plate coated with carbon black placed in the path of the front reveals regular markings caused by spherical, divergent detonation. These marks are similar to interference patterns, and thus suggest that several waves are in fact propagated. This effect may be due to "microdetonations" originating at hot spots scattered more or less regularly in front of the combustion wave. This hypothesis can be tested by simulating a series of such microdetonations; explosions are initiated in an inert gas, either at a series of points (producing spherical waves) or by parallel wires (producing cylindrical waves). In this way, structures can be generated which exactly resemble those found in detonations. As they move through the medium, each microdetonation coalesces with its neighbour, at an angle of incidence which varies with total distance travelled. At a certain degree of incidence, a third wave is produced at the junction of the first two which may even move ahead of the original front. Rather than study this phenomenon in all its complexity, we shall consider the simpler case of the collision between a planar incident wave and a reflected wave. As we shall see, at a certain degree of incidence, a third wave can be formed.

Consider a planar shock moving through a medium which encounters a wall at an angle to its direction of propagation, forming a wedge shape (although this is just one possible case). If the slope of the wedge is sufficiently steep (Fig. 7.8) a reflected wave, often called a "transverse wave", will be generated and the reflection is said to be "regular". However, if the slope is smaller (Fig. 7.9), point P in Figure 7.8 detaches from the wall in accordance with the conservation conditions for mass, momentum, and energy. A third wave appears, PM, known as the Mach stem, which travels faster than the incident wave PI, and point P becomes a triple point. In this case, the reflection is known as "Mach reflection". The medium (zone 3 in Figure 7.10) through which only the PM wave has passed is separated from the medium (4 in Figure 7.10) through which waves PI and PR have passed by a slip stream, or slip line, shown in Figure 7.10 as SP. The local fluid flow velocities near point P are rather low. As a first approximation, the system can be thought of as being the result of the transverse wave PR reflecting off the slip stream SP, since the slope of SP is indeed greater than that of the wall. This very rough approximation reduces the cases shown in Figures 7.9 and 7.10 to that in Figure 7.8.

If the side walls are coated with carbon black, then the passage of a detonation groups the carbon black into lines whose regularity depends on the extent

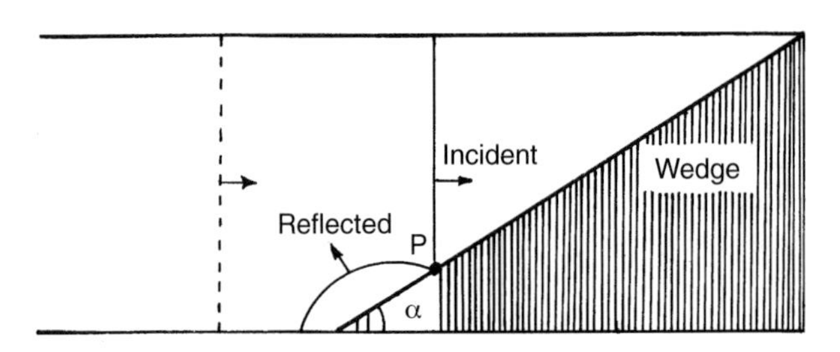

Figure 7.8 "Regular" reflection of a shock wave (case where the angle α is large enough).

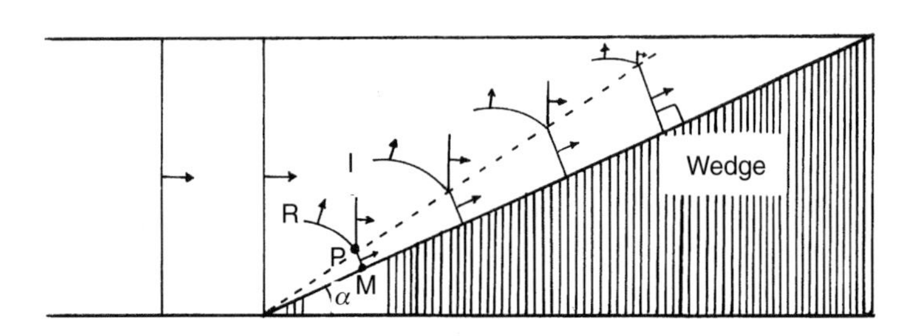

Figure 7.9 "Mach reflection" (case where the angle α is small enough) with the appearance and displacement of triple point P.

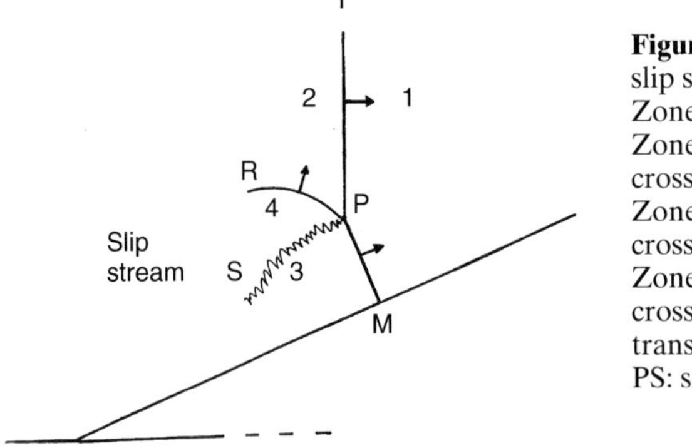

Figure 7.10 "Mach reflection" and slip stream.
Zone 1: initial medium.
Zone 2: medium which has been crossed by incident wave PI.
Zone 3: medium which has been crossed by the Mach stem PM.
Zone 4: medium which has been crossed by incident wave PI and the transverse wave PR.
PS: slip stream.

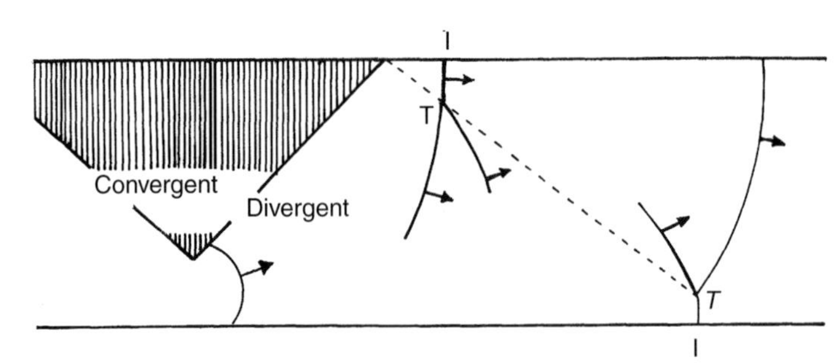

Figure 7.11 Configuration which will produce the three-dimensional structure of Figure 7.10, relating to collision between transverse waves.

to which the detonation is structured. They consist of triple points or rather of the points on the slip stream SP close to the triple points P, and reveal the multidimensional structure. This said, the phenomena, although similar, are more complex for a detonation than for a pure shock in the absence of reaction.

Experimentally, a one-dimensional detonation can be generated which is initially stable, with no transverse wave, and suitably oriented. This is achieved by using a convergent-divergent system such as that shown in Figure 7.11. A Chapman-Jouguet detonation is produced upstream of the convergent section. Having passed the convergent/divergent section, the detonation has the desired properties. It is initially reflected at I on the upper wall of the tube. The situation is in fact analogous to that in Figure 7.9 with the upper wall of the divergent section in Figure 7.11 acting as the wall of the tube in Figure 7.9 and the wall of the tube in Figure 7.11 taking the place of the wedge in Figure 7.9. A triple point is produced which follows the path shown by the dashed line. The detonation is then reflected again, this time off the lower wall. Whilst the phenomenon is complex, a detailed analysis shows that following each collision with a transverse wave, the incident wave and the Mach wave swap over.

7.8 DETONATIONS IN CONDENSED MEDIA

Depending on their nature, particle size, physical characteristics and confinement conditions, many solids or solid mixtures can either deflagrate or detonate; although clearly the solid must store large amounts of chemical energy. Solid propellants are solid multiphase mixtures mainly consisting of a fuel and an oxidant which must deflagrate without detonating. In contrast, explosives are unstable compounds, whose sudden, exothermic decomposition can result in a detonation. To produce a detonation, the amount of chemical energy stored by a compound does not have to be great. However, although it must be possible to release this energy very rapidly following appropriate ignition, such as that provided by an impact, a hot wire, an electric current, etc. In a somewhat arbitrary way, it is normal to distinguish between primary and secondary explosives. The former can only detonate (insofar as the transition from deflagration to detonation occurs too quickly for anything other than detonation to be detected), whilst the latter can, depending on the conditions and especially on the ignition technique, deflagrate or detonate.

Detonation in solid or liquid condensed media (or phases), although much more complex, nevertheless resembles detonations in gases. The principles are the same, but the higher densities mean that the local pressures produced may be much greater, and may rise to 10^4–10^6 bar. Thus, it is no longer possible to consider the gases produced as behaving ideally. Selected values of detonation speed, pressure and temperature, are given in Table 7.3 below.

TABLE 7.3 CALCULATED AND MEASURED VALUES OF DETONATION SPEED AND CALCULATED VALUES OF THE TEMPERATURES AND PRESSURES ATTAINED (After Mader[6])

Compound	Speed ($m \cdot s^{-1}$)		T (K)		p (10^{-5} bar)	
	Experiment	Calculated	Experiment	Calculated	Experiment	Calculated
Liquid **nitrométhane**	6290	6397	3380	3359	1.41	1.28
Liquid **trinitroglycerine**	7580	7406	3470	3905		2.23
Trinitrotoluene :						
Solid, $\rho = 1.64\ g \cdot cm^{-3}$	6950	7008		3143	1.90	2.01
Solid, $\rho = 1.061\ g \cdot cm^{-3}$	5254	5245		3417	1.10	0.82
Solid, $\rho = 0.732\ g \cdot cm^{-3}$	4200	4450		3264	0.59	0.44

The calculated values given in Table 7.3 are the averages of those determined using different equations of state. The calculated and experimental values given for trinitrotoluene highlight the high dependency on initial density, which is itself dependent on the degree of compaction.

The transition from deflagration to detonation can only occur if the mass exceeds a critical value which itself depends upon many parameters, including the geometry. For a given geometry, the surface area-to-volume ratio must not be too great; for example, the ratio for a sphere of radius r is $3/r$. If r is too small then the value of $3/r$ may drop below the critical limit at which the heat lost by the surface is greater than that produced inside the volume. The critical mass is very small for primary explosives such as lead nitride, $Pb(N_3)_2$, and larger for secondary explosives such as trinitrotoluene, $CH_3C_6H_2(NO_2)_3$. The former are often used to ignite the latter, as described in more detail by Quinchon et al.[7].

6. Mader C.L. (1979) *Numerical Modeling of Detonations*, University of California Press.
7. Quinchon J. et al. (1987) *Les poudres, propergols et explosifs*, t. I, *Les explosifs*. Lavoisier.

7.9 SUPERSONIC FLOW WITH PREMIXED COMBUSTION

In the above discussion, the propagation of a detonation in a fluid at rest was studied by writing the flow equations (7.3), (7.4), (7.5) and (7.6) referred to a coordinate system moving with the detonation wave. These are obviously the same equations as those which govern a flow in which a premixed combustion zone is stabilised perpendicularly to the flow. These equations can thus be used to obtain information relating to the phenomena occurring in this zone. This approach was pursued in chapter 5, using the Hugoniot equation to study deflagration, and particularly to determine whether the pressure increased or decreased across the deflagration. However, this was not sufficient and additional equations were required to describe the mass and heat transfer phenomena which play a crucial role in deflagration (more so than in detonation) and which could not be neglected. On the other hand, in the case of stabilised flames in high velocity flows, heat and mass transfer by molecular diffusion play a less important role and equations (7.3), (7.4), (7.5) and (7.6) are once again suitable, and can be used to describe stabilised combustion in these flows. This is particularly applicable to combustion in supersonic flow, which is of specific practical interest in the design of hypersonic aircraft engines.

Equations (7.3), (7.4), (7.5) and (7.6) relate to the case of uniform, continuous flow in a flow tube of constant cross-sectional area and thus can be applied directly to the case of a combustion zone stabilised normally to the flow in a combustion chamber of constant cross-sectional area. The fact that combustion occurs over a very short distance (which is also the case for a detonation wave) or over a finite distance is of no real consequence, especially if the cross-sectional area of the combustion chamber remains constant over this distance.

Figure 7.5 can therefore be reconsidered in terms of combustion in a flow as opposed to combustion propagation. Equally, that which has been called detonation should be called combustion in supersonic flow, since the flow is supersonic relative to the combustion. In the strong detonation region, a subsonic combustion follows a planar shock, whereas in the region of weak detonations supersonic combustion occurs, with the flow velocity associated with the combustion remaining above the speed of sound.

Supersonic combustion, in the absence of shock, has been investigated in flows using experiments involving guided flows in tunnels. Models of supersonic-combustion ramjets have been tested in wind tunnels. The combustion zone which can actually be stabilised in the flow is thick, and is neither a wave nor really normal to the flow. However, in a combustion chamber of uniform cross-sectional area, the characteristics of the burnt gases appear to correspond reasonably well with those of a point below T on the adiabatic for the burnt gases H in Figure 7.5.

Instead of plotting Hugoniot curves and Rayleigh lines on pressure/volume diagrams such as the one shown in Figure 7.5, a method which is frequently used is the Mollier diagram for burnt gases in chemical equilibrium. The plot consists of an enthalpy-entropy diagram on which isotherms and isobars can be drawn. The points previously indicated in Figure 7.5 (called J and K) on the adiabatic H for the burnt gases can be plotted and verify, in both representations, the Rayleigh and Hugoniot equations.

The values which must be determined are those of the variables in the final system; shock, detonation, or stabilised supersonic combustion, for a given initial system, which in practice are the parameters p_f, T_f, ρ_f, V_f, h_f and s_f. Two series of relationships are used to determine these values; the first combines the state equation with thermodynamic equations which, for an ideal gas, are:

$$p = \rho r T \tag{7.32}$$

$$\mathrm{d}h = c_p \mathrm{d}T \tag{7.33}$$

$$\mathrm{d}s = c_p \frac{\mathrm{d}T}{T} - r\frac{\mathrm{d}p}{p} \tag{7.34}$$

and by integrating (7.33) between 0 K and T and (7.34) between a reference state p_r/T_r and the state p/T:

$$h = c_p\, T + h_0 \tag{7.33'}$$

$$s = c_p \ln\left(\frac{T}{T_r}\right) - r \ln\left(\frac{p}{p_r}\right) + s_r \tag{7.34'}$$

where c_p, h_0, s_r, T_r and p_r are constants, and more complicated equations if the gas cannot be considered as ideal. The second series combines the conservation equations which have already been described, but are written slightly differently:

$$\rho_f V_f = \rho_1 V_1 \tag{7.35}$$

$$p_f + \rho_f V_f^2 = p_1 + \rho_1 V_1^2 \tag{7.36}$$

$$h_f + \frac{V_f^2}{2} = h_1 + \frac{V_1^2}{2} = h + \frac{V^2}{2} \tag{7.37}$$

where the subscripts 1 and f have the same meaning as in section 7.3. Hence, for an ideal gas, and if c_p is the same before and after shock and detonation:

$$h_f(\text{at } T_f) - h_{f,0}\,(\text{at } 0\text{ K}) = c_p T_f \quad \text{and} \quad h_1\,(\text{at } T_1) - h_{1,0}\,(\text{at } 0\text{ K}) = c_p T_1$$

$$h_f - h_1 = c_p(T_f - T_1) + h_{f,0} - h_{1,0}$$

or, again by setting $(h_{1,0} - h_{f,0}) = q$ (the heat released by the reaction at 0 K, which is positive for a detonation and zero for a shock without reaction):

$$h_f - h_1 = c_p(T_f - T_1) - q \tag{7.38}$$

Combining equations (7.35) and (7.36) gives:

$$p_f + \frac{\rho_1^2 V_1^2}{\rho_f} = p_1 + \rho_1 V_1^2 \tag{7.39}$$

which is the Rayleigh equation. Combining equations (7.35) and (7.36) gives:

$$h_f - h_1 = \frac{V_1^2}{2} - \frac{1}{2}\left(\frac{p_1 + \rho_1 V_1^2 - p_f}{\rho_1 V_1}\right)^2 \tag{7.40}$$

which is the Fanno equation and contains the unknown p_f (although clearly the equation could have involved V_f if p_f had been eliminated).

In practice a value of p_f is chosen, and ρ_f deduced from (7.39), then T_f from the state equation, then h_f, then s_f (from equations (7.32), (7.33) and (7.34) for an ideal gas), which yields the Rayleigh curve in the plane h/s of Figure 7.12 (which is a straight line in the plane $p/(1/\rho)$ of Figure 7.5).

Again by choosing a value for p_f, a value for h_f is obtained from (7.40) then, for an ideal gas, T_f from (7.33′), followed by s from (7.34′), which gives the Fanno curve in the plane h/s of Figure 7.12. Equations (7.38) and (7.40) show that the Fanno plot depends on the heat released by combustion, which is not the case for the Rayleigh curve.

The general profiles of the Rayleigh and Fanno curves are shown in Figure 7.12, taking as a basis for the latter, as above, equation (7.40), with V_f, eliminated (by choosing a value of p_f, and not the converse). An example of the numerical calculation of a plot such as this is given in the worked examples at the end of the chapter.

At the limiting case, neither a shock nor a reaction would occur, and the initial and final systems would be identical. This implies that for $q = 0$, both curves pass through the representative point of the initial system h_1/s_1. In the case of a detonation, the Rayleigh curve, for which the equation contains no term q for the heat released on combustion, still passes through this point, although the Fanno curve does not.

Since the representative point of the final state for a given value of p_f has to satisfy equations (7.39) and (7.40), then it must be common to the two curves. There are therefore three possible solutions: the two curves may have two points in common (J and K in Figure 7.12, which are analogous to points J and K of Figure 7.5), a single common point when they are tangents to each other, or zero when they do not intersect. In this way the three possibilities shown in Figure 7.5 are thus represented, depending on the slope of the Rayleigh curve, since the problem in question is the same.

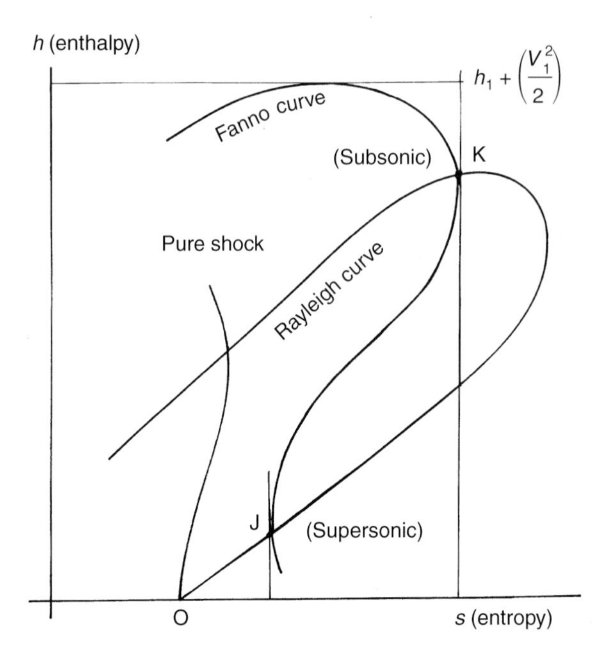

Figure 7.12 Mollier diagram with the Rayleigh and Fanno curves shown for the case where there are two solutions represented by points J and K (J for weak detonations and K for strong detonations). It can be shown that at points J and K, the tangents to the Fanno curve are vertical. The diagram is drawn, as indicated in section 7.9 of this chapter, by eliminating the final flow velocity V_f. The figure also shows the Rayleigh curve corresponding to a shock without reaction.

On the Fanno plot, the entropy passes through a maximum at point K and through a minimum at point J. Since the enthalpy at point K is greater than that at point J then the flow velocity is lower since the quantity $(h + V^2/2)$ is the same across the transition, and the pressure is greater. If the initial flow is supersonic, i.e. if V_1/a_1 is greater than one, a point such as K will correspond to what has been termed a strong detonation and thus, in this section, to combustion with shock. On the other hand, the solution given by point J, where h_f is at its smallest value and V_f at its greatest, is the solution in the absence of shock, considered here, where combustion occurs in supersonic flow.

In fact, the situation for the two curves is very similar when V_1/a_1 is smaller than 1, i.e. within the flammability limits. A point analogous to point T in Figure 7.5 can be identified, above which there is the analogue of point J and below it the analogue of point K. However, the entropy at this point K is found to be less than that for the initial flow, thus invalidating this solution.

Consider again the case where V_1/a_1 is greater than unity. The relative positions of the two curves in Figure 7.12 can change depending on the value of V_1. If V_1/a_1 is reduced, then at a certain value of V_1 the two curves will be tangential at one single point. This point corresponds to the double solution $V/a = 1$ of the Chapman-Jouguet regime, given by point T in Figure 7.5. If V_1/a_1 continues to fall, then the two curves will no longer intersect, which then corresponds to the case where the Rayleigh line in Figure 7.5 no longer intersects the Hugoniot adiabatic. There is no longer a solution for a stabilised combustion of the type considered in this section.

WORKED EXAMPLES

■ **1)** Using relationship (7.31) to obtain the temperature T_f after detonation, the product pv can be derived using the equation of state of the products then, using the Hugoniot of the products, the separate values of the pressure and density, hence the speed of sound for the conditions in question (from expression (7.10). Finally, relationship (7.13) will give the detonation speed in the Chapman-Jouguet state.

■ **2)** Numerical tables are available which give the conditions of the post-shock state for a given initial state. For the relationships given in section 7.4 of this chapter, these tables provide the ratios of the pressure before and after shock, the temperatures, and the flow velocities.

Consider air at 272 K and at a pressure of 0.689 bar, flowing at Mach number 2 and passing into a normal shock. Taking c_p/c_v to be equal to 1.4 and the "average molar mass" of air as 29 g, expression (7.2) gives:

$$a_1 = \left(\frac{1.4 \times 8.314 \times 272}{0.029}\right)^{1/2} = 330.45 \text{ m·s}^{-1}$$

$$V_1 = 2a_1 = 660.9 \text{ m·s}^{-1}$$

and the tables give $V_1/V_2 = 2.667$, from which:

$$V_2 = 247.8 \text{ m·s}^{-1} < V_1$$

Obviously, the equations from section 7.4 could have been used instead of the tables. Expression (7.17) would have given a T_2/T_1 value of 1.6875 which, with (7.2), would give $a_2 = 429.27$ m·s^{-1}. Equation (7.19) could then be used to obtain $Ma_2 = 0.58$, which is the same as the value of V_2/a_2 obtained with V_2 as determined above.

■ **3)** Consider the possible symmetry (achieved by interchanging the subscripts 1 and 2) or non-symmetry of expressions (7.16), (7.17), (7.18) and (7.19) and of the physical significance of this possibility. $Ma_1 < 1$ implies that $s_2 < s_1$, which makes the problem non-symmetric, since this condition contradicts the Second Law of Thermodynamics.

■ **4)** What happens to the calculations in section 5 of this chapter if a state equation other than the ideal gas equation is used, for example $p(v - b) = rT$ or another ?

■ **5)** Show that if the Chapman-Jouguet condition holds, then this implies that the Hugoniot curve is a tangent to both the Rayleigh curve passing through the origin and to the isentrope (pv^{γ} = constant for an ideal gas), and regardless of whether the process is a detonation (point T of Figure 7.5) or a deflagration (with the analogue of point T for deflagration but with the tangent from point I going towards the right rather than to the left. See section 7.9 above).

■ **6)** Construct, point-by-point, the Rayleigh and Fanno curves for the slightly idealised case of an ideal gas system. As in section 7.9 of this chapter, start by assuming an initial system, for example:

$$p_1 = 1 \text{ bar} \quad T_1 = 1000 \text{ K} \quad \text{and} \quad V_1 \text{ such that } Ma_1 = 2$$

Take the pressure, p_1, of the unburnt gases to be the reference pressure, and similarly for the temperature, i.e. T_1, of the unburnt gases. Thus, the lower limit of integration in the calculation of the entropy is s_1, again the value for the unburnt gases.

The "average molar mass" of air is:

$$r = \frac{8.314 \times 10^3}{29} = 287 \text{ SI}$$

and $c_p = 10^3 \text{ J·kg}^{-1}\text{·K}^{-1}$, which is slightly less than 7 $\text{cal·mol}^{-1}\text{·K}^{-1}$, the normal value for a diatomic gas.

A first Fanno curve can be plotted using an initial value for the heat, q, released by combustion i.e. $0.3 \times 10^6 \text{ J·kg}^{-1}$.

Continuing as described in section 7.9, the following values are found:

For p_f (bar)	1.3	1.5	2.0	3.0	4.0	5.0
$10^6\,(h_f - h_1)$ ($J{\cdot}kg^{-1}$)	0.3838	0.4371	0.5614	0.7715	0.9304	1.038
$(s_f - s_1)$ ($J{\cdot}kg^{-1}{\cdot}K^{-1}$)	249.53	246.26	246.63	256.53	259.85	250.2

The table clearly shows (and indeed more so than the visual representation in Figure 7.13) that s passes through a maximum at around the fifth value calculated and through a minimum at around the second value. Repeating this procedure for the two other initial values of q, i.e. of 0.5×10^6 and 10^6 $J{\cdot}kg^{-1}$ yields two other Fanno curves whereas the Rayleigh curve is unique, since it is independent of q.

For $q = 0.5 \times 10^6$ $J{\cdot}kg^{-1}$, the Rayleigh curve no longer appears to intersect the Fanno curve, a result which becomes very clear for $q = 10^6$ $J{\cdot}kg^{-1}$. The significance of this result is that there are no solutions, and that stabilisation (and hence supersonic stabilised combustion) cannot be reached for the values of p_1, T_1, Ma_1, c_p and q in question, or alternatively, that q is too large for the condition $Ma_1 = 2$ to be imposed with the values of p_1, T_1 and c_p chosen for the initial system.

Consider Figure 7.13 again. If the locus of points corresponding to the same p_1 value is produced, e.g. for 5 bar, then isobars are plotted. Similarly, isotherms can be plotted for given temperature values T_1. All points of intersection between an isotherm and an isobar define a pair of values of p_1 and T_1 corresponding to a pair of h_f and s_f values.

■ **7)** A numerical application of combustion in supersonic flow for uniform cross-sectional area (discussed in section 7.9).

This application obeys the following classical equations:

$$\dot{m} = \rho V = \rho_1 V_1 \qquad \dot{m}V + p = d_1 = \rho_1 V_1^2 + p_1$$

$$h + \left(\frac{V^2}{2}\right) = h_1 + \left(\frac{V_1^2}{2}\right) (= h_{t,1} \text{ which defines the subscript } t)$$

which for an ideal gas gives:

$$p = \rho RT \quad h = c_p T + h_0 \quad h_1 = c_p T_1 + h_{1,0}$$

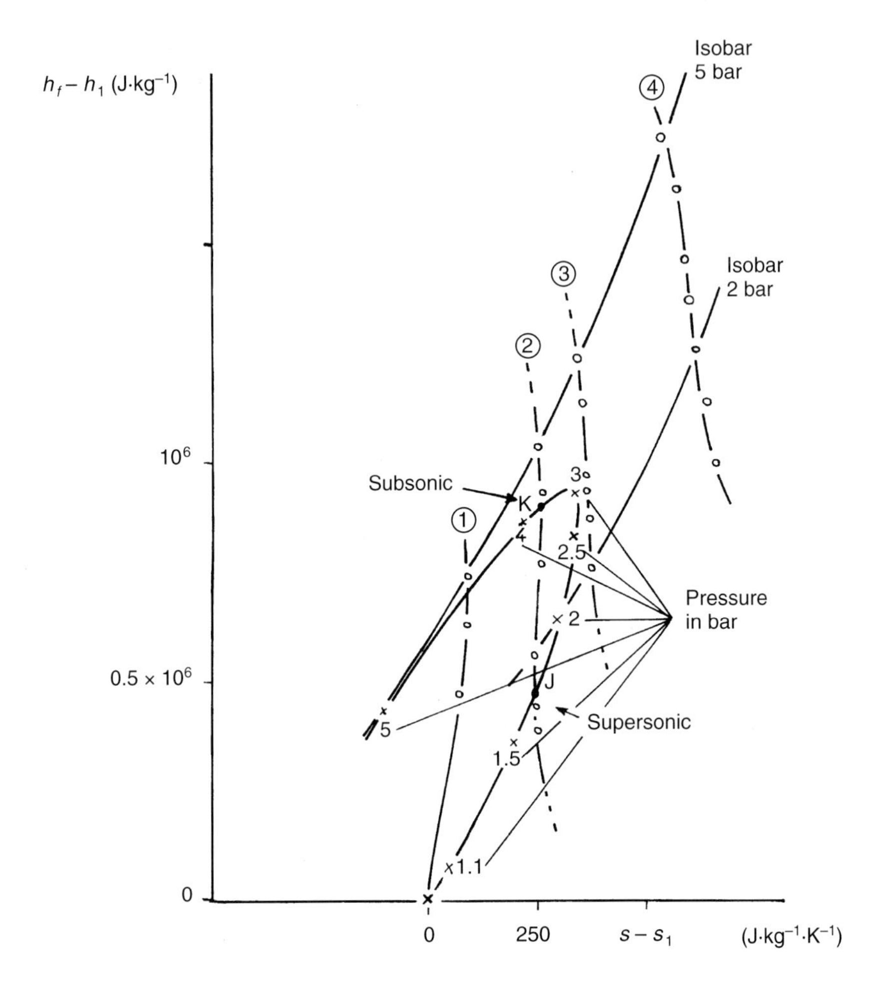

Figure 7.13 Mollier diagram relating to the numerical application of worked example 6 showing, as in Figure 7.12, the curve relating to shock without reaction, i.e. for $q = 0$ (curve 1), the Rayleigh curve, and a Fanno curve with the two solutions J and K (curve 2) for which $q = 0.3 \times 10^6$ J·kg^{-1}. Also shown are two other Fanno curves for which $q = 0.5 \times 10^6$ and 10^6 J·kg^{-1} (curves 3 and 4), which do not intersect the Rayleigh curve, as well as the 2 and 5 bar isobars. Neither the 2.5 and 1.5 bar isobars, nor the isotherms have been plotted in order to avoid overcrowding the figure. Curves 2, 3 and 4 are of the same general shape as the Fanno curve of Figure 7.12 but are less pronounced because of the numerical values chosen. It is worth noting that points J and K correspond well with the extreme values of entropy.

where R, c_p, h_0 and $h_{1,0}$ are constants. The system can be solved graphically by setting:

$$x = \frac{\gamma^2 - 1}{\gamma^2}\left(\frac{\dot{m}}{d_1}\right)^2 (c_p T_{t,1} + h_{1,0} - h_0), \text{ which is known}$$

$$y = \frac{\gamma + 1}{\gamma}\frac{\dot{m}}{d_1}V$$

$$z = R\left(\frac{\dot{m}}{d_1}\right)^2 T$$

By eliminating p:

$$\dot{m} = \rho V \qquad \dot{m}V + \rho\, RT = d_1 \qquad h + \frac{V^2}{2} = c_p T_{t,1} + h_{1,0}$$

or:

$$\dot{m}V + \frac{\dot{m}}{V}RT = d_1 \quad \text{and} \quad c_p T + h_0 + \frac{V^2}{2} = c_p T_{t,1} + h_{1,0}$$

where:

$$\frac{\dot{m}}{d_1}\left(V + \frac{RT}{V}\right) = 1 \quad \text{and} \quad c_p T + \frac{V^2}{2} = c_p T_1 + h_{1,0} - h_0$$

The last two expressions, when combined with the definitions of x, y and z, lead with $c_p/R = c_p/(c_p - c_v) = \gamma/(\gamma - 1)$ to:

$$\frac{\dot{m}}{d_1}\left[\frac{d_1}{\dot{m}}\frac{\gamma}{\gamma+1}y + \left(\frac{d_1}{\dot{m}}\right)^2 z\,\frac{\gamma+1}{\gamma}\frac{\dot{m}}{d_1}\frac{1}{\gamma}\right] = 1 \tag{a}$$

and:

$$\frac{c_p}{R}\left(\frac{d_1}{\dot{m}}\right)^2 z + \frac{1}{2}\left(\frac{\gamma}{\gamma+1}\right)^2\left(\frac{d_1}{\dot{m}}\right)^2 y^2 = \left(\frac{d_1}{\dot{m}}\right)^2 \frac{\gamma^2}{\gamma^2 - 1}x \tag{b}$$

(a) multiplied by $(\gamma + 1)y$ gives:

$$\gamma^2 y^2 - \gamma(\gamma + 1)y + (\gamma + 1)^2 z = 0 \tag{a'}$$

and (b) by $(\gamma - 1)(\gamma + 1)^2$ gives:

$$z = \frac{\gamma}{\gamma+1}\left[x - \frac{1}{2}\frac{\gamma-1}{\gamma+1}y^2\right] \tag{b'}$$

Eliminating z between (a') and (b') gives:

$$y - \frac{y^2}{2} = x \qquad (c)$$

which is a quadratic equation in y with two roots if x is less than or equal to 1/2 (Fig. 7.14), since x is itself defined by the conditions at the entrance to the reactor, and once known, p can be found from:

$$p = d_1 - \dot{m}V = d_1\left(1 - \frac{\gamma y}{\gamma + 1}\right)$$

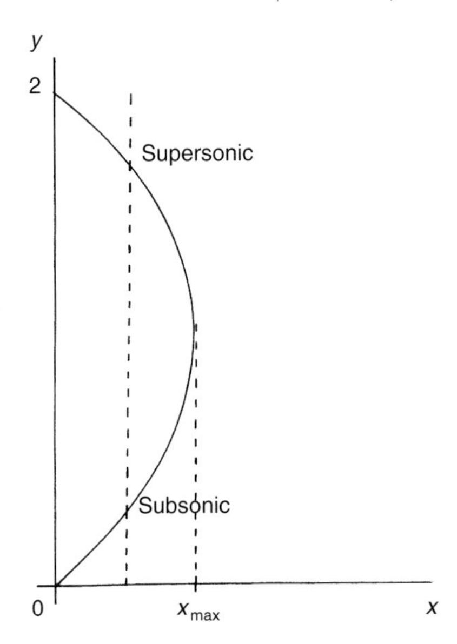

Figure 7.14 Representation of equation (c) of worked example 7 showing the two possible values of x for a given value of y.

The larger of the two roots corresponds to the highest V and to supersonic flow, with the lower value corresponding to subsonic flow. This can be shown by calculating the speed of sound in the burnt gases, $(\gamma RT/M)^{1/2}$, and comparing it with V. The double root found for $x = 1/2$ gives:

$$\frac{1}{2} = \frac{\gamma^2 - 1}{\gamma^2}\left(\frac{\dot{m}}{d_1}\right)^2 (c_p T_1 + h_{1,0} - h_0)$$

which corresponds with the Chapman-Jouguet condition. Setting $\dot{m} = \rho_1 V_1$ gives an equation in V_1 which allows V_1 to be calculated as a function of T_1, ρ_1, p_1,$(h_{1,0} - h_0)$, c_p and R.

CHAPTER 8
FLAME IGNITION

8.1 THE IGNITION PHENOMENON

So far we have considered a wide variety of flames, laminar or turbulent, premixed or diffusion, stable or propagating without really looking into how they are initially ignited or lit. Whilst the concept of self-ignition was studied in chapter 4 (by considering how a gaseous mixture evolves as a function of different values of its initial temperature and pressure), in practice we usually seek to ignite the mixture at a single point or in a small volume of the medium, and the term ignition or inflammation rather than self-ignition is used. What are the actual processes involved?

Ignition is achieved through supplying energy, over a given period of time, to a given volume of a medium consisting of a premixed blend of fuel and oxidant. The energy is often provided by a spark, a small jet of hot gas, or even by a suitably-focused laser pulse. In each of these cases the conditions are different and far from straightforward; ionised species are created by a spark as a result of a highly complex and poorly understood process, whereas jets of hot gas (of limited duration) may carry with them reactive species and cause motion within the gas. A laser pulse is perhaps the simplest way to supply the required energy, although care must be taken to ensure that the light energy is absorbed, and serves only to increase the internal energy of the molecules in the gas medium, without producing imbalances in the populations of the vibrational energy levels of the molecules present in the gas.

When describing and calculating the ignition process, any energy supplied must be included in a simple, yet realistic manner. There are two models commonly used to satisfy this requirement. The first considers that initially there is a zone of given size (a small sphere of known radius for example) in the homogeneous medium whose temperature is higher than the surrounding mixture, and models the subsequent evolution of the medium. The second considers that in this zone of known size a certain amount of heat per unit time (or power) is released over a period extending from the initial time up to a given time later. The temperature of the zone will increase and ignition may ultimately occur. The second model more closely simulates what actually happens, at least in the case of a spark or laser source, while the first model can be considered as being the limiting case of the second which assumes that an infinite amount of power is introduced over a very short time (and is thus more schematic and less realistic).

Whether the ignition conditions correspond to the first or to the second model, it is possible to describe approximately how ignition will proceed; the heat deposited in the ignition zone (either at an initial time or over a certain time period) first of all diffuses into the surrounding medium through thermal conduction. The heated zone may be expanded if the degree of heating is great enough. At the same time, as soon as the temperature of the heated zone is high enough, chemical reactions begin between the fuel and oxidant, and they require a certain amount of time to raise the temperature of the medium sufficiently. If this temperature increase is high enough, and occurs quickly enough compared with the temperature drop caused by the diffusion of heat to the surroundings, then the ignition zone will remain hot, and a flame is established which propagates towards the surroundings in the form of a deflagration.

Solutions are required to the following questions concerning the ignition process:

If we start by considering that ignition is initiated by a hot zone, which assumes the deposition of an infinite amount of energy in an infinitely short time, under what conditions will ignition lead to flame propagation? The important parameters in this problem are the size of the heated zone and its temperature and, more specifically, when will they be large enough to ignite the mixture for a given set of temperature and pressure conditions?

Moving on to consider ignition by heat deposition, there are two additional parameters, namely the power (instantaneous) of the heat deposition and its duration. On the other hand, the temperature of the hot zone is no longer an entry in the problem. Once again we must determine the range in which the three parameters (size, power and duration) can be varied whilst still leading to successful ignition.

The amount of energy required to ignite a flame under given conditions is one of the results that any calculation would be expected to provide. At the end of this chapter we shall see that normally only very small amounts of energy are needed, of the order of a few millijoules, which might seem surprising. However, one should bear in mind that a flame under normal conditions is extremely thin (typically around 0.3 mm). If it were possible to create in the medium a small sphere with a radius of just 1 mm, and if the temperature of this sphere were raised to the maximum flame temperature, i.e. around 2000 K, then it is practically certain that a flame would propagate. The energy contained in such a sphere at this temperatures would be approximately

$(4/3) \times \pi \times 10^{-9} \times 1000 \times 2000 \times 0.2$, or about 1.6 millijoules (since the c_v for air is about 1000 joules per kilogram, and its density at 2000 K is about 0.2 kilograms per cubic metre). This figure of 1.6 millijoules must, however, be increased if ignition is initiated under conditions approaching the limits for flame propagation (as defined in chapter 5).

These ignition-related problems are quite similar to those investigated in 1940 by Frank-Kamenetzkii and considered in Chapter 4, although Frank-Kamenetzkii studied very slight temperature increases in a zone of finite size, whereas here we are concerned with considerable increases in temperature. Frank-Kamenetzkii's approach can therefore only be applied when the medium is already very close to the conditions required for self-ignition.

The phenomena associated with heat transfer and chemical reactions described above are also accompanied by acoustic effects. Every time a small volume of gas expands it generates pressure waves which create infinitely small perturbations in the pressure and velocity fields (as discussed in chapter 7). These perturbations propagate at the speed of sound from the ignition zone towards the periphery. In general, however, these pressure perturbations move away quickly enough from the ignition zone for ignition to be considered as taking place under substantially isobaric conditions. Thus, strictly speaking, the phenomena of pressure waves and ignition are not coupled. However, this statement is not true when ignition occurs in a finite medium: pressure waves reflecting from the walls may return at exactly the right moment to interact with the start of the chemical reactions, depending on the size of the combustion zone to be ignited.

The possibility of another, very specific, case of coupling has been demonstrated recently. When the temperature distribution in the ignition zone is not homogeneous and has certain characteristics, an interaction between the pressure waves and the chemical reactions at the point of ignition may lead to the propagation of a detonation rather than a deflagration. We shall return to this case at the end of the chapter.

The theoretical solutions to ignition-related problems, even when simple, have associated difficulties which are too involved to be covered by the scope of this book. Hence, the remainder of this chapter will focus primarily on a description of the results of numerical calculations, together with a discussion of their physical significance. These results are taken from recent work by Champion et al.[1], He and Clavin[2], for sections 8.2 and 8.3, and by Warnatz and his co-workers: Maas and Warnatz[3], He and Clavin[4] and Zeldovitch et al.[5], for section 8.4.

---

1. Champion M., Deshaies B., Joulin G., Kinoshita K. (1988) *Combustion and Flame*, 65, p. 319.
2. He L., Clavin P. (1992) *Combustion and Flame*, 93, 4, p. 391.
3. Maas U., Warnatz J. (1991) Dynamics of deflagrations and reactive systems: flames. *Progress in Astronautics and Aeronautics*, 131, AIAA.
4. He L., Clavin P. (1992) *24th Symposium (Int.) on Combustion*. The Combustion Institute.
5. Zeldovitch Ya. et al., *Astronautica Acta*, 15, p. 971.

8.2 HOT SPOT IGNITION

8.2.1 Results of a typical calculation

The problem of ignition induced by a hot spot of given initial size is difficult to study in an analytical manner, although accurate results can be obtained using numerical calculations. It is even possible to take detailed chemical reactions into consideration if they are fully understood, which is the case for the combustion of hydrogen in oxygen. Equally, differences in the coefficients of diffusion of the various species, as well as their variations with temperature, can also be included. As mentioned above, the model assumes that the pressure remains constant and that any gas expansion is neglected. The equations to be integrated are then simply those dealing with chemical species and energy.

These equations are deduced from the general equations given in chapter 3, with modification made to allow for ignition occurring with spherical symmetry, hence:

$$\frac{\partial}{\partial t}(\rho Y_i) = \frac{1}{r^2}\frac{\partial}{\partial r}(r^2 j_i) + \rho w_i \qquad i = 1, \dots n \tag{8.1}$$

replaces 3.12, and 3.14 is replaced by:

$$\frac{\partial}{\partial t}(\rho u) = \frac{1}{r^2}\frac{\partial}{\partial r}\left(r^2\left(j_q + \sum_{i=1}^{n} h_i j_i\right)\right) \tag{8.2}$$

The Y_i terms are the mass fractions of the various species, for which j_i are the diffusional fluxes which are functions of the radial concentration gradients, and which are controlled by w_i, the rates of reaction. j_q is the conductive heat flux and u the internal energy, from which the temperature can be deduced.

The numerical solution to this system of equations does not present any specific problems other than those resulting from the chemical reaction terms, w_i, mentioned in chapter 2.

Figures 8.1a and b show the calculated changes in radial temperature profiles for the ignition of an H_2—O_2 mixture containing 94% hydrogen (molar), i.e. fuel-rich, and at atmospheric pressure. Figure 8.1a shows the temperature changes for a spherical pocket of gases, with an initial radius of 1.4 cm, burnt at 1300 K, which is the maximum adiabatic temperature corresponding to this mixture. The temperature at other points in the mixture is 298 K. Ignition does not occur and the temperature then decreases at the centre of the pocket. On the other hand, ignition does occur in the case shown by Figure 8.1b, which plots the changes in a pocket of initial radius 1.45 cm, with the same initial temperature values. After an initial decrease, the temperature rises to reach 1300 K, with the hot zone progressively extending.

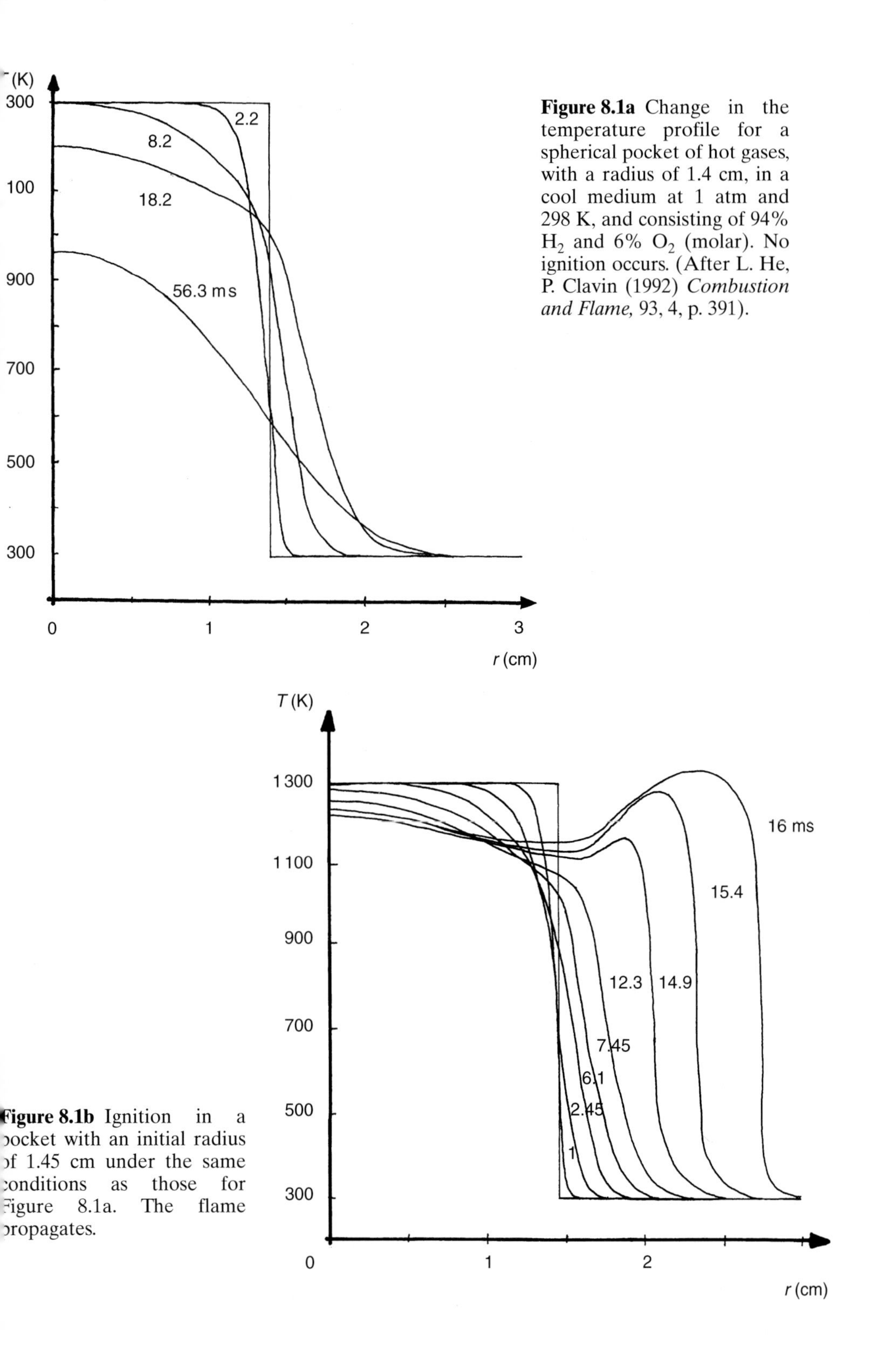

Figure 8.1a Change in the temperature profile for a spherical pocket of hot gases, with a radius of 1.4 cm, in a cool medium at 1 atm and 298 K, and consisting of 94% H_2 and 6% O_2 (molar). No ignition occurs. (After L. He, P. Clavin (1992) *Combustion and Flame,* 93, 4, p. 391).

Figure 8.1b Ignition in a pocket with an initial radius of 1.45 cm under the same conditions as those for Figure 8.1a. The flame propagates.

For a spherical pocket of gas at a given temperature and of known composition under given conditions, the result of these calculations is that ignition only occurs if the radius of the pocket r_a is greater than a certain critical value. Figure 8.2a illustrates this critical radius as a function of mixture composition (for fuel-rich mixtures) and Figure 8.2b for fuel-lean mixtures. The critical radius is generally very small (1 mm) except when the mixture is close to the limits for flame propagation, as in the case shown in Figure 8.1. Note also that the flame thickness e_L in the propagation regime increases as these limits are approached, but that the critical radius for ignition increases to a much greater extent.

The temperature of the hot zone is also important, but in a very non-linear manner, since below the self-ignition temperature, studied in chapter 3, the limiting radius is infinite and drops quite quickly as a function of increasing temperature in the ignition pocket, to level off at a value approximately the same as the value of the radius given in the preceding figures when the temperature exceeds the adiabatic temperature corresponding to the end of combustion.

8.2.2 Physical explanation

A physical explanation of this critical radius is possible by taking into consideration the fact that a spherical flame (containing burnt gases at its centre) could, **in theory**, be stationary and not propagate if the size of the reaction zone was exactly the same as the critical radius. Indeed, if we restrict our approach to considering the simplified representation which involves a single reaction and with Lewis numbers equal to unity, then applying Fourier's law for j_q, equations (8.1) and (8.2) reduce to a single equation, for example the following equation for energy:

$$\left.\begin{array}{l} \left(\dfrac{\lambda}{\rho c_p}\right)\dfrac{1}{r^2}\dfrac{\mathrm{d}}{\mathrm{d}r}\left(r^2\dfrac{\mathrm{d}T}{\mathrm{d}r}\right)+\dfrac{q}{c_p}w=0 \\ \text{making the assumptions that} \left\{\begin{array}{l}\text{at } r=0 \quad T=T_{\text{ad}} \\ \text{at } r=\infty \quad T=T_0\end{array}\right. \end{array}\right\} \tag{8.3}$$

where $T_{\text{ad}} = T_0 + q/c_p$ is the adiabatic temperature at the end of combustion (at constant pressure) of the mixture (defined in chapter 1, section 1.2.3) in question, for an initial temperature T_0. The non-stationary term has not been included since only the stationary regime is of interest, and λ/c_p is assumed to be constant.

Note that once again this is not the same as the problem investigated by Frank-Kamenetzkii since the limiting conditions are different. Furthermore, T is not assumed to be close to T_0, nor are the reactants assumed to be consumed.

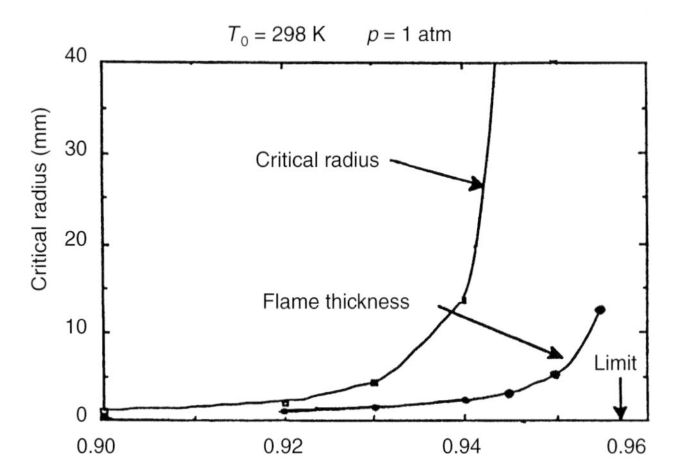

Figure 8.2a Critical radius as a function of mole fraction of H_2 near the propagation limit, on the fuel-rich side. The mixture is again at 1 atm and 298 K. Note that the critical radius is generally of the same order of magnitude as the thickness of the laminar flame propagating through the medium in question, but becomes much greater near the propagation limit of this laminar flame (as defined in chapter 5). Same reference as for Figure 8.1.

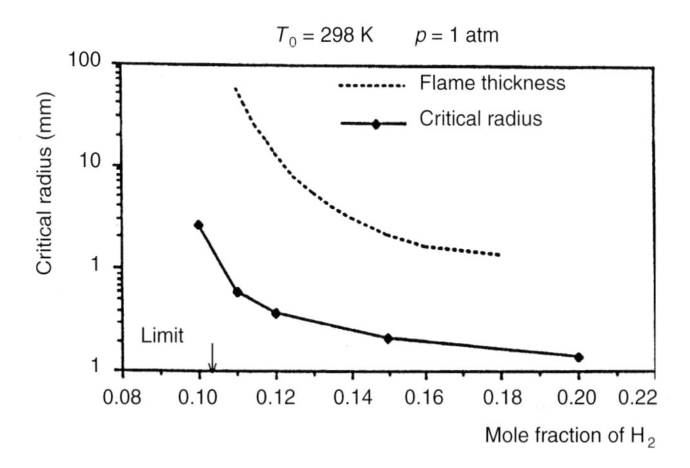

Figure 8.2b Critical radius near the extinction limit, from the fuel-lean side. The critical radius is at all points smaller than the laminar flame thickness. Same reference as for Figure 8.1.

Taking the Arrhenius law for w:

$$w = kY_{Ox}Y_K \exp -\left(\frac{T_A}{T}\right)$$

where, because the Lewis number is equal to unity, Y_{Ox} and Y_K are functions of T alone, and for a fuel-lean mixture, Y_K becomes equal to zero if $T = T_{ad}$ since all the fuel is consumed.

By defining $\theta = (T - T_0)/(T_{ad} - T_0)$, which is always less than one, $\beta = T_A(T_{ad} - T_0)/T_{ad}^2$, and $\alpha = (T_{ad} - T_0)/T_{ad}$, then the following is obtained:

$$w(\theta) = kY_{Ox}Y_K \exp\left(-\frac{T_A}{T_{ad}}\right) \exp\left[\frac{\beta(\theta - 1)}{1 + \alpha(\theta - 1)}\right]$$

where Y_{Ox} and Y_K are functions of θ, the equation then becomes:

$$-\frac{1}{r^2}\frac{d}{dr}\left(r^2\frac{d\theta}{dr}\right) = \frac{k}{(\lambda/\rho c_p)} \exp\left(-\frac{T_A}{T_{ad}}\right)\frac{Y_K Y_{Ox}}{T_{ad} - T_0} \exp\left[\frac{\beta(\theta - 1)}{1 + \alpha(\theta - 1)}\right] \tag{8.4}$$

Now assume that both β and k are very large, such that $w(\theta)$ is very small except when $\theta = 1$ ($T = T_{ad}$). This assumption, which implies that the chemical reactions occur over a surface where $\theta = 1$, is often made in order to simplify the calculations.

In this case, except for the surface where $\theta = 1$, which is a sphere with a radius of, say, r_f, equation (8.4) reduces to:

$$\frac{d}{dr}\left(r^2\frac{d\theta}{dr}\right) = 0 \tag{8.5}$$

The temperature profile θ thus has the form shown in Figure 8.3 where, if $r < r_f$, then $\theta = 1$ (which is the solution of equation (8.5)). The slope of this temperature profile exhibits a discontinuity at $\theta = 1$, where chemical reaction takes place.

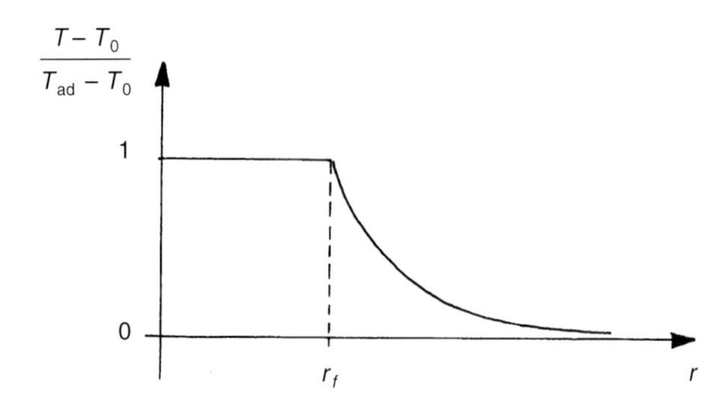

Figure 8.3 Temperature profile for a stationary spherical flame (unstable).

In order to calculate r_f the details of what happens at this discontinuity must be studied. An asymptotic expansion is set for θ and r in terms of 1/β:

$$\theta = 1 + \frac{\theta_1}{\beta} + ..$$

$$r = r_f + \frac{\eta}{\beta} + \ldots$$

θ_1 (η) is required, and equation (8.4) thus becomes (neglecting terms in 1/β and higher):

$$-\frac{d^2\theta_1}{d\eta^2} = \Lambda\,\theta_1 \exp\theta_1 \qquad (8.6)$$

with $\Lambda = (\rho k c_p/\beta^2\lambda) \exp(-T_A/T_{ad})\, Y_{K,0} Y_{Ox,f}/(T_{ad} - T_0)$ (where the subscripts 0 and f denote the states before and after combustion).

Equation (8.6) can be integrated to:

$$-\frac{d\theta_1}{d\eta} = \left(2\,\Lambda \int_0^{\theta_1} \theta_1 \exp\theta_1\, d\theta_1\right)^{1/2}$$

The identification of $d\theta/dr)_{r_f-\varepsilon} = d\theta_1/d\eta)_\infty$ gives r_f.

And $1/r_f = (2\Lambda)^{1/2}$ since it is known that $\int_0^\infty \theta_1 \exp\theta_1\, d\theta_1 = 1$

Thus r_f may have a well-determined value which is stationary, unique and a function of the characteristics of the chemical reaction and of the thermal diffusivity.

However, this theoretical solution is not confirmed in practice since it is not stable, and indeed it can be demonstrated that a slightly smaller flame evolves by diminishing (and eventually extinguishing) whereas a slightly larger flame evolves by growing such that the whole medium is progressively ignited.

The value of r_f determined previously for the stationary spherical flame therefore corresponds to the critical radius for ignition (for the case where the hot pocket is at the adiabatic temperature T_{ad} corresponding to the above calculation).

This theoretical approach to hot spot-induced ignitions leads to the concept of the existence of a critical zone size i.e. the minimum size required for ignition to occur. We have seen that the rate of energy deposition in the initial hot zone was assumed to be infinite; the presence of a critical size for the hot zone thus implies that the amount of energy deposited must have a certain minimum yet finite value for ignition to be successful.

8.3 IGNITION THROUGH ENERGY DEPOSITION

In practice, a flame is ignited by introducing energy, at a finite power, in as small a region as possible of the gaseous medium. The following questions are raised by this practical observation; firstly, can the ignition zone be very small, or is there a critical radius (as above) and, secondly, is there a minimum quantity of energy which must be deposited before ignition is achieved, or must this energy be transferred over a given period of time? The answer to the first question is that ignition can be induced even by an extremely localised (pinpoint) deposition, since even in this case the energy deposited diffuses radially to occupy a gradually-increasing zone. Moreover, it is unwise to consider that the time over which energy deposition occurs does not matter. In fact, even if the ignition time is extended, if the power of the deposition is too low it may not be sufficient to compensate for diffusional heat losses away from the ignition zone, and consequently the temperature in this zone would not reach a high enough level to initiate chemical reaction. This is indeed what will be seen to be the case.

The existence of a spherical flame which is in theory stationary and non-propagating (such as that described in the previous section) is also possible so long as energy is supplied inside this sphere. If the energy supply is assumed to be constant with time and applied at a pinpoint location (at $r = 0$), the calculation from the previous chapter can be reconsidered, changing the condition at $r = 0$ to one for a given heat flux.

Hence, based on a similar treatment to that of the previous section, where r_f^0 is the previous solution when the power of the energy source Q_s is zero, we find that:

$$\frac{r_f}{r_f^0} = \exp\left\{-\frac{\dfrac{\varphi}{r_f^0}}{\dfrac{2r_f}{r_f^0}}\right\} \tag{8.7}$$

where φ is proportional to the power of the source:

$$\varphi = \frac{Q_s \dfrac{T_A(T_{ad} - T_0)}{T_{ad}^2}}{4\pi\lambda(T_{ad} - T_0)}$$

Where T_A is the activation temperature of Arrhenius' Law. The existence of a stationary solution r_f ($Q_s \neq 0$) then depends on the existence of solutions to this implicit equation and it can be shown that no solution exists if $\varphi/r^0_f > 2/e$. If this is not the case, there are two solutions, and since one is stable and the other unstable (which can be demonstrated) the situation is thus entirely different. The stable solution, if achieved, can prevent ignition. The natural conclusion is that ignition inevitably requires a minimum power, such that $\varphi > (2/e)r^0_f$, to allow no stable solution to exist.

If this minimum power is provided continuously, then the energy provided does not need to be supplied to a volume of a critical size, (instead it is supplied to a pinpoint location), although this corresponds to infinite energy and hence is not possible in practice. Even when this minimum power is provided, the energy deposition will in practice cease after a certain time, corresponding to a certain amount of deposited energy. Will ignition occur, or not, as a function of this amount of deposited energy?

Numerical calculations provide the accurate answer to this question, although as yet these have only been performed for certain simple cases.

Figure 8.4a shows an example of the result of a calculation of the changes in flame radius for various ignition times at a given power. It is immediately apparent that under these conditions ignition does not occur (the flame stops growing) if the time during which energy is deposited is too short. There is, therefore, a certain amount of energy which must be deposited to ensure ignition.

However this amount of energy is not a constant which is independent of deposition time. Figure 8.5 shows the amount of energy which must be deposited ($E_s = Q_s \cdot t$) as a function of deposition time. In general, the energy required increases with deposition time. A minimum energy level does, however, exist for certain curves for values of β (as defined in the previous section) which are not too large; i.e. for cases in which the activation temperature is not too high, there is no advantage in igniting with high power over a short period of time since this would result in a much larger increase in the local temperature than is required for a moderate activation temperature.

Looking again at Figure 8.4a, it can be seen that the energy deposition time must be such that the flame radius exceeds a certain critical value, similar in size to the critical radius r_f^0 defined in the case when no energy is deposited in section 8.2 (since the latter was chosen to make the flame radius dimensionless). Thus instead of considering a minimum ignition energy, a minimum ignition radius is again of more relevance, although in this case the radius does not correspond to the amount of energy deposited but to the size of the flame resulting from the energy deposition after a certain energy deposition duration. In the case where ignition occurs, the temperature profiles are similar to those shown in Figure 8.4b. The temperature may rise greatly at the energy deposition point and is, at least according to the theory, initially infinite!

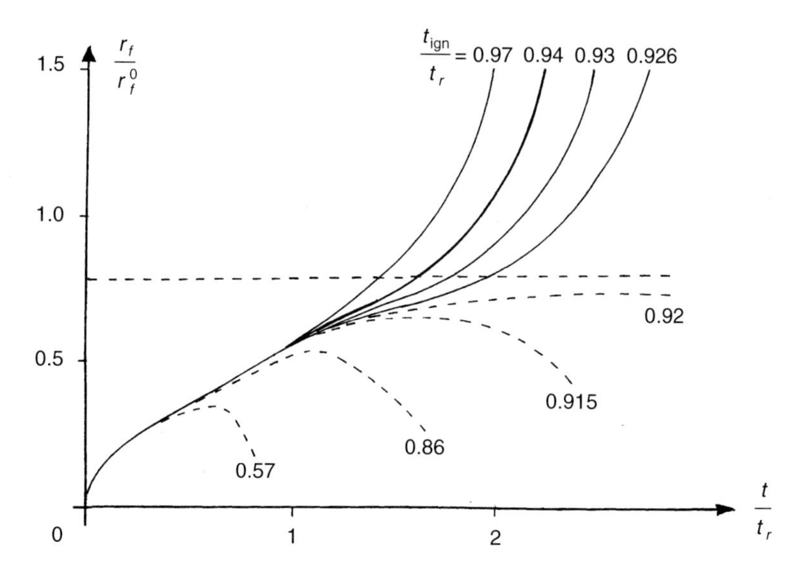

Figure 8.4a Calculated growth in flame radius with energy supplied for various times. The reference time t_r used to make t dimensionless is:

$$t_r = \frac{c_p(T_{ad})\rho(T_{ad})(r_f^0)^2}{\lambda(T_{ad})}$$

which is the characteristic time for the evolution of the unstable spherical flame introduced in section 8.2. The ignition times are also taken with respect to this time. After Champion, Deshayes, Joulin and Kinoshita (1988), *Combustion and Flame,* vol. 65, p. 319.

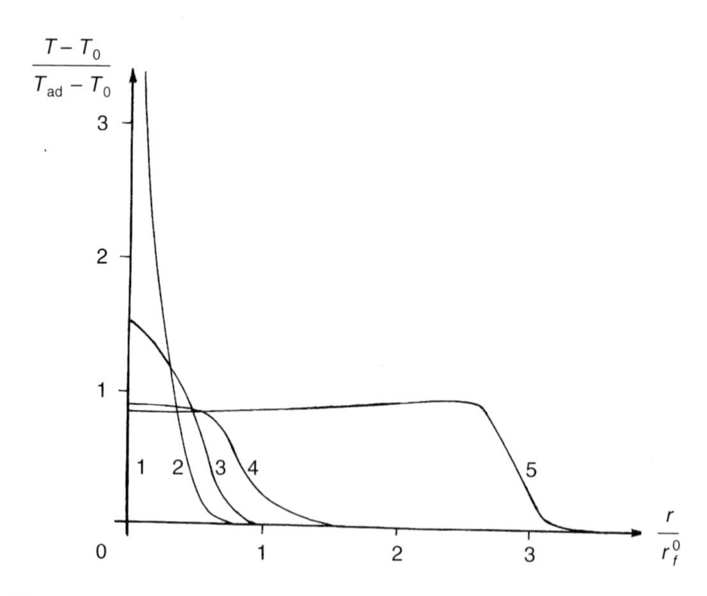

Figure 8.4b Calculated temperature profiles during (successful) ignition with energy deposition.

1: $t/t_r = 0$ 2: $t/t_r = 0.06$ 3: $t/t_r = 0.11$ 4: $t/t_r = 0.34$ 5: $t/t_r = 1.25$.
Same reference as for Figure 8.4a.

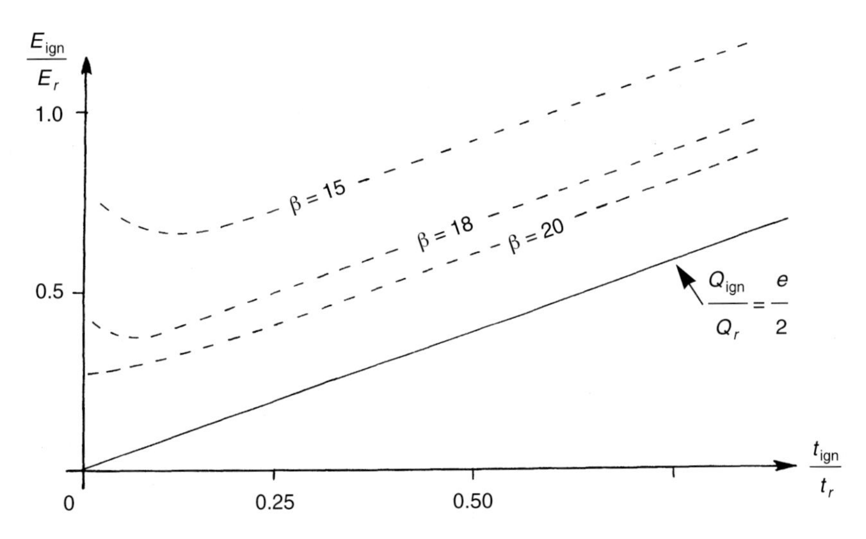

Figure 8.5 Energy required for ignition as a function of deposition time (made dimensionless).
The reference energy is that contained in the unstable spherical flame introduced in section 8.2 and given by:

$$E_r = \frac{4\pi\lambda(T_{ad})T_{ad}^2 r_f^0 t_r}{T_A}$$

the reference power is $Q_r = E_r/t_r$. Same reference as for Figure 8.4.

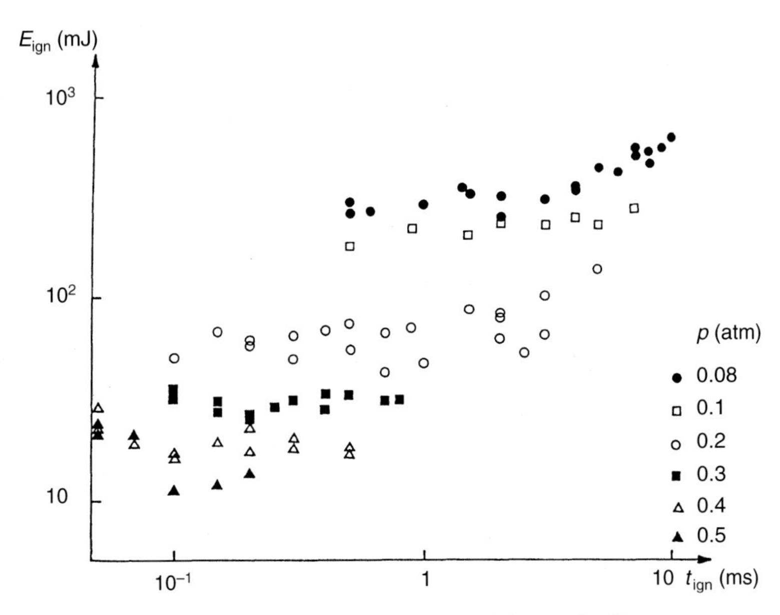

Figure 8.6 Experimental results for the minimum ignition energy required to ignite a propane-air mixture with an equivalence ratio of 0.8. The energy deposition is by a quasi-pinpoint spark provided by two very thin electrodes. Same reference as for Figure 8.4.

What then, in practice, are the energies required for ignition? Figure 8.5 indicates that they are of the same order of magnitude as the amount of energy E_r used to make the energy for ignition dimensionless. In the case of a propane-air mixture with an equivalence ratio of 0.8 at room temperature and 0.5 bar, this energy is calculated to be approximately 20 mJ (involving an estimation of the global activation temperature) and the reference time t_r is approximately 10 ms. Figure 8.6 shows, in values which have not been made dimensionless, the result of attempts to ignite the above mixture at various pressures, with the minimum ignition energies shown as a function of ignition duration (with ignition initiated by a spark). All the experimental points, when plotted using dimensionless parameters such as in Figure 8.5, are found to lie around the minimum of the curve corresponding to $\beta = 18$.

8.4 IGNITION WITH CONSECUTIVE PRESSURE PERTURBATIONS

Any localised release of energy in a gaseous medium inevitably leads to a pressure variation at the point where the energy is produced which then propagates through the rest of the medium. This is easy enough to explain, and is due to the fact that if a small volume of gas is heated and its temperature is increased. Then the resulting expansion in volume pushes against the surrounding fluid.

If, hypothetically, the surrounding fluid is assumed to resist all displacements, then the expansion would be prevented and the pressure would rise greatly in the small fluid volume under consideration until it reached the value calculated by applying $pV = nRT$ for a constant volume V. In reality, the pressure rise is much smaller in the heated volume since it can expand, altering the pressure of the surrounding fluid and setting it in motion.

Pressure perturbations (in a fluid or a solid) propagate in the form of waves and, when the amplitude of these waves is very small, their velocity is equal to the speed of sound for that medium (which can be calculated to be the square root of the partial derivative of the pressure with respect to the density, the entropy being taken as constant: $(\partial p/\partial \rho)_s$. A perturbation which is not infinitely small transforms, after travelling a certain distance, into a shock wave which may then propagate at a supersonic velocity. A more detailed explanation of this phenomenon was given in chapter 7.

Clearly, these pressure perturbations are stronger if the energy is deposited more rapidly. For example, if the size of the deposition zone divived by the deposition time is defined as being the characteristic rate of the deposition process, then the pressure perturbations become greater as the ratio of this velocity to the speed of sound in the medium increases. However, even if these perturbations are strong, they move away quickly (since the speed of sound is approximately 300 m/s under normal conditions) from the zone where the energy is deposited and thus from where the temperature is highest and where ignition, if it is to occur, will do so. In most cases, the time required for ignition to occur is long enough to allow the pressure (or shock) wave to travel a long way from the point where the energy was deposited, and hence the pressure at this point will have had time to return to substantially the value it had before energy deposit occurred.

If, however, ignition occurs very rapidly then coupling can take place with the pressure perturbation. Indeed, on one hand the rates of the chemical reactions are directly dependent on pressure (as seen in chapter 2), and on the other hand (more importantly) they are dependent on the temperature; the pressure perturbation also being accompanied by a temperature perturbation. If the pressure wave is an isentropic "pulse" then the isentropic compression is known to produce a temperature rise which can be determined as a function of γ alone (the ratio of the specific heat capacities of the gas) and, if a shock wave is produced, the resulting temperature rise can also be calculated using the classic equations (depending in addition on the propagation velocity of the shock; see chapter 7). Thus the pressure perturbation increases the rate of reaction on passing a given point and may, if ignition occurs over a very short time period, trigger the ignition of the medium.

An example of this type of coupling can be obtained by calculation. The procedure involves performing a numerical integration of the complete balance equations, and must include not only the equations for the chemical species and the energy but also those for the momentum balance (Navier-Stokes equations).

Figure 8.7a shows temperature profiles obtained during the ignition of a hydrogen-oxygen mixture in a cavity where the equivalence ratio of the medium is initially unity, with a uniform temperature of 1150 K, and a pressure of 1 bar. This experiment shows ignition occurring as a function of time: the elevated temperature zone grows and the maximum temperature rises to 2900 K in 2×10^{-5} s. The figure clearly shows the temperature perturbation caused by the shock wave which moves away from the ignition zone very quickly. The flame propagates away from the ignition zone much more slowly and in this case there is no coupling between ignition and the pressure wave. In contrast, Figure 8.7b shows temperature profiles for an identical situation except that the initial pressure is 2 bar. In this case, the pressure wave provides a greater degree of heating and the elevated temperature zone is much closer to the pressure wave, even managing to catch up with it after a certain time, to produce a very steep temperature front which ultimately propagates more quickly than the pressure wave itself! The ignition-pressure coupling has in fact produced a detonation wave.

The previous examples illustrate how coupling can occur if the initial temperature of the medium is high enough (in fact even if no hot zone is present at $t = 0$ the mixture could have ignited spontaneously, but over a much longer

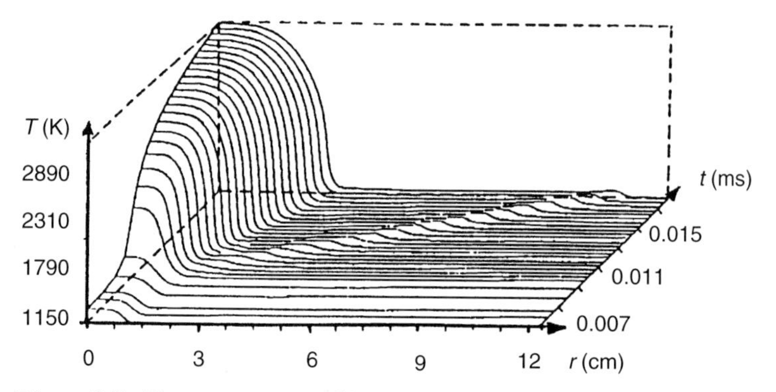

Figure 8.7a Temperature profiles calculated during the "normal" ignition of an H_2—O_2 mixture initially at a sufficiently high temperature ($p_0 = 1$ bar, equivalence ratio = 1). At the initial instant a spherical temperature distribution is assumed and defined by:

$$T(r, 0) = 1150 + 150 \times \exp(-(r/1\text{ mm})^2)$$

(After Maas U. and Warnatz J. (1991) Dynamics of deflagrations and reactive systems: flames. *Progress in Astronautics and Aeronautics*, Vol. 131, AIAA).

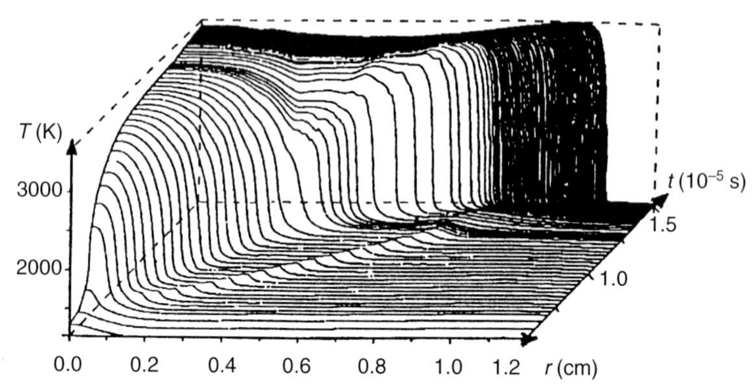

Figure 8.7b Temperature profiles calculated during the ignition of a detonation of the mixture described above. The conditions are exactly the same, except that the initial pressure is now 2 bar. Same reference as for Figure 8.7a.

time) and if the pressure perturbation is large enough. The fairly high initial temperature ensures that the rate of reaction will be highly dependent on the temperature increase caused by the pressure wave.

Would it be possible to initiate a detonation wave, simply by local heating, in a medium at a much lower temperature? The answer, as demonstrated through calculations by Y. Zeldovitch and co-workers in the 1970s is "Yes", although a much larger heating zone is required whose temperature distribution falls off with a shallow gradient towards the periphery.

An example of the result of a calculation of this type of phenomenon is shown in Figures 8.8a and 8.8b, which plot the temperature and pressure profiles respectively at various times after ignition. As Figure 8.8a indicates, the profile of the initial temperature is a very large pocket of radius 1.5 cm within which the temperature rises from 450 K at the outside to 1350 K at the centre. The calculation in this case corresponds again to the combustion of an H_2—O_2 mixture with an equivalence ratio of unity but for a global reaction.

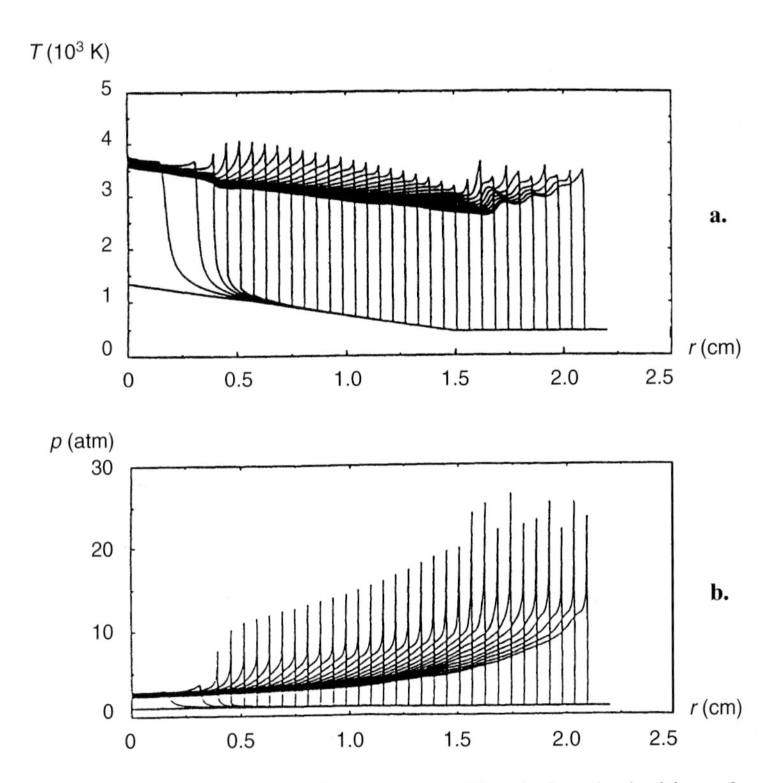

Figure 8.8 Temperature and pressure profiles during the ignition of a detonation in a medium at room temperature, produced by a hot zone of controlled temperature gradient. Results taken from L. He and P. Clavin (*24th Symposium (Int.) on Combustion*, The Combustion Institute, 1992). The curves are plotted from points taken at 1.3 ms intervals after the initiation.

The rise in temperature to around 3600 K at the centre can be clearly seen, followed by the formation of a very steep temperature front which propagates very quickly. Figure 8.8b, showing the pressure peaks, confirms that detonation does take place, and it is also possible to verify that once the pressure peak increases to a radius of 0.6 cm it has the same height as the pressure detonation.

Zeldovitch offers an explanation as to what then happens, based again upon a consideration of the rate of reaction under the ignition conditions. An induction delay is defined for the combustion, τ_i (related approximately to the inverse of the rate of reaction) and which is strongly temperature-dependent, especially when the H_2—O_2 mixture is at a temperature of about 1300 K. Imagine that the medium is at 1350 K; the reactions will occur spontaneously after τ_i (1350) but, next to this zone, in the medium at 1300 K, the reactions will occur (also spontaneously, i.e. not influenced by what is happening nearby) but slightly later, at τ_i (1300), and further away still, where the temperature is 1250 K, they will take place slightly later still. Thus the impression is given that the reaction zone propagates from the point of highest temperature (in this case at 1350 K) towards lower temperature regions, and at a velocity of $(d\tau_i/dx)^{-1}$, i.e. $(d\tau_i/dT \cdot dT/dx)^{-1}$. Zeldovitch called this velocity the propagation velocity of spontaneous self-ignition, since it is very different from the burning velocity of a flame as well as the velocity of propagation of the detonation. It is dependent mainly on the sensitivity of the induction period (hence roughly on the inverse of the rate of reaction) at the temperature in question.

The condition required for a detonation to form, and hence for coupling to occur between the ignition and pressure wave effects, is that this spontaneous velocity must have a value approaching the speed of sound. If the spontaneous velocity is very high compared to the speed of sound then self-ignition will occur virtually throughout the whole volume. On the other hand, if this velocity is small then quasi-isobaric deflagration will occur with a pressure wave which moves away very rapidly. Of course, for detonation (and indeed for deflagration) to occur, the maximum temperature (at the centre) must be high enough for the induction delay to be shorter than the time required for heat conduction to lower the temperature.

However, it is not certain that the detonation wave produced under these conditions can propagate far enough to reach the medium at room temperature. Indeed, Figures 8a and 8b show the first signs that the temperatures and pressures during the detonation are perturbed when the detonation reaches a radius of 1.5 cm, where the temperature is only 450 K. When the radius reaches 1 cm, or even slightly before, the maximum pressure deviates slightly from the Von Neumann pressure. Another calculation, identical except that the temperature of the surrounding medium is only 360 K, with the same temperature gradient and same central temperature of 1350 K, clearly indicates a sharp drop in maximum pressures and temperatures above a radius of 1.5 cm. This is also found for different, greater values of the temperature gradient. As a result, the detonation slows down, the reaction zone thickens and the velocity of the maximum temperature zone reduces compared to that of the pressure wave; this progressively leads to a pressure wave which moves away rapidly from the deflagration wave. It appears, therefore, that if the temperature of the medium is too low, then a stable detonation wave cannot be initiated in this way.

This problem of initiation of detonation resulting from a hot zone—or indeed from a pre-established deflagration—is not yet completely resolved, as stated in chapter 7. The exact conditions which favour detonation initiation are not known quantitatively. However, this chapter has hopefully removed the veil of mystery which surrounds the physical and chemical mechanisms, to reveal that the processes involved are essentially interactions between self-ignition and pressure waves.

CHAPTER 9

COMBUSTION OF LIQUIDS AND SPRAYS

9.1 GENERAL POINTS

In numerous practical applications, the fuel supplied to an engine or burnt in a furnace is a liquid. Petrol for internal combustion engines, diesel for Diesel engines, heavy fuel oil for industrial burners, and kerosene for turbo-jets are all liquid fuels. In terms of storage, liquids offer an important practical advantage in that the volume required for liquids is much less than that for gases. Rocket engines are powered by fuel/oxidant combinations which are either both liquids, for example dinitrogen peroxyde N_2O_4 (an oxidant) and UDMH (unsymetrical dimethylhydrazine) used in the first-stage engine of the Ariane rocket, or the oxidant alone is a liquid, such as the liquid oxygen used in cryotechnical oxygen-hydrogen engines where the hydrogen vaporises before being injected into the combustion chamber.

Even the simplest of all flames, a candle flame, is fed by liquid fuel. The candle wax melts to produce a liquid which is soaked up by the wick before vaporising and burning.

However, the candle demonstrates yet another aspect of the combustion process commonly-encountered in practical applications in that the wax does not burn in liquid form. Instead it is vaporised by the heat of the flame, usually decomposing to a certain extent in the process, and is then drawn towards the gaseous oxidant at which point combustion occurs. There are practically no examples, except perhaps liquid monopropellants, (in which the fuel and oxidant are combined in the same compound), where exothermic chemical reaction takes place in the liquid phase.

Consider, as an example, a small amount of alcohol heated in a spoon and then lit. Once combustion is established, a stable gaseous flame forms above the surface of the liquid. The heat produced by the flame acts in several ways. It heats up the surface of the liquid by conduction (and often by radiation) causing the liquid to vaporise, and then it heats these vapours as they approach the flame. In addition, the flame heats the air diffusing towards the flame, such that a reaction zone is sustained between the fuel and oxidant vapours in exactly the same way as a classical gas-phase diffusion flame. This situation is shown schematically in Figure 9.1.

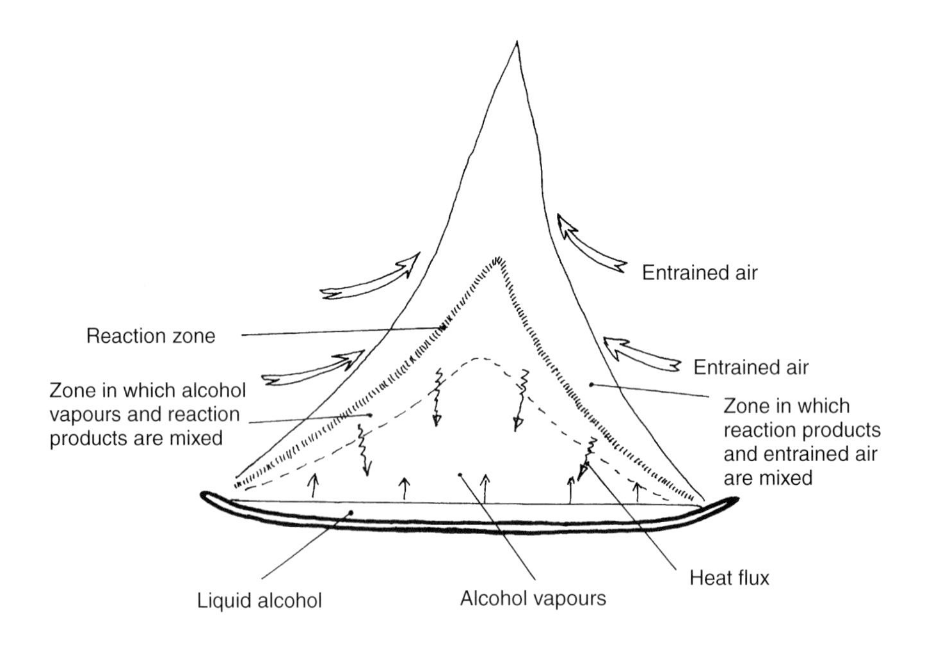

Figure 9.1 A simple experiment demonstrating liquid combustion (or, how to flambé brandy!)

Since the flame is self-sustaining close to the liquid surface, then in practical applications, the best way to increase the amount of "combustion per unit volume" is to maximise the air-liquid contact area. This is the reason why all burners which operate with liquid fuel inject the fuel into the combustion area in the form of a spray of droplets. The same principle is applied when two propellants (fuel and oxidant) are both injected as droplets into a combustion chamber. As the two sprays mix, the droplets of the least volatile product burn, with a flame close to their surface, in the vapours produced by the more volatile product.

The exception to this type of situation is encountered in fires (usually caused accidentally) involving fuel storage. The simple experiment shown in

Figure 9.1 is recreated on a larger scale when a layer of combustible liquid spreads out over the ground or across water or in a tank which then burns directly with a flame which extends over the surface of the fuel. The situation is a little more complicated, but the principle is the same as that for a burning droplet.

The combustion of a spherical drop of liquid fuel in a gaseous oxidising medium is therefore a typical situation of general interest. Numerous experiments have been carried out on this subject over the years using either a drop held by a very thin silicon fibre, a drop in freefall, or even a large porous sphere saturated with liquid. The situation was studied theoretically for the first time around forty years ago (Godsave[1] and Spalding[2]), building on theories developed at the turn of the century relating to droplet vaporisation (Stefan and Kelvin).

In this chapter, we will start by considering the combustion of a single drop, before moving on to a more complete description of sprays. We will discuss the combustion of a spray of droplets reacting with the gas in which they are distributed (fuel droplets in an oxidising gas, or vice-versa), in addition to the combustion of a jet of drops in atmospheric air. The first case is similar to a premixed flame and the second to a diffusion jet flame. Flames in sprays have been studied much less, both experimentally and theoretically, than flames burning purely in the gas phase, and for this reason we will concentrate on a discussion of the theory, since a great deal of progress remains to be made in relation to the experimental aspects in most cases.

9.2 COMBUSTION OF AN ISOLATED DROP

We have chosen to consider the combustion of a drop of fuel in an oxidising atmosphere, although clearly the reverse situation could equally well have been taken as an example. The first case to be considered will be the idealised one of a spherical drop of fuel in a gas at rest and unaffected by gravitational forces, such that the flame geometry remains spherical. This case will then be extended to that of a drop moving with respect to the surrounding gas, since this is closer to a real situation.

1. Godsave G.A.E. (1953) *4th Symposium on combustion*, The Combustion Institute, 818-830.
2. Spalding D.B. (1953) *4th Symposium on combustion*, The Combustion Institute, 847-864.

9.2.1 Quasi-stationary theory of the combustion of a spherical drop

The simplest case to study is not in fact that of a drop whose diameter reduces as it burns, but rather that of a drop which is fed with fuel at its centre by a very narrow tube, such that its diameter remains constant. With this arrangement, the problem is perfectly stationary, and if the drop is assumed to be spherical then the flame and the drop can be described using just one spatial variable, namely distance from the centre of the drop.

Assume that this problem had been solved and that a flow rate of liquid, $\dot{m}$, had been determined which was sufficient to maintain a constant drop diameter. It could then be assumed that this flow rate corresponds exactly to the reduction in the drop radius which would be observed if the supply was stopped. The radius r_{drop} of the fuel-deprived drop could then be calculated directly as : $\dot{m} = -\,\mathrm{d}(4\pi\rho_L r^3_{\text{drop}}/3)/\mathrm{d}t)$, where ρ_L is the density of the liquid. Thus:

$$\frac{\mathrm{d}}{\mathrm{d}t}(r_{\text{drop}}) = -\frac{\dot{m}}{4\pi r^2_{\text{drop}}\rho_L} \tag{9.1}$$

This is true only if the rate of reduction of radius is very small, such that the drop passes through a series of quasi-stationary states as it shrinks. Let us assume this to be true for our theoretical approach, and compare the result with experiment at the end.

The theoretical representation and solution of this problem are not particularly difficult provided that the following simplifying assumptions are made :

- the pressure of the gases surrounding the drop is constant and their velocity low,
- the drop is at uniform temperature, including its surface,
- the c_p for the gas is constant irrespective of its composition and temperature,
- an infinitely rapid chemical reaction of the type $K + \nu Ox \rightarrow P$ occurs, i.e. the fuel K evolving off the drop and the oxidant Ox cannot co-exist,
- the coefficient of molecular diffusion for Ox and K and the thermal diffusivity coefficient are all equal to D, and ρD can be assumed to be a constant.

Let us turn our attention to the gas surrounding the drop, and write the balance aerothermochemical equations of mass and energy for a spherical shell enclosing the drop at a distance r from the centre. Figure 9.2a shows the expected profiles of Y_{Ox}, Y_{K}, and T as a function of r. Since the reaction is very fast, Y_{Ox} is zero from the surface of the drop to the reaction surface, and Y_{K} is zero at all points on the outside of this reaction surface; in exactly the same way as for the diffusion flame described in chapter 5.

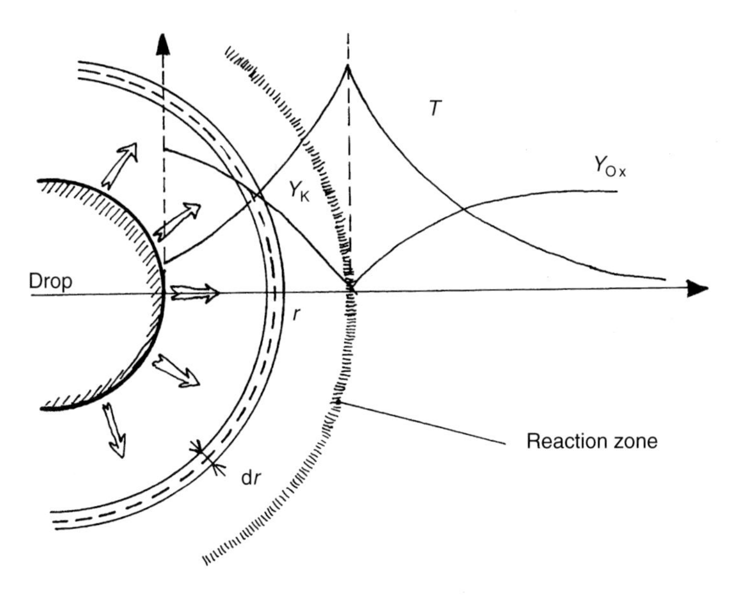

Figure 9.2a Schematic representation of the spherical flame surrounding a drop of liquid fuel.

The following equations are obtained:

- total mass balance: $\frac{d}{dt}(4\pi r^2 \rho v) = 0$ (v is the radial gas velocity)
- mass balance for Ox:

$$\frac{d}{dr}(4\pi r^2 \rho v Y_{Ox}) = \frac{d}{dr}\left(4\pi\rho D r^2 \frac{dY_{Ox}}{dr}\right) - \nu w M_{Ox} \,.\, 4\pi r^2$$

- mass balance for K:

$$\frac{d}{dr}(4\pi r^2 \rho v Y_K) = \frac{d}{dr}\left(4\pi\rho D r^2 \frac{dY_K}{dr}\right) - w M_K \,.\, 4\pi r^2$$

- energy balance:

$$\frac{d}{dr}(4\pi r^2 \rho v T) = \frac{d}{dr}\left(4\pi\rho D r^2 \frac{dT}{dr}\right) + \left(\frac{q}{c_p}\right) .\, w \,.\, 4\pi r^2 \qquad (9.2)$$

w is the molar rate of reaction per unit volume, as described for a gaseous diffusion flame (see chapter 5 and the discussions therein concerning q and w). M_{Ox} and M_K are the molar masses of Ox and K respectively.

Since the single chemical reaction occurs infinitely rapidly, a new variable Z can be defined: $Z = Y_{Ox}/\nu M_{Ox} - Y_K/M_K$, which will satisfy an equation containing no reaction term. Also, since Y_{Ox} and Y_K cannot simultaneously be non-zero, Z will represent $Y_{Ox}/\nu M_{Ox}$ if positive and $-Y_K/M_K$ if negative.

Hence the equations for Y_{Ox} and Y_K can be replaced by an equation for Z, namely:

$$\frac{d}{dr}(4\pi r^2 \rho v Z) = \frac{d}{dr}\left(4\pi\rho D r^2 \frac{dZ}{dr}\right)$$

i.e. on setting $\dot{m} = 4\pi r^2 \rho v$, $\quad \dot{m}Z - 4\pi r^2 \rho D \dfrac{dZ}{dr} = \text{a constant}$ (9.3)

Equally, it can easily be shown that $Z_T = c_p T/q + Y_{Ox}/\nu M_{Ox}$ satisfies the same equation as Z:

$$\dot{m}Z_T - 4\pi r^2 \rho D \frac{dZ_T}{dr} = \text{a constant} \tag{9.3'}$$

Advancing the solution further requires introducing boundary conditions, although they are not immediately apparent.

The easiest boundary conditions to obtain are those at infinity, far from the droplet, since the values of Y_{Ox} and the temperature are known. These will be denoted $Y_{Ox,\infty}$ and T_∞, and are the values for the pure air in which the gas is burning. At the surface of the drop the conditions of interest do not include the values of T or Y_K (referred to as T_{drop} and $Y_{K,drop}$), but instead the mass and heat fluxes. Indeed, the heat flux serves to vaporise the liquid (and only this since the drop is at uniform temperature and the liquid thus arrives heated to its surface temperature and no more):

$$4\pi\rho D r_{drop}^2 \frac{dT}{dr} = +\frac{\dot{m}L_v}{c_p} \tag{9.4}$$

where L_v is the latent heat of vaporisation per unit mass, i.e. Δh of vaporisation.

On the other hand, the total flux (through convection and diffusion) of the fuel vapour which evolves from the surface into the gas phase is $\dot{m}$, equal to the flow rate of liquid fuel fed continuously into the centre of the drop:

$$\dot{m}Y_{K,\,drop} - 4\pi\rho D r_{drop}^2 \left(\frac{dY_K}{dr}\right)_{drop} = \dot{m} \tag{9.5}$$

These conditions allow the constants in equations (9.3) and (9.3′) to be fixed, giving:

$$\dot{m}Z - 4\pi r^2 \rho D \frac{dZ}{dr} = -\frac{\dot{m}}{M_K} \tag{9.6}$$

and:

$$\dot{m}Z_T - 4\pi r^2 \rho D \frac{dZ_T}{dr} = -\dot{m}\frac{c_p T_{drop}}{q} - \frac{\dot{m}L_v}{q} \tag{9.6'}$$

Equations (9.6) and (9.6′) may now be integrated to find Z and Z_T as a function of r. For (9.6), for example, we obtain:

$$\dot{m}\left(Z + \frac{1}{M_K}\right) = 4\pi r^2 \rho D \frac{dZ}{dr}$$

Whence:

$$\frac{dZ}{\left(Z + \dfrac{1}{M_K}\right)} = \frac{\dot{m}}{4\pi\rho D}\frac{dr}{r^2}$$

which integrates to give:

$$Z = -\frac{1}{M_K} + \left(\frac{Y_{Ox,\infty}}{\nu M_{Ox}} + \frac{1}{M_K}\right)\exp\left(-\frac{\dot{m}}{4\pi\rho D r}\right) \tag{9.7}$$

Using the boundary condition at infinity:

$$Y_{K,\infty} = 0 \quad \text{and so} \quad Z(\infty) = \frac{Y_{Ox,\infty}}{\nu M_{Ox}}$$

Similarly, we can obtain:

$$Z_T = \frac{c_p T_{drop}}{q} - \frac{L_v}{q} + \left(\frac{c_p(T_\infty - T_{drop})}{q} + \frac{Y_{Ox,\infty}}{\nu M_{Ox}} + \frac{L_v}{q}\right)\exp\left(-\frac{\dot{m}}{4\pi\rho D r}\right) \tag{9.7'}$$

These equations yield the profiles for Y_{Ox}, Y_K and T as functions of r, surrounding the drop, as functions of T_{drop}, the surface temperature, and of $\dot{m}$. However these last two quantities are not known.

However, we have not made full use of all the physical conditions that we know. At the surface of the drop, the process of vaporisation implies some form of a relationship between T_{drop}, $Y_{K,drop}$ and the pressure around the drop, since the partial pressure of K is equal to the saturated vapour pressure at the temperature T_{drop}. If $Y_{K,drop}$ were equal to one, i.e. if no combustion product could diffuse to the surface of the drop, this condition could be written simply as $T_{drop} = T_{vap}(p)$, the boiling temperature of the drop at the pressure in question. In reality, $Y_{K,drop}$ is slightly less than unity and thus T_{drop} is slightly less than $T_{vap}(p)$. Nevertheless, if the equation for vaporisation equilibrium is written:

$$p_s(T_{\text{drop}}) = p \frac{\dfrac{Y_{\text{K,drop}}}{M_{\text{K}}}}{\dfrac{Y_{\text{K,drop}}}{M_{\text{K}}} + \dfrac{(1 - Y_{\text{K,drop}})}{M_{\text{P}}}} \tag{9.8}$$

where the ratio on the right-hand side corresponds to the mole fraction of the fuel vapour at the surface, then the additional equation is obtained which enables $Y_{\text{K,drop}}$ and T_{drop} to be found, since the function $p_s(T)$ is known.

We now have all the elements required for the calculation of $\dot{m}$, an unknown for the system. Moreover, the two first-order differential equations (9.3) and (9.3′) are linked at three boundary conditions, namely the values of Z_T and Z at infinity, and the expression (9.8). Thus, there is one equation too many which is why $\dot{m}$ can be determined, expressed either as a function of $Y_{\text{K,drop}}$ or as a function of T_{drop}. In the former case, $\dot{m}$ is calculated from (9.7) expressed at $r = r_{\text{drop}}$, which gives:

$$\frac{\dot{m}}{4\pi\rho D_{\text{drop}} r_{\text{drop}}} = \ln \left(\frac{\dfrac{1}{M_{\text{K}}} + \dfrac{Y_{\text{Ox},\infty}}{\nu M_{\text{Ox}}}}{\dfrac{1}{M_{\text{K}}} - \dfrac{Y_{\text{K,drop}}}{M_{\text{K}}}} \right) \tag{9.9}$$

and in the latter case, equation (9.7′) is used, which may be written as applying to the surface of the drop, yielding:

$$\frac{\dot{m}}{4\pi\rho D_{\text{drop}} r_{\text{drop}}} = \ln \left(\frac{c_p(T_\infty - T_{\text{drop}}) + \dfrac{Y_{\text{Ox},\infty} q}{\nu M_{\text{Ox}}}}{L_v} + 1 \right) \tag{9.9′}$$

In both expressions, T_{drop} and $Y_{\text{K,drop}}$ can be found independently of $\dot{m}$ by using equation (9.8) and by equating (9.9) to (9.9′).

Based on expressions (9.7) and (9.7′), $Y_{\text{Ox}}(r)$, $Y_{\text{K}}(r)$ and $T(r)$ can be plotted. Indeed if a transformed variable z is defined such that $z = \exp(-\dot{m}/4\pi\rho D r)$, which is equal to 1 at $r = \infty$, and equal to $z_{\text{drop}} = \exp(-\dot{m}/4\pi\rho_{\text{drop}} D_{\text{drop}} r_{\text{drop}})$ at $r = r_{\text{drop}}$ (where D_{drop} is the value of D at the surface of the drop), then the Y_{Ox}, Y_{K} and T profiles consist of a series of straight lines as a function of z, and are plotted in Figure 9.2b.

The radius of the reaction surface, r_F, is easily determined, since it is the point where $Y_{\text{K}} = Y_{\text{Ox}} = 0$ and thus where $Z = 0$, giving:

$$\ln \left(\frac{\dfrac{Y_{\text{Ox},\infty}}{\nu M_{\text{Ox}}} + \dfrac{1}{M_{\text{K}}}}{\dfrac{1}{M_{\text{K}}}} \right) = \frac{\dot{m}}{4\pi\rho_{\text{drop}} D_{\text{drop}} r_F}$$

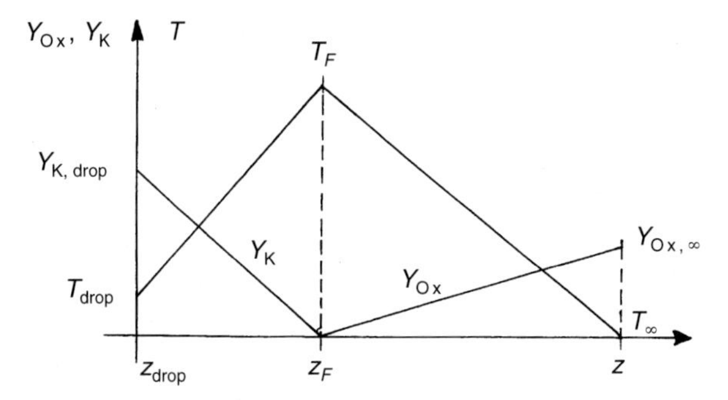

Figure 9.2b Temperature, Y_{Ox} and Y_K profiles as a function of the reduced coordinate z, around the drop.

and, by taking into account (9.9′), the ratio r_F/r_{drop} is obtained independently of $\dot{m}$:

$$\frac{r_F}{r_{drop}} = \frac{\ln\left(1 + \dfrac{c_p(T_\infty - T_{drop}) + \dfrac{Y_{Ox,\infty}q}{\nu M_{Ox}}}{L_v}\right)}{\ln\left(1 + \dfrac{Y_{Ox,\infty}M_K}{\nu M_{Ox}}\right)} \tag{9.10}$$

The maximum temperature, T_F, at the reaction surface is deduced using (9.7′) and (9.7) with $Z = 0$ and $Z_T = c_pT_F/q$:

$$T_F = T_{drop} - \frac{L_v}{c_p} + \frac{T_\infty - T_{drop} + \dfrac{L_v}{c_p} + Y_{Ox,\infty}\dfrac{q}{c_p\nu M_{Ox}}}{1 + \dfrac{Y_{Ox,\infty}M_K}{\nu M_{Ox}}}$$

$$= \frac{T_\infty + \left(T_{drop} - \dfrac{L_v}{c_p} + \dfrac{q}{c_p}\right)\dfrac{Y_{Ox,\infty}}{\nu M_{Ox}}}{1 + \dfrac{Y_{Ox,\infty}M_K}{\nu M_{Ox}}} \tag{9.11}$$

9.2.2 The d^2 rule

Finally, from equations (9.1) and (9.9), it is possible to deduce the reduction in the diameter of the drop by assuming that T_{drop} is known:

$$\frac{dr_{drop}}{dt} = -\frac{\dot{m}}{4\pi r_{drop}^2\rho_L} = -\frac{\rho_{drop}}{\rho_L}\frac{D_{drop}}{r_{drop}}\ln(1 + B)$$

where $B = (c_p(T_\infty - T_{\text{drop}}) + Y_{\text{Ox},\infty}q/\nu M_{\text{Ox}})/L_v$ is known as the "transfer parameter". Since r_{drop} appears in the denominator position of the right-hand term, then the derivative of the square of the radius is a constant. If the drop diameter is denoted by d_{drop}, the so-called "d^2" law is obtained:

$$d_{\text{drop}}^2 = d_{\text{drop}}^2(t=0) - Kt, \quad \text{where} \quad K = 8D_{\text{drop}}\frac{\rho_{\text{drop}}}{\rho_L}\ln(1+B) \quad (9.12)$$

which predicts a linear reduction in the square of the diameter d_{drop} as a function of time.

The plots in Figures 9.3 show the results of experiments performed by S. Okajima and S. Kumagai for two types of drop.

The linear drop in d_{drop}^2 predicts reasonably well what actually occurs. In fact, the rule is not valid initially because the drop must be heated to the uniform temperature T_{drop} before it can be said that all the heat it receives serves to heat it. The drop's rate of evaporation is obviously much reduced during this period. The rule also does not hold at the end of the drop's lifetime when its radius is very small, since $\mathrm{d}r_{\text{drop}}/\mathrm{d}t$ is very large at this point and the quasi-steady-state approximation can no longer be applied.

However, the numerical values obtained from (9.12) as well as from (9.10) and (9.11) are not obviously in agreement with experiment. The shortcoming of the numerical calculation is the fact that ρD was assumed to be constant, as was c_p, and it is difficult to decide what their approximate values should be since ultimately they depend largely on the temperature and composition of the gases, which vary around the drop.

Equally, note that the ratio r_F/r_{drop}, given by equation (9.10), and assumed to be constant in the context of the quasi-steady-state theory, varies greatly as the drop burns. This is apparent from Figure 9.3 which also plots d_F, the flame diameter. The latter does not decrease in anything like the same way as the diameter of the drop; even increasing initially as the flame establishes around the drop! This is well illustrated by the example of a drop of heptane, for which theory suggests that $r_F/r_{\text{drop}} = 15$ (approximately); in fact this ratio varies between 7 and 17 during combustion, demonstrating that the quasi-steady-state approximation is indeed only an approximation and that it is more valid for determining the drop radius than for other quantities.

All of these non-steady-state effects may be calculated and indeed have been by Sirignano and co-workers[3], although their study goes beyond the scope of this book.

3. Sirignano W.A. et al. (1983) *Progress in Energy and Combustion Sciences*, 9, 291–322.

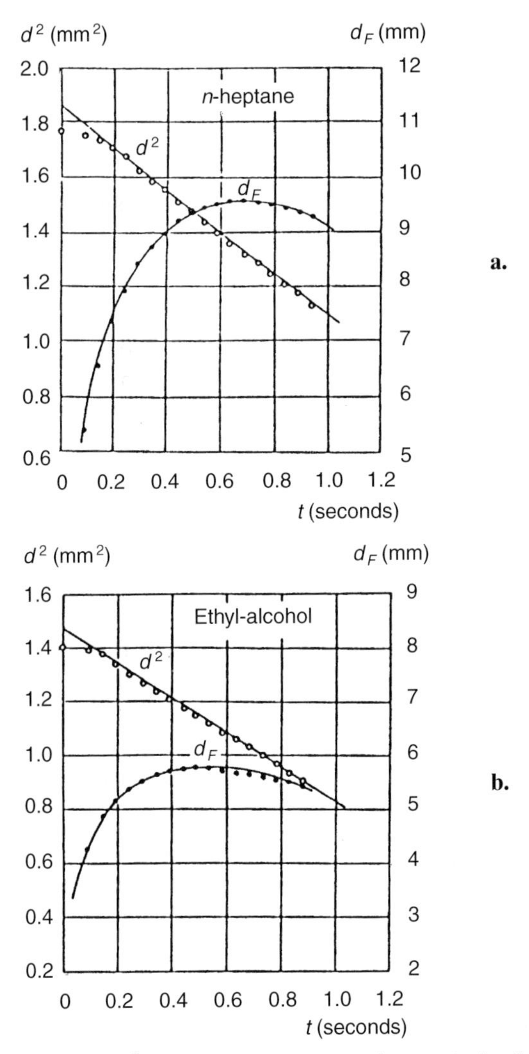

Figure 9.3 Experimental "d^2" laws for different drops. d_F is the flame diameter (After Okajima S. and Kumagai S. (1978) *15th Symposium (Int.) on Combustion*. The Combustion Institute. 402–407). The "d^2" rule clearly holds. unless the quasi-steady-state approximation concerning the ratio d_F/d_{drop} is not valid.

9.2.3 Combustion of a drop in motion

When a drop moves with respect to the surrounding gas, the spherical symmetry of the flow around it is broken. The flame adopts an elliptical form around the drop, as illustrated in Figure 9.4a.

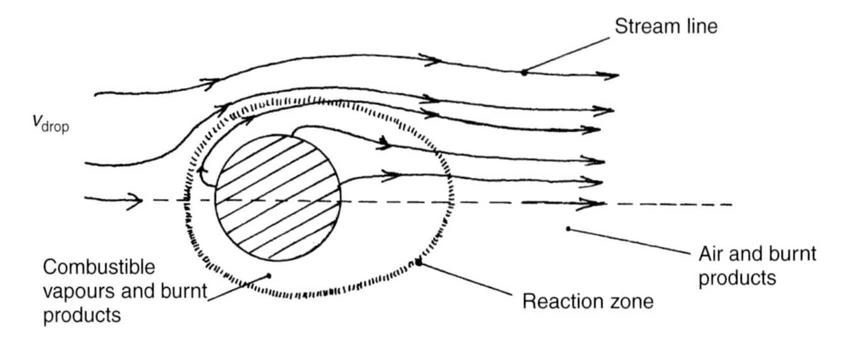

Figure 9.4a. The reaction surface surrounding a drop which is in motion with respect to the air.

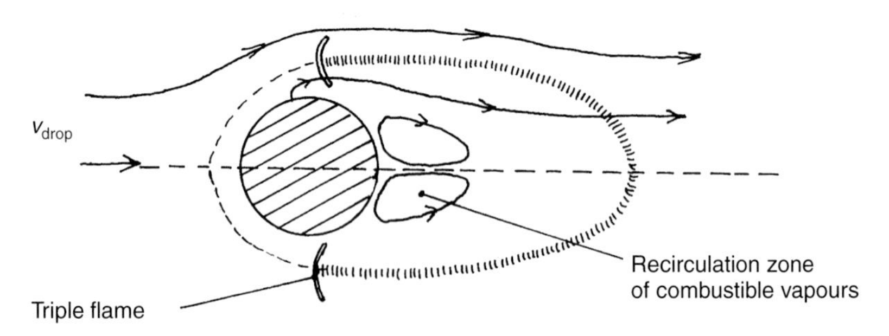

Figure 9.4b Flame extinguished at the stagnation point (facing the flow). A triple flame (see chapter 4) develops at the point where the flame is held, all around the drop. A toric recirculation zone forms behind the drop.

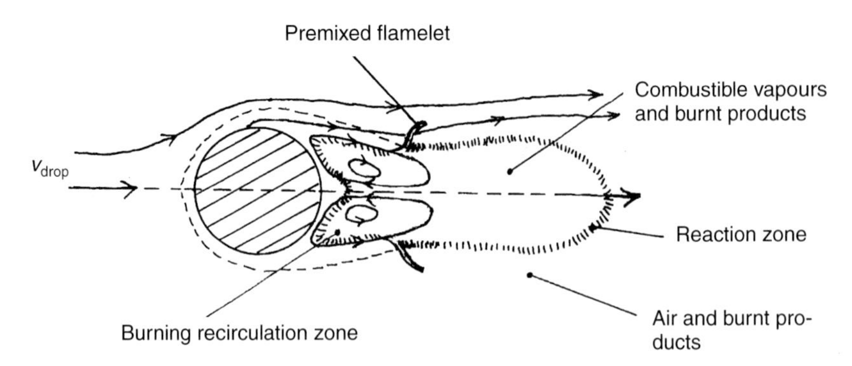

Figure 9.4c At a higher velocity, the flame may establish in the wake of the drop. It is stabilised by the recirculation behind the drop, which consists of a fuel-rich mixture, and on which a small premixed flame is held, followed by a diffusion flame.

A new theoretical treatment is required for this case, which assumes that the velocity of the drop is low. In fact, a Peclet number is defined for the drop compared with the gas: $Pe_{drop} = \rho_{drop} r_{drop} v_{drop} / D_{drop}$, and when Pe_{drop} is not too large the result is that the "d^2" law is still valid, but with a different coefficient :

$$d^2_{drop} = d^2_{drop}\,(t = 0) - K't$$

with: (9.13)

$$K' = K\,(1 + 0.3\,Re^{1/2}_{drop}\,Pr^{1/3})$$

Experiments have shown that the domain within which this equation is valid may extend to Peclet number values Pe_{drop} of the order of one hundred, a result not predicted by the theory. Re_{drop} is a Reynolds number for a drop, and Pr the Prandtl number ($Re_{drop} = \rho_{drop} v_{drop} r_{drop} / \mu_{drop}$, $Pr = \mu_{drop} c_{p,\,drop} / \lambda_{drop}$ equal to μ_{drop} / D_{drop} in this case).

However, even if the equation still holds, it only relates to the integral of the vaporisation around the drop. Clearly in this case the vapour flow rate per unit surface area, which leaves the drop at each point on it, depends on the position of this point. Vaporisation is greatest on the "leading side" of the drop than on the "trailing side"

As the drop's velocity in relation to the air increases further, it may reach a velocity at which the flame extinguishes on the leading side. It is then stabilised at a certain distance from the stagnation point, and continues to burn through and into the wake of the drop. In this case, the flame structure is no longer that of a pure diffusion flame (in which a reaction zone separates a gas rich in fuel from a gas rich in oxidant). In a more extreme case, at an even higher relative velocity, the flame burns in the wake of the drop only, and a premixed flame may form slightly ahead of the diffusion flame. These possibilities are shown in Figures 9.4b and 9.4c. Above a very high velocity, the flame is blown out, which can also be expressed by writing that if the Damkölher number of the drop ($d_{drop} / v_{drop} \tau_c$, where τ_c is a characteristic chemical time associated with propellants combustion) is too small then combustion cannot be sustained.

9.3 COMBUSTION OF A SPRAY OF DROPLETS

9.3.1 General points

The droplets forming a spray may burn in two different ways. The first case is analogous to that of a premixed flame, where the droplets (selected here arbitrarily as being the fuel) are dispersed in a gaseous medium (the oxidant) such as air, to form a premixed medium which is homogeneous on a macroscopic scale, although clearly heterogeneous when considered on a scale close to the size of an individual droplet. In a spray such as this, theoretically extending over an infinite volume, it is conceivable that a flame zone could propagate from droplet to droplet following its initiation at a certain point or across a plane. This type of situation is frequently encountered in practice, and many accidents have been caused by sprays of droplets igniting, or by the analogous case of suspensions of solid particles in air (e.g. coal dust or even flour). Let us try to investigate the structure of the flame zone, its burning velocity and thickness, and the conditions under which propagation is possible. Note, however, that complete solutions to the problems that we shall consider are not available at the current time.

We shall start by considering this case of a spray burning in an oxidising gas, before moving on to a second case, which is more similar to a diffusion flame, established between a flow of fuel droplets surrounded by an inert gas or fuel gas on one side, and an oxidising gas on the other.

9.3.2 Combustion in premixed sprays: the various types of flame

When studying a "flame" propagating in a premixed spray, the first point to consider is whether the drops are small enough and volatile enough to completely vaporise before the reaction zone present in the flame reaches them. If this is the case, then the situation being studied is that shown schematically in Figure 9.5. The flame consists of a preheating zone within which the drops are heated and completely vaporised, and a reaction zone when the gases (a blend of fuel and oxidant) are heated further to the point of reaction, and which is located at the rear of the flame. Thus, except for one minor modification, the picture is that of a premixed type flame in the gas phase.

This type of flame is encountered when the time required to vaporise the drops is less than the time required for the premixed flame to cross the diame-

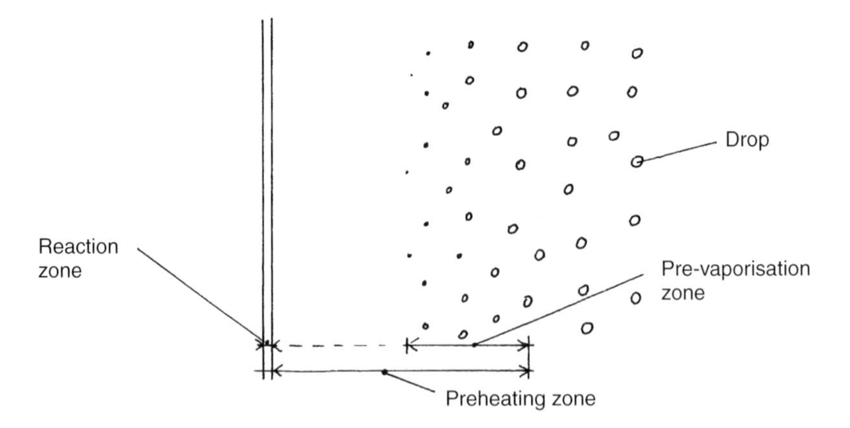

Figure 9.5 Schematic representation of a pre-vaporisation flame.

ter of the drop. Thus, as a rough approximation, $d_{\text{drop}}^2/K)/(d_{\text{drop}}/S_L)$ must be less than a certain value, where K is the constant in the "d^2 law" and S_L the burning velocity of a premixed laminar flame burning the vapours of the propellant combination being studied. Hence, the diameter of the droplets, d_{drop}, must be small compared with K/S_L. S_L is proportional to $(D/\tau_c)^{1/2}$ (where D is the diffusivity in the gas surrounding the drop and τ_c a time characteristic of the chemistry; see chapter 5) and K is proportional to D. Then, K/S_L is proportional to $(D/\tau_c)^{1/2}$, i.e. to the thickness of the premixed laminar flame propagating at a velocity S_L. Hence, the result is that the drops must be fairly small relative to the characteristic thickness e_L of the corresponding premixed laminar flame. The exact value of the ratio $\mathrm{d}_{\text{drop}}/e_L$ is difficult to determine through a dimensional reasoning of the type which has just been discussed, although clearly this value may become greater still as $\rho_{\text{drop}}/\rho_L \ln(1 + B)$ increases because this factor directly affects K. In other words, if the pressure is higher, $\rho_{\text{drop}}/\rho_L$ will be greater at higher gas pressures, so the drops can be larger and, if the liquid is very volatile or releases large amounts of heat on burning in air, then B will be higher and again the drops can be larger. In practice, with hydrocarbon droplets, a diameter of less than 10 microns is required for the flame to have the structure described in Figure 9.5.

When the droplets diameter is larger than this value they cannot burn quickly enough to prevent a diffusion flamelet from surrounding them while they vaporise. However, in determining the exact structure of the flame, it is important to consider whether or not the spray is sufficiently diluted to allow each drop to be surrounded by a diffusion-type "flamelet", i.e. can the droplet burn alone or in a group with other droplets? The key parameter in answering this question is the ratio of the average distance between two drops in the spray to the radius which the flamelet surrounding each drop would have if the drops were far enough apart to burn independently (and which was termed r_F in section 9.2).

The importance of this parameter was demonstrated for the first time by Kerstein and Law[4], in the form $n^{1/3}r_F$, where n is the number of drops per unit volume (and thus $n^{-1/3}$ is the average spacing between drops). If the arrangement of the drops in the spray is assumed to be uniform, only two possibilities could result, depending on whether the value of this parameter were greater or less than a certain critical value, close to unity. The drop distribution may, however, be random and a theory specially adapted to this situation (the "percolation" theory) can be applied to show that three categories of spray must be considered:

a) If $n^{1/3}r_F < 0.41$, the spray is not very dense and although certain drops are surrounded by a flamelet, there is still a certain probability that flamelets

---

4. Kerstein A. and Law C.K. (1982) *19th Symposium (Int.) on Combustion*, The Combustion Institute, 961-969.

will surround groups consisting of two, three or even more drops. In this case, "group combustion" is said to be occurring. Of course, as $n^{1/3}r_F$ tends to zero then every drop can be assumed to burn with its own "individual" flame.

b) If $n^{1/3}r_F > 0.73$, the spray is very dense, to the extent that rather than diffusion flamelets surrounding groups of drops it is the groups of drops which surround the flamelets! The flamelets surround pockets of oxidising gas which mix progressively with the combustion products as these products are formed.

c) Between these two extremes, i.e. for $0.41 < n^{1/3}r_F < 0.73$, both situations may be present and hence a diffusion flamelet of infinite length may be formed, which surrounds neither a group of drops nor a pocket of gas. This regime is referred to as the "percolation combustion" regime.

These three types of combustion in a premixed spray are shown in Figures 9.6a, b, and c.

Consider an infinite, isotropic spray in the process of burning, i.e. a spray which has been ignited at a number of random points dispersed throughout its volume rather than a spray in which a "flame" propagates in a specific direction. As time passes, the three combustion regimes identified above succeed one another.

If the spray is initially dense enough such that $n^{1/3}r_F > 0.73$, then the first regime to establish will be characterised by the presence of gas pockets (Fig. 9.6b), followed by the percolation combustion regime (Fig. 9.6c) since as the drop diameter decreases then so does r_F. Finally, groups of drops surrounded by diffusion flamelets will be visible (Fig. 9.6a), but only if the spray is sufficiently fuel lean to allow all the fuel drops to burn. In fact, if the spray is fuel-rich, then combustion will cease, leaving behind drops which are partially vaporised and only partially burnt (and which will probably then vaporise in the burnt gases away from the actual flame zone) and hence it is in a percolation combustion regime that combustion will finish.

If, on the other hand, a distinct flame propagates through the spray, the three types of structure will be formed within the flame front from upstream to downstream (or only two if the mixture is fuel-rich). This is shown, for example, in Figures 9.7a and b, for the case of a spray which is fuel lean and initially dilute or dense, respectively.

Note that the overall equivalence ratio of the spray may be independent of the parameter $n^{1/3}r_F$ and, indeed, even if $n^{1/3}r_F$ is large (when there are a large number of drops per unit volume) the spray may be fuel-lean overall, if the density of the gaseous oxidant is high, when the pressure is quite high.

To conclude this description of flame propagation in a spray, particular mention must be made of a specific feature of sprays, as shown schematically in Figure 9.7. At the very front of the flame zone, the drops which enter the zone must be ignited since they are heated by the high temperatures which exist at the centre and back of the flame. The drop is not ignited uniformly

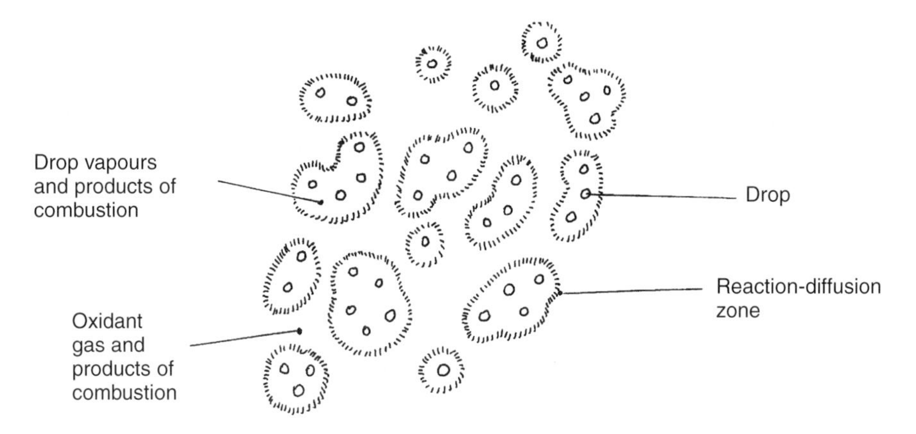

Figure 9.6a Group combustion.

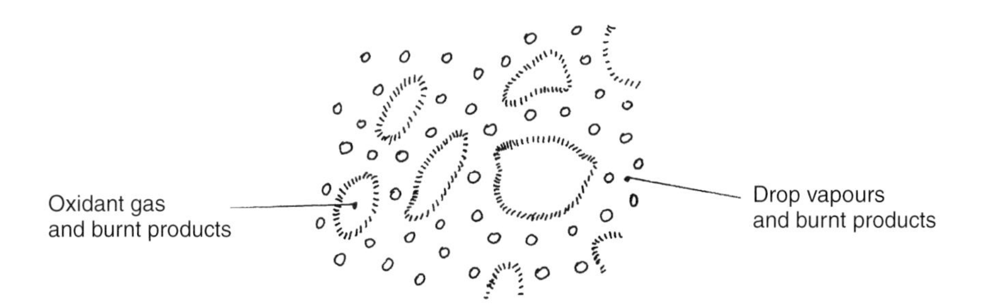

Figure 9.6b Combustion in pockets.

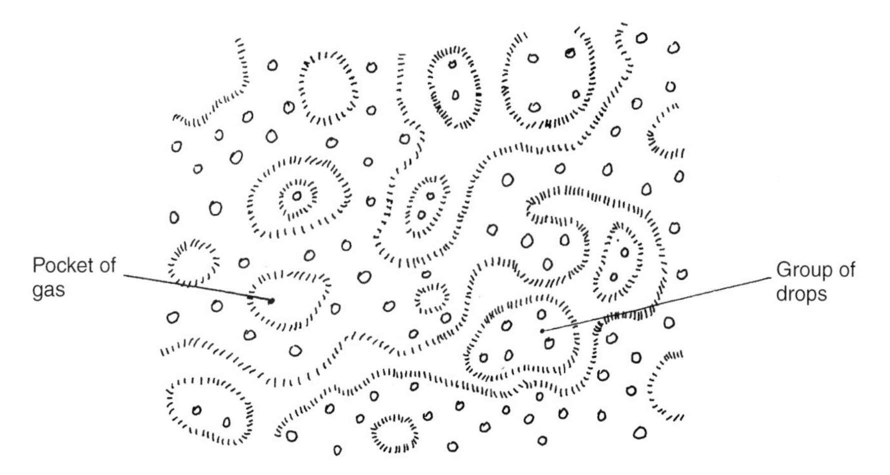

Figure 9.6c Percolation combustion.

over its entire spherical surface but preferentially at the side facing the oncoming flame zone. When a fairly large drop is ignited under these conditions, it is possible to observe the propagation of a "triple flamelet", very similar to that stabilised in the attachment zone of a diffusion flame lifted off a burner, or to that which forms near a drop in quite a strong air flow (see Figure 9.4b). This situation is shown schematically in Figure 9.7 for various drops as they ignite. The diffusion flamelets surrounding the drops, and lengthened by the

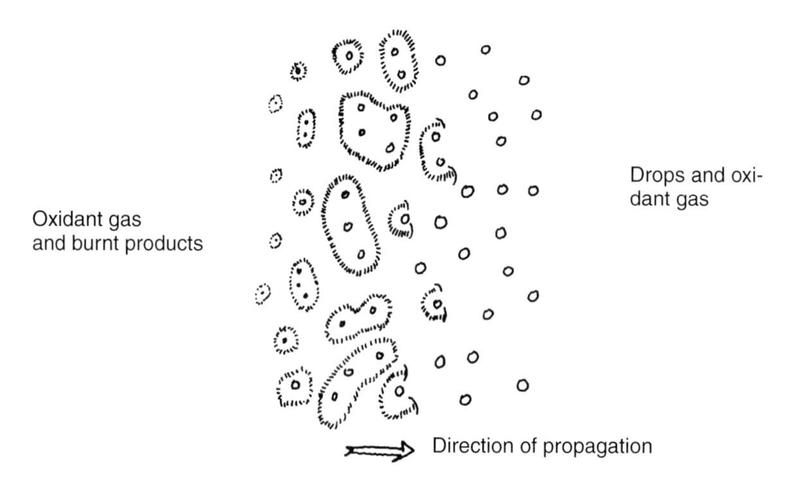

Figure 9.7a Flame propagation in a dilute spray in fuel-lean conditions.

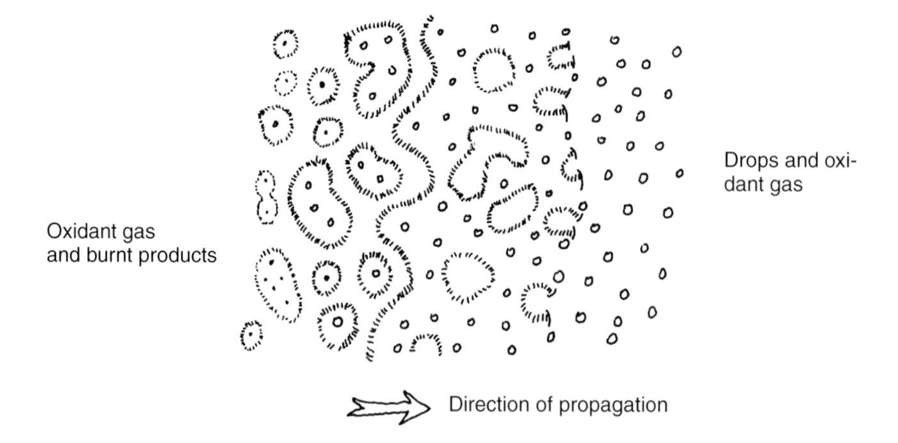

Figure 9.7b Flame propagation in a dense spray in fuel-lean conditions.

premixed flames, are shown which, if the spray is dense enough, can extend from one drop to the next.

Finally, the preceding discussions can be summarised by the diagram shown in Figure 9.8, as a function of the two dimensionless parameters d_{drop}/e_L and $n^{1/3}r_F$ (remembering that r_F is proportional to d_{drop}).

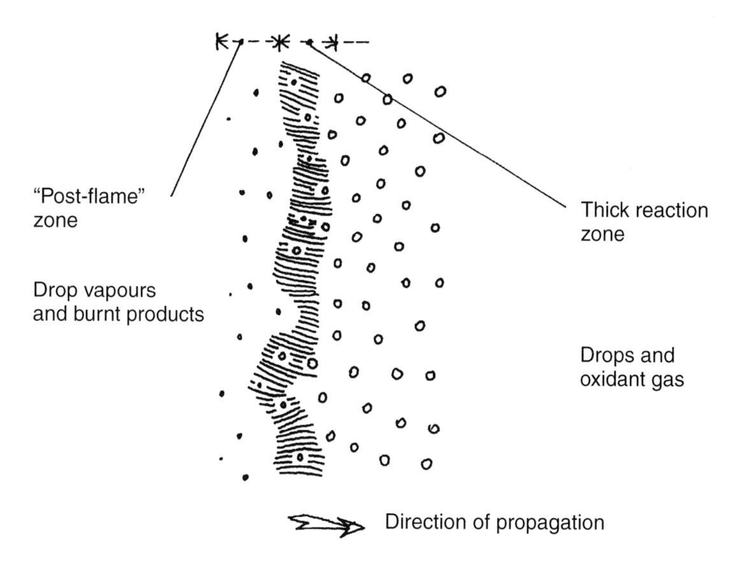

Figure 9.7c Propagation of a "distributed combustion" flame in a spray rich in drops.

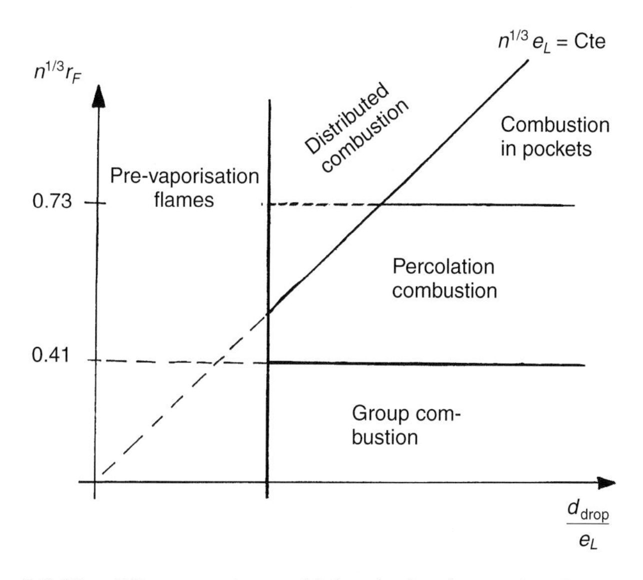

Figure 9.8 The different regimes which exist for the combustion of a spray.

In fact, drop pre-vaporisation does not really depend on whether there are many drops or not. However, once the drops have vaporised, a more homogeneous gas-phase medium results (and one which is richer in fuel) if there had been a large number of drops than if there had not. In addition, the flame structure in the gas phase is only likely to resemble a perfectly premixed flame at the top of the diagram. For low values of $n^{1/3}r_F$, local fluctuations in equivalence ratio will probably be large enough to disrupt the reaction zone. Small diffusion flamelets may even be observed if these fluctuations lead to the formation of localised gas pockets in which the equivalence ratio is less than unity at certain points and greater than unity elsewhere. This also depends on the overall equivalence ratio of the spray.

The assumption that the chemical time scale τ_c is sufficiently low and the global activation temperature for combustion is quite high has been made implicitly throughout the preceding discussion. In fact, it is on the basis of this condition alone that the reaction zones of the diffusion flamelets, of the pre-vaporisation flame and of triple flamelets, can be taken to be thin. This condition is also required to enable r_F to be defined accurately and hence for the definition of the three regimes relating to sprays composed of large drops.

If this were not the case, we would have to allow for the characteristic thickness of these reaction zones. In cases where the thickness is too large compared to the separation between drops, the burning spray would become a medium containing reactions distributed homogeneously in space, with no small-scale structure other than the random presence of drops. Let this thickness be characterised by e_L, the premixed flame thickness for the same propellants (which depends on the chemistry through τ_c, on the diffusivity of the gases and of the heat). In this case, the condition that $n^{1/3}e_L$ is greater than a critical value corresponds to a straight line passing through the origin in the plane $(n^{1/3}r_F,\ d_{\text{drop}}/e_L)$, since $n^{1/3}e_L = n^{1/3}r_F \cdot e_L/\beta d_{\text{drop}}$, where β is the ratio r_F/d_{drop} which can be found from equation (9.10). Thus, in this case, the upper part of the diagram in Figure 9.8 would correspond to this combustion which could be called "distributed". A schematic representation of the structure of such a flame is given in Figure 9.7c.

9.3.3 Rate of combustion and flame velocity in premixed sprays

Consider a premixed spray consisting of drops which are large enough not to vaporise completely before burning, and where there are very few drops. In this case, each drop will burn surrounded by a diffusion flamelet. This is the limiting case for the spray shown in Figure 9.6b. Moreover, if the drops are all of the same size, it is then easy to calculate the rate of combustion of the drops per unit volume of spray, based on the information deduced about the rate of combustion of a single drop.

The rate per unit volume w_G is, by definition, the mass of the drops consumed per unit time per unit volume of spray, and is therefore given by $w_G = n\mathrm{d}(4\pi\rho_L r^3_{\text{drop}}/3)/\mathrm{d}t$, where n is the number of drops per unit volume, ρ_L the density of the liquid forming the drops, and r_{drop} the radius of a drop:

$$w_G = 4\pi\rho_L n r^2_{\text{drop}} \frac{\mathrm{d}r_{\text{drop}}}{\mathrm{d}t} = 2\pi\rho_L n r_{\text{drop}} \frac{\mathrm{d}r^2_{\text{drop}}}{\mathrm{d}t} = \frac{\pi\rho_L}{2} r_{\text{drop}}(-K)n$$

where K is the constant in the d^2 rule given by equation (9.12).

With $\rho_G = 4\pi r^3_{\text{drop}}\rho_L n/3$, where ρ_G is the mass per unit volume of the drops in the spray, the following expression is finally obtained which shows that the combustion rate is proportional to $\rho_G^{1/3}$

$$w_G = -\left(\frac{3}{32}\right)^{1/3} \pi^{2/3} K(\rho_L n)^{2/3} \rho_G^{1/3} \tag{9.14}$$

This formula may be used, as for a gas-phase mixture, to calculate the combustion time for a given mixture, through applying a mass balance. Indeed this can be written, for a spatially homogeneous spray, as :

$$\frac{\mathrm{d}\rho_G}{\mathrm{d}t} = w_G = -K'\rho_G^{1/3} \tag{9.15}$$

with $K' = (3/32)^{1/3}\pi^{2/3}K(\rho_L n)^{2/3}$, a constant. Integration of (9.15) simply gives the decrease in ρ_G as a function of time.

Equation 9.15 is not satisfactory from two points of view. Firstly, it does not take into account interactions between diffusion flamelets, which is understandable since it is assumed that each drop burns independently of the others. The equation should therefore be modified to take into consideration flamelet structures such as those shown in Figure 9.6. In addition, however, the equation also implicitly assumes that the droplets ignite spontaneously: in fact w_G is non-zero as soon as ρ_G has a non-zero value and becomes greater still as ρ_G increases. No allowance is therefore made for the phenomenon of spray ignition as a function of spray temperature.

Research work is currently underway in an attempt to model these two effects satisfactorily; however, the absence of sufficiently detailed experimental results have not yet allowed the models to be verified. At the present time, only the trends in the rates of reaction can be reasonably predicted, as a function of ρ_G, n, and T. Figures 9.9a and b show the relationship obtained in a recent study by Borghi and Loison[5]. Figure 9.9a indicates in particular how the effect of n is to reduce the rate of reaction compared to equation (9.14); and experiment has indeed confirmed that when two drops burn too close to each other, their diameter reduces relatively more slowly. The model used clearly confirms that n is involved in the form (dimensionless) of the Kerstein-Law parameter (in this case in the form $nr_F^{1/2}$ since the model considers one plane only). Moreover, Figure 9.9b illustrates the effect of the temperature of the medium; spray ignition is dependent on the temperature being high enough. Temperature increases in the spray prior to combustion result in increases in the rate of reaction from zero up to values corresponding to self-ignition, which is the case illustrated by Figure 9.9a. The dimensionless quantity representing this effect is very probably T/T_F, where T_F is the maximum temperature in the diffusion flamelet surrounding a drop; and is simulated in the model by the presence of a certain number of "hot particles" in the medium. These two results, although realistic, are nonetheless derived from a model. A solid theory able to quantify these two phenomena does not exist at the present time.

When studying a flame propagating in a premixed spray, a knowledge of rates of reaction such as these can allow an estimate to be made of the burning velocities of the flame, as well as its thickness. Indeed, by comparing this case with a premixed gaseous flame, the velocity S_{spray} can be considered as being related to a thermal diffusivity coefficient D_{spray} and to an average reaction time (with a constant K_s):

$$S_{\text{spray}} = \frac{K_s D_{\text{spray}}^{1/2}}{\tau_G^{1/2}} \tag{9.16}$$

where τ_G can be defined such that:

$$\tau_G^{-1} \rho_G(0) = \int_0^{\rho_G(0)} w_G \, d\rho_G \tag{9.16'}$$

However, the exact value of D_{spray} is not known–is it the diffusivity coefficient of the gas between the drops, D, or should it be determined in another way? Similarly, the flame thickness in the spray, e_{spray}, should be given roughly (with a constant K_e) by:

$$e_{\text{spray}} = K_e D_{\text{spray}}{}^{1/2} \tau_G{}^{1/2} \tag{9.17}$$

The knowledge of formulae such as (9.14), or improvements which take into account $n^{1/3} r_F$ and T/T_F, are thus of particular interest in this field, although very few experimental or theoretical results are available.

5. Borghi R. and Loison S. (1993) *22nd Symposium (Int.) on Combustion*, The Combustion Institute, 1541–1547.

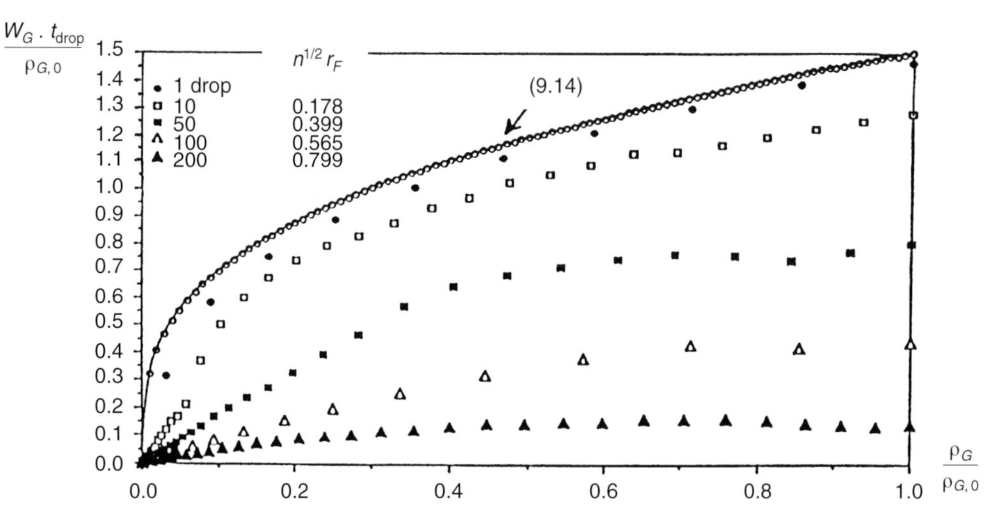

Figure 9.9a Effect of spray density on rate of combustion. The rate of combustion is made dimensionless by the combustion time t_{drop} of an isolated drop with the same characteristics and the same initial diameter, and by the initial mass of the drops per unit volume of spray. The plots were calculated (Borghi R. and Loison S. (1993) *24th Symposium (Int.) on Combustion*, The Combustion Institute, 1541–1547) by a numerical simulation in which the drops were dispersed in a plane, such that the percolation parameter was $r_F n^{1/2}$. Formula (9.14) is also plotted, and the difference between this and the simulation for a single drop is due to the numerical precision of the calculation.

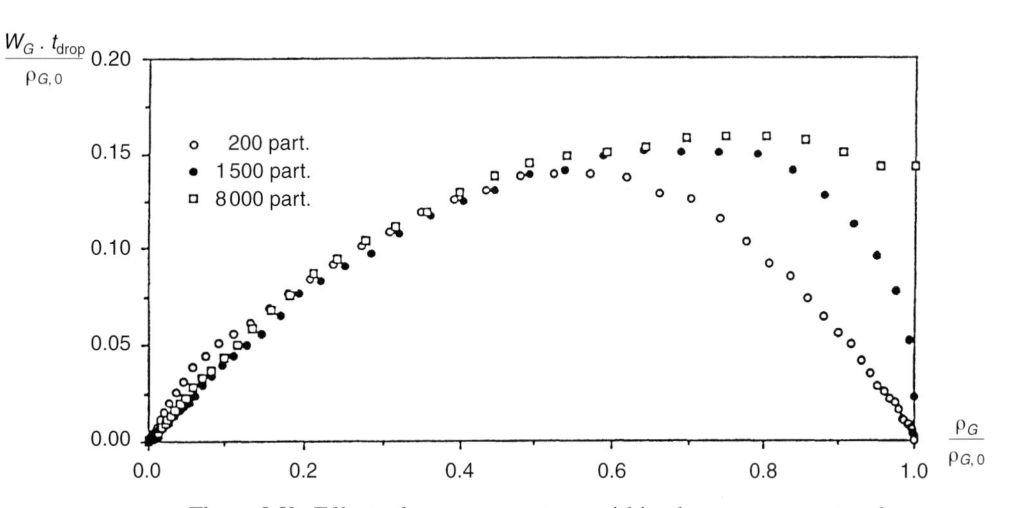

Figure 9.9b Effect of gas temperature within the spray on rate of combustion. The numerical simulation (the same as for Figure 9.9a) represents this temperature by the presence, initially, of "hot particles" which diffuse into the gas separating the drops. The initial number of these particles is indicated in the key. The spray studied in this experiment corresponds to an initial $r_F n^{1/2} = 0.799$.

9.4 DIFFUSION JET FLAMES WITH DROPLETS

9.4.1 Description

Diffusion jet flames play an important role in many industrial applications, one of which is the industrial burner. Jets of droplets, sometimes very dense, are sprayed inside a current of air, mix with the air and are then ignited. The droplets are formed at the mouth of the burner from a liquid jet, and obviously cannot instantaneously produce a homogeneous mixture within the surrounding air. On the axis of the liquid injector, a spray of fuel droplets is produced inside a zone of fuel vapour which is in turn surrounded by a flow of air (the oxidant). The two flows will gradually mix together and burn. The situation is therefore analogous to that of a diffusion jet flame, but with droplets present in addition to the reacting gases. One would therefore expect to see a flame zone establish, consisting of a stationary reaction zone separating a fuel-rich medium from a fuel-lean medium, and a flame length which is dependent on the velocity of the jets.

Figure 9.10a shows the expected situation schematically, distinguishing the different regions of the flame and the shape of the transversal profiles of the gas temperature and of the mass fractions of the gases and the drops of liquid. Figure 9.10 b gives an example of a series of values measured in a flame of this type. The profiles are radial, at a set distance from the injector, and indicate the composition, temperature and gas velocity.

The overall structure of this type of flame closely resembles the structure of a diffusion flame burning with only gas-phase reactants. The most notable difference is highlighted by Figure 9.10a; the mass fraction of the fuel in the gas phase, Y_v, increases initially before reducing with distance from the flame axis. This is due to the fact that the gas-phase fuel is initially in the form of droplets, which must first be vaporised before diffusing and ultimately burning. If $Y_v + Y_l$ had been plotted, where Y_l is the mass fraction of the liquid fuel in the mixture, then this quantity would have behaved in exactly the same way as the mass fraction of the fuel in an entirely gas-phase diffusion flame.

It should be noted, however, that the definition of Y_v, Y_{O_2} and Y_l for sprays is not quite as general as that of Y_K and Y_{Ox} for gas-phase fuels. Indeed, in the latter case, Y_K and Y_{Ox} may exist locally in as small a volume as desired, provided it is much larger than that occupied by a molecule. Conversely, Y_v, Y_{O_2} and Y_l in a spray cannot be defined for such small volumes; their definition requires that the volume in question should be much greater than the size of a droplet. Since the size of one droplet is approximately 10–100 μm, this volume cannot be much smaller, in practice, than 1 mm^3.

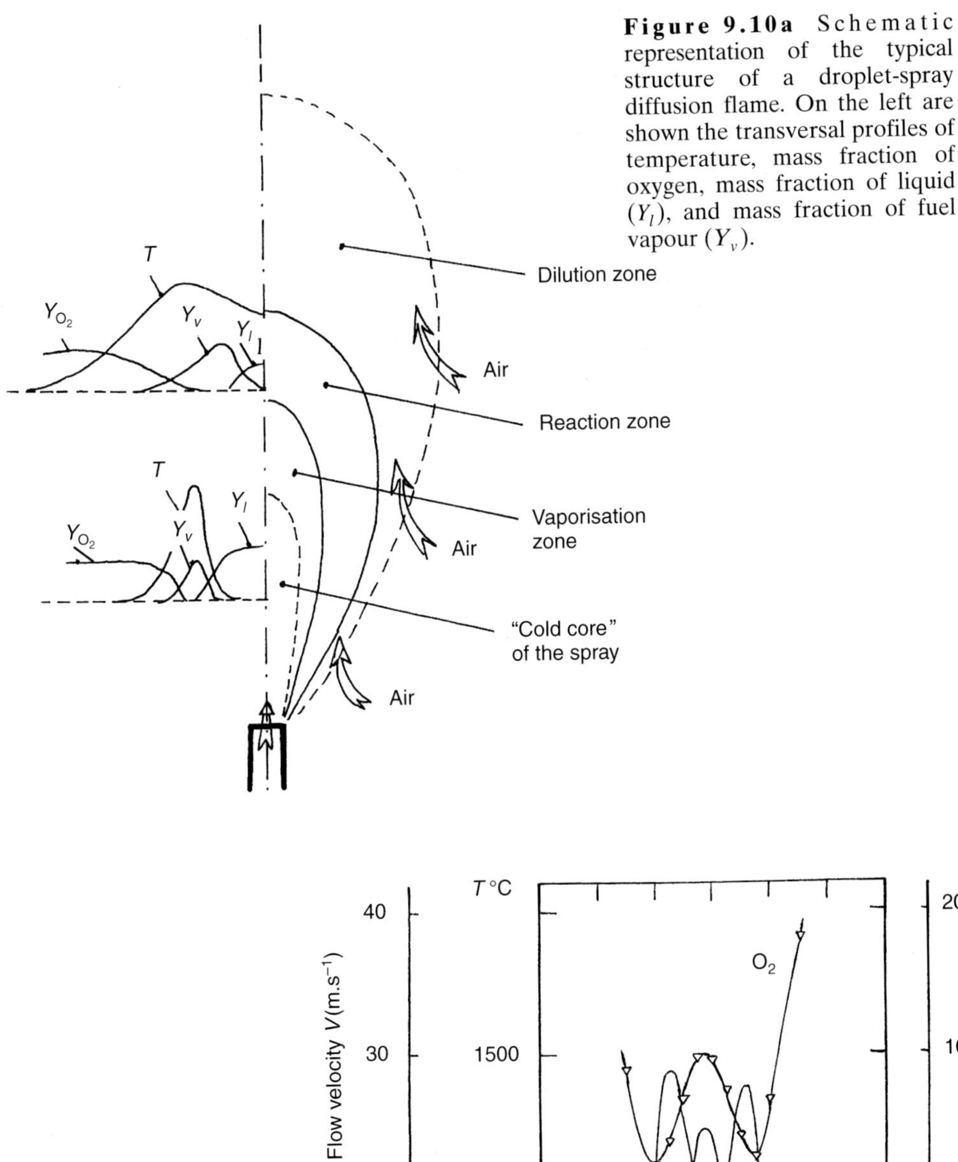

Figure 9.10a Schematic representation of the typical structure of a droplet-spray diffusion flame. On the left are shown the transversal profiles of temperature, mass fraction of oxygen, mass fraction of liquid oxygen, mass fraction of liquid (Y_l), and mass fraction of fuel vapour (Y_v).

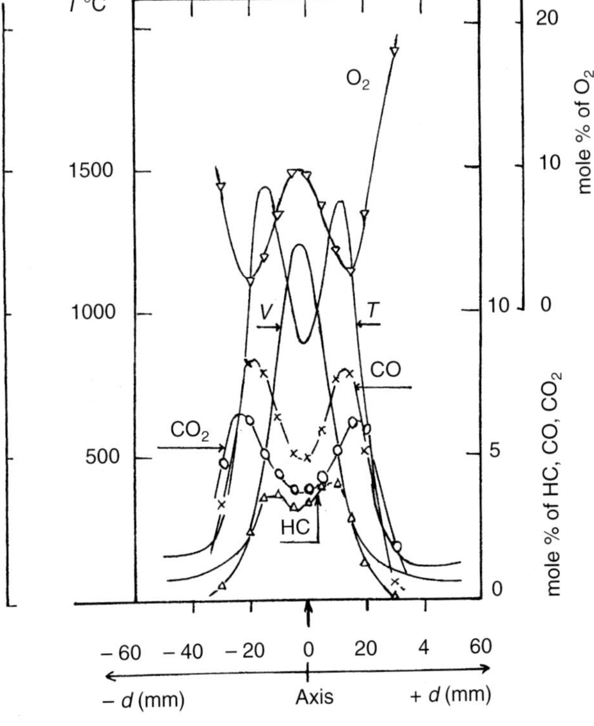

Figure 9.10b Measurements taken in a droplet-spray diffusion flame. Radial profiles obtained for velocity (V), temperature, and mole fractions (as percentages) of O_2, CO, CO_2, and unburnt hydrocarbons (HC) (After Faeth G.M. (1983) *Progress in Energy and Combustion Sciences*, 9, 1–76).

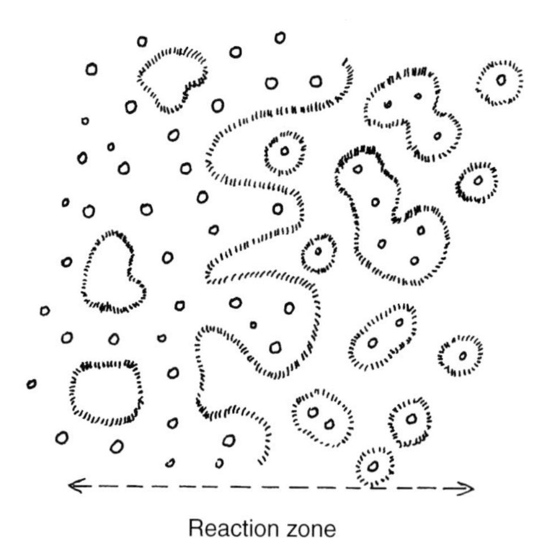

Figure 9.11a Local structure of a droplet-spray diffusion flame, far from the attachment region.

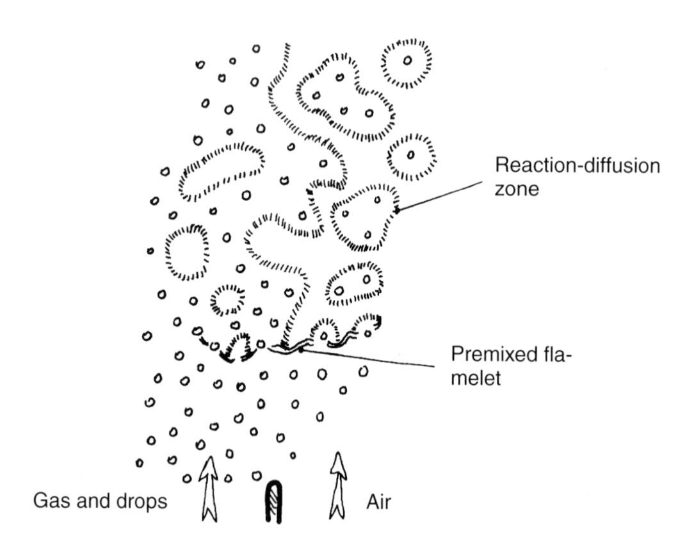

Figure 9.11b Local structure, in the attachment region, of a droplet-spray diffusion flame.

If the structure of the flame is analysed in more detail, through considering more closely a volume of several cubic millimetres centred on the line referred to as the external limit of the spray in Figure 9.10a, then the picture revealed would be something like that shown in Figure 9.11a. On the droplet jet side the medium has a high droplet density, with a few oxidising gas pockets, and surrounded by a diffusion flamelet. On the air side, the medium has a lower droplet density, with what droplets there are surrounded either individually or in groups by a diffusion flamelet. In fact, this description is not greatly different from that which would describe the structure of a premixed flame propagating in a spray when observed on a small scale (percolation flame type). The difference compared with the situation shown in Figure 9.6c, is that there is now clear anisotropy in the structure; the number of drops per unit volume decreases from left to right, from the jet of drops towards the entrained air, which was not necessarily so in Figure 9.6c.

The stabilisation of a droplet-spray diffusion flame is probably very similar to that of entirely gaseous diffusion flames and there is no reason to doubt this. In particular, a triple flame is likely to occur in a spray, where the central "diffusional" reaction zone is attached to a droplet-rich premixed flame on one side and to a droplet-poor premixed flame on the other. This situation is shown schematically in Figure 9.11b, where the diffusion flamelets (partially surrounding the droplets) and the premixed flamelets, rich or lean, are present in the premixed spray flames. There are, of course, more droplet-rich premixed flames on the rich side (on the left) than on the poor side (the air side on the right). These premixed spray flames have, as stated in the previous section, a burning velocity and it is this velocity which is responsible for stabilising droplet-spray diffusion flames.

9.4.2 "Modelling" droplet-spray diffusion flames

A spray of droplets can be considered as being a continuous medium. It is therefore possible to define a liquid mass fraction in a "fluid particle", which is simply the total mass of liquid divided by the total mass of liquid plus gas. As stated above, this approach requires the fluid particle to be "large" enough to contain several thousand drops. This definition can also be extended to enable it to be applied locally, but only so long as it is accepted that it is valid only in a statistical sense, i.e. in defining the mass fractions and all the necessary quantities as averages of a large number of flows which are identical on a macroscopic level, but all different on the scale of the drops.

For this new "continuous medium", balance equations can again be written, as in chapter 3, to try to calculate the structure and conditions for the diffusion flame of interest (as at the end of chapter 3, section 3.4). The development of these equations and the discussion of their various terms goes beyond the scope of this book, although the demonstration of the physical problems which arise is clear. Two types of problem are apparent, the first is due to the fact that the droplets are not in general moving at the same velocity as the gases, even though they are in the same "fluid particle". This is also true for the molecules of different gases and may be represented, as we have seen, by diffusion fluxes and coefficients. However, the use of a coefficient of diffusion for droplets is less well justified theoretically than for gases. The second problem concerns obtaining the rate of reaction in a fluid particle since this can no longer be calculated by analysing the chemical reactions occurring in a homogeneous medium. This is because the fluid particle has a heterogeneous nature inherent of a medium containing drops, but also high heterogeneities in the concentrations of gas-phase fuel and oxidant, with possibly diffusion or premixed flamelets (as shown in Figures 9.6, 9.7 and 9.11). Here, the problem is quite similar to that of turbulent combustion where, in a mean turbulent flame, very fine flamelets can exist which are in continuous motion. Under these conditions, the equations for the rates of reaction in spray flames must be closer to those of the "Eddy Break-Up" type (see chapter 6) than to those of an Arrhenius type. Only when the spray has a very low droplet density or when the reactions are very slow can the use of Arrhenius-type formulas be justified.

The approach which considers the spray to be like a gaseous medium, with a coefficient of diffusion for the drops and for the gases, was termed the *locally homogeneous flow approach* (LHF) by Faeth[6], one of the pioneers of these studies. Its strengths and weaknesses can be highlighted by studying a spray of drops which do not burn but simply vaporise. This experiment has been conducted under certain conditions, for drops of freon in air, with the conclusion that the LHF hypothesis is only valid for very small drops, of a size no greater than a few microns (and also depending on the density of the drops themselves). A modification to this approach involves retaining the Eulerian balance equations for the gas and the drops which make up the spray, but to consider the gas and drops as having two different velocities, with momentum being exchanged between the two types of flow (which are intimately mixed, as are the flows of different chemical species). This approach yields some interesting results but introduces some complex problems when drops of several sizes, hence of several possible velocities, are present in the spray. To overcome the difficulties arising from this latter conditions, another approach has been developed which applies Lagrangian balance equations to the liquid phase (in fact to each drop in this phase), with a statistical simulation of the trajectories of a certain number of drop samples.

---

6. Faeth G.M. (1983) *Progress in Energy and Combustion Sciences*, 9, 1–76.

Figures 9.12a and 9.12b compare the measurements made in a jet of droplets (which evaporate but do not burn) with the theoretical calculations produced from an LHF approach and a statistical Lagrangian simulation. Freon droplets were used in this experiment with average diameters of 38 or 50 μm. Figure 9.12a shows the mass fraction of the freon (liquid plus gas) along the jet axis while Figure 9.12b gives the temperature of the gases along the *x*-axis of the jet. The differences between the LHF approach and the measurements can be seen (the latter probably involving a certain experimental error also) as can the apparent usefulness of the Lagrangian stochastic simulation.

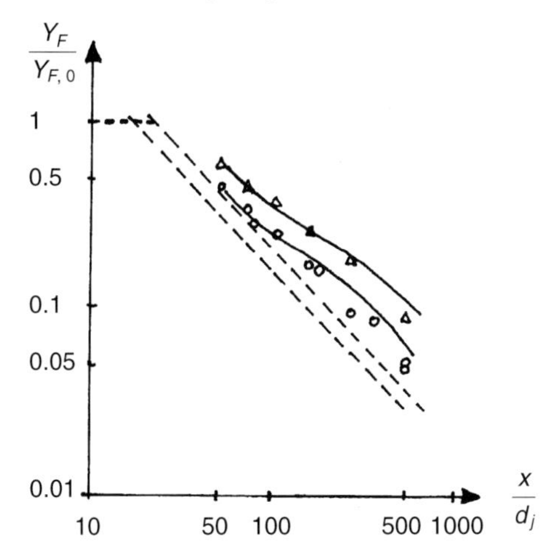

Figure 9.12a Axial distribution of the mass fraction of freon along the axis of a cold jet of freon droplets in air at room temperature. The broken lines are the results of a calculation made using the LHF hypothesis, and the full lines those obtained using a Lagrangian calculation for initial diameters of: ○ – d_O = 38 μm, and Δ – d_O = 50 μm respectively (After Solomon A.S.P. and Faeth G.M. (1985) *Journal of Heat Transfer*, 107, 679).

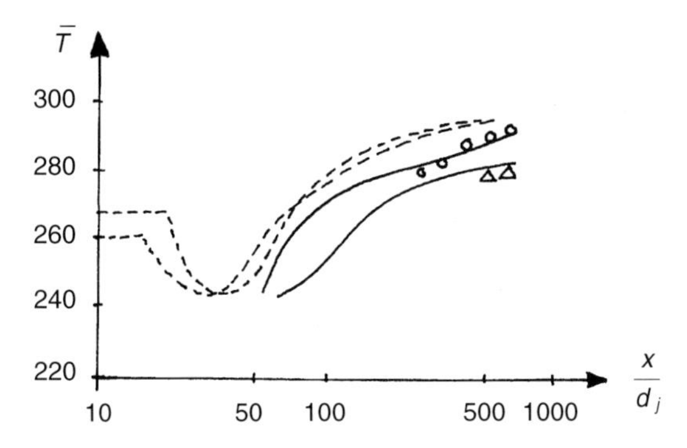

Figure 9.12b Axial temperature distribution of the gases in a jet of freon droplets. *T* initially decreases due to vaporisation. Same reference as for Figure 9.12a.

In order to calculate the rate of vaporisation of the droplets, which is important in determining their size at any instant, and which is also involved in the calculation of their trajectories, the what should by now be familiar d^2 law is used with some modifications made to allow for the convection of the gases surrounding the drop. This is achieved simply by differentiating equation (9.12):

$$\frac{\mathrm{d}(d_{\mathrm{drop}})}{\mathrm{d}t} = -\frac{K}{2d_{\mathrm{drop}}}$$

but with K varying along the trajectory of the drop as a function of its temperature and its velocity, and as a function of those of the gases encountered. It is thus possible, without any particular difficulties, to allow for the fact that the drops are of various sizes, on one hand because they are not identical on injection and on the other since they do not all follow the same trajectory. At the present time, calculations involving statistical Lagrangian simulations of the liquid phase are widely used even though they present certain difficulties when the spray is dense. However, to describe such methods in the detail they deserve would be to go beyond the scope of this book.

CHAPTER 10

POLLUTANT EMISSIONS IN COMBUSTION REACTIONS

10.1 GENERAL POINTS

For various reasons, combustion reactions can release products known as "pollutants" which are dangerous for man and the environment either by causing direct or indirect health problems, by destroying vegetation and crops, damaging buildings, or causing both small- and large-scale modifications to atmospheric and climatic conditions. Generally, two main types of noxious emission are classified; firstly those limited both in terms of their life and scale, and for which the effects are non-cumulative; and secondly those which act on a large scale, whose effects are cumulative, and for which it is often difficult to predict and quantify their consequences in the future. Natural regulation systems (physical, chemical, or biochemical) can often repair the damage caused by a polluting event when the source of the emission belongs to the former case, but not for the latter.

In order not to stray from the context of this book, namely combustion, only physico-chemical mechanisms relating to the emission of pollutants will be discussed, together with any consequences and possible remedial action. The spreading of pollutants will be treated only briefly and we shall pass over matters relating to their possible environmental degradation as well as their economic and social cost.

The principal causes of pollution produced by combustion are:

- those resulting, whether combustion is complete or not (and thus in addition to CO_2 which is almost always present), from the presence of impurities in the fuel, particularly sulphur and nitrogen, as well as radioactive elements contained in the fuel and released with the burnt gases, such as ^{14}C entrained by CO_2,
- those caused by incomplete combustion arising from the attainment of chemical equilibria and irrespective of the equivalence ratio of the initial mixture, hence even for an equivalence ratio of unity (as shown for CO in Figure 10.1),

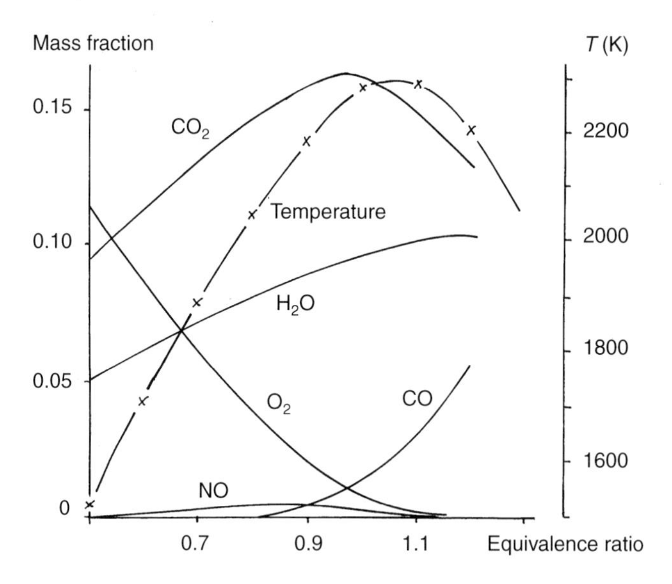

Figure 10.1 Effect of equivalence ratio (calculated) on the equilibrium reached by adiabatic combustion for a given volume, of propane-air mixtures, and with an initial temperature and pressure of 298 K and 1 bar. The amount of CO at equilibrium increases with increasing equivalence ratio (same figure as that shown in Figure 1.2).

- those which are again due to incomplete combustion but result from the malfunctioning of the combustion apparatus itself, (burners, engines, etc.); such as misfiring, insufficient residence time, incorrect mixing of fuel and oxidant, etc.

Various pollutants will now be considered in turn, grouping into sections those which it is convenient to treat together.

10.2 POLLUTION BY CO_2, CO, ALDEHYDES, SMALL HYDROCARBONS, ALKENES, PEROXIDES, TETRAETHYL LEAD AND PAN

10.2.1 CO_2

The complete combustion of any carbon-containing compound automatically leads to the production of CO_2. In the case of hydrocarbons, proportionally more CO_2 is produced as the C/H ratio increases. The crucial step in this emission, i.e. the reaction:

$$CO + OH^{\bullet} \rightarrow CO_2 + H^{\bullet} \tag{10.1}$$

has been clearly identified, since the termolecular reaction:

$$CO + O^{\bullet} + M \rightarrow CO_2 + M \tag{10.2}$$

is far too slow to be of importance.

It is impossible to prevent CO_2 from being emitted by the combustion of these compounds. Moreover, CO_2 emissions continue to rise and the total volume emitted by natural phenomena and human activity has become so great that natural regulatory chemical and biochemical processes cannot prevent the concentration of CO_2 in the air from increasing. Although CO_2 is non-toxic, it plays a fundamental role in the thermal balance of the Earth's atmosphere, and contributes greatly to the "greenhouse effect". At the present time, it is still difficult to predict the short- or long-term consequences of this greenhouse effect.

Slightly less than half the CO_2 produced by the combustion of fossil fuels stays in the atmosphere. Thus, even in the eighties the total volume of CO_2 in the atmosphere was increasing at a rate of about 0.5 per cent per year[1]. Dramatic, yet one-off events such as the burning of the oil wells in Kuwait have only a minor effect on a planetary scale; in fact, the emissions of CO_2 following the Gulf War only accounted for three per cent of the total world-wide production of CO_2 generated by the combustion of fossil fuels[2].

---

1. Prather M.J., Logan J.A. (1994) *25th Symposium (International) on Combustion*, 1513–1525.
2. Ferek R.J., Hobbs P.V., Herring J.A., Laursen K.K., Weiss R.E., Rasmussen R.A. (1992) *J. Geophys. Res.* 97, 14483–14489.

10.2.2 CO

Carbon monoxide (CO), on the contrary, is quite toxic and starts to become dangerous at mixing ratios above 50 ppm, which is quite small. CO is produced by the incomplete combustion of carbonaceous compounds in engines and burners, for a variety of reasons: when the overall or local equivalence ratio is too high, if the residence time in the combustion zone is too short, or from attaining equilibria such as:

$$C + CO_2 \rightleftharpoons 2CO \tag{10.3}$$

which is favoured by high temperatures and low pressure. The reaction

$$CO_2 + H_2 \rightleftharpoons CO + H_2O \tag{10.4}$$

is also favoured by high temperatures, but is independent of pressure. As might be predicted, the amount of CO released increases with the equivalence ratio of the mixture (Figure 10.1). Soots, although difficult to oxidise, can also produce CO through the reaction:

$$C_n + OH^\bullet \rightarrow CO + C_{n-1} + H^\bullet \tag{10.5}$$

with the CO then able to yield CO_2. The rate constant is high, but this reaction is usually slow since the OH concentration is low. The OH radical is found in many flames, particularly in the plume (*cf.* Fig. 1.3) and can also be formed in exhaust gases by the reaction:

$$H_2O + M \rightarrow H^\bullet + OH^\bullet + M \tag{10.6}$$

which has a high activation energy and thus, as a result, its rate drops greatly if the temperature falls.

As a general rule, since CO is produced as part of an alkane oxidation mechanism and is then more or less completely oxidised in turn to CO_2, its concentration passes through a maximum during combustion.

For the combustion of methane, one of the paths leading to CO production is that giving formaldehyde, HCHO, beginning by the creation of methyl radicals, for example by the following reaction:

$$CH_4 + O_2 \rightarrow CH_3^\bullet + HO_2^\bullet \tag{10.7}$$

and followed by:

$$CH_3^\bullet + O_2 \rightarrow CH_3O_2^\bullet \tag{10.8}$$

at least so long as the temperature is not too high. By reacting with CH_3, HO_2 or another CH_3O_2 radical, CH_3O_2 can lead to CH_3O, which in turn leads to formaldehyde, HCHO, and to methanol, CH_3OH, e.g. through the sequence:

$$CH_3O^\bullet + CH_4 \rightarrow CH_3OH + CH_3^\bullet \tag{10.9}$$

and:

$$CH_3O^\bullet + O_2 \rightarrow HCHO + HO_2^\bullet \tag{10.10}$$

At higher temperatures, CH_3O_2 is unstable and, under these conditions, there is evidence to suggest that CH_3 reacts with O_2 to give CH_2OOH, the decomposition of which gives OH and, as at lower temperatures, HCHO. After this the attack of O_2 or OH on HCHO yields the CHO radical which can lead to CO via:

$$CHO^{\bullet} + O_2 \rightarrow CO + HO_2 \tag{10.11}$$

Combustion is responsible, more or less directly, for approximately 40 per cent of all CO emissions. As a result of the greater industrialisation of the northern hemisphere compared with the south, and carbon monoxide's short life in the atmosphere (approximately 2.5 months), its concentration is higher in the northern hemisphere than in the southern hemisphere, i.e. approximately 120 ppb in the first compared with 50 in the second[3].

For more information on the combustion chemistry of hydrocarbons, the reader may wish to consult the book Combustion Chemistry[3].

10.2.3 Aldehydes, alkanes, alkenes, peroxides, tetraethyl lead and PAN

Aldehydes, like CO, are known reducing agents. Their presence in exhaust gases is therefore also an indication of incomplete combustion and in this case is due to insufficient residence time in the engine's combustion chambers. Whilst aldehydes are not toxic as such, their odour is unpleasant. Alkanes of low molar mass and alkenes are also products of incomplete combustion. Combustion gases also contain peroxy radicals which have been exhausted from the combustion zone before being able to continue to propagate the mechanism. In engines, it is often at the walls that incomplete combustion occurs, due to the wall acting as a heat sink (hence creating lower temperatures) and by promoting the recombination of radicals which would otherwise be chain propagators. The pollutants produced by a diesel engine differ from those from a spark-ignition engine. Indeed, a well-tuned diesel engine should emit less CO and unburnt compounds, but more smoke and aldehydes. Clearly an engine in poor condition, whose pistons and valves leak, pollutes much more than a well-maintained engine.

The incomplete combustion of alkanes, in the presence of halogenated impurities, may also produce, in addition to CO, compounds such as CH_3Cl and CH_3Br. This is not the only cause of their presence in the atmosphere, although it is responsible for 10 to 50 per cent[4]. Having said this, atmospheric chlorine, combined or uncombined, is produced mainly from carbon tetrachloride CCl_4 and chloro-fluro-carbons in processes unrelated to combustion.

---

3. Gardiner W.C., ed. (1984) *Combustion Chemistry*, Springer-Verlag.
4. Novelli P.C., Masarie K.A., Tans P.P., Lang P.M. (1993) *Science*, 263, 1587–1590.

Despite increasingly stringent laws, petrols for spark-ignition engines containing lead-based additives are still available on garage forecourts. One of the most common additives is tetraethyl lead $(C_2H_5)_4Pb$, which is included to avoid the knocking produced by undesirable self-ignition in the cylinder head (see chapter 4). The lead reacts with impurities in the fuel, notably sulphur, to produce lead sulphide.

Peroxyacetyl nitrate (usually referred to simply as PAN) is an eye-irritant and is toxic to vegetation. Present even in areas remote from human activity, it is concentrated mainly in large cities and particularly those which combine heavy industry with a very sunny climate. It is not, strictly speaking, produced during combustion but rather as its consequence via the reactions shown below. The mechanism which produces this pollutant, and others of the same type, is complex but relatively well understood, and involves impurities, unburnt compounds, NO_2, chlorinated compounds etc. which are all present in the atmosphere around centres of industrialisation. These compounds can, through chemical and occasionally photochemical reactions, lead to the formation of free radicals, including OH, which then undergo the following reactions:

$$OH^{\bullet} + RH \text{ (various unburnt compounds)} \rightarrow R^{\bullet} \text{ (including } R'CO^{\bullet}) + H_2O \quad (10.12)$$

$$R'CO^{\bullet} + O_2 + M \rightarrow R'COO_2^{\bullet} + M \quad (10.13)$$

$$R'COO_2^{\bullet} + NO_2 + M \rightarrow R'COO_2NO_2 \text{ (PAN if } R' = CH_3) + M \quad (10.14)$$

$R'CO$ can also be produced by the incomplete oxidation of alkanes or from ozonide decomposition; with ozonides themselves a product of a combination reaction between alkenes and ozone (occasionally involving a photochemical aspect). Laboratory simulations indicate that PAN and ozone appear during the photooxidation of hydrocarbon—NO_2—air mixtures, which is why PAN is formed more readily in sunny, high traffic density cities (and thus affects Los Angeles to a greater extent than London).

PAN is stable enough in cool air and if the NO_2/NO ratio is high. Indeed, it has been shown that quite large quantities of NO_2 are trapped in the lower atmosphere in the form of PAN. This compound can therefore, by means of atmospheric circulation, transport NO_2 from cool regions to warmer regions. Useful additional details on this subject can be found in the book by Finlayson-Pitts and Pitts, Jr[5].

5. Finlayson-Pitts B.J., Pitts J.N., Jr (1986) *Atmospheric Chemistry*, Wiley.

10.3 POLLUTION BY SOOTS AND POLYAROMATICS

Polyaromatics are compounds containing fused aromatic rings, usually C_6 rings, but may also contain other types of rings. Figure 10.2 shows six compounds selected from the many members of this family of compounds, and which have all been detected in exhaust gases. As a general rule (which can be easily verified) the H/C ratio in these molecules decreases as the number of rings increases. The ratio is equal to 1 for benzene, C_6H_6, hence for one ring, 8/10 for naphthalene, $C_{10}H_8$, which has two rings, 10/14 for anthracene and phenanthrene, $C_{14}H_{10}$, both containing three rings, 10/16 for pyrene and fluoranthene, $C_{16}H_{10}$, both with four rings, and 12/20 for benzo(a)-pyrene, $C_{20}H_{12}$ which contains five (and which is a mutagen and carcinogen), and so on. Following this sequence through to its conclusion would end with graphite, which contains a very large number of rings and hence a H/C ratio approaching zero.

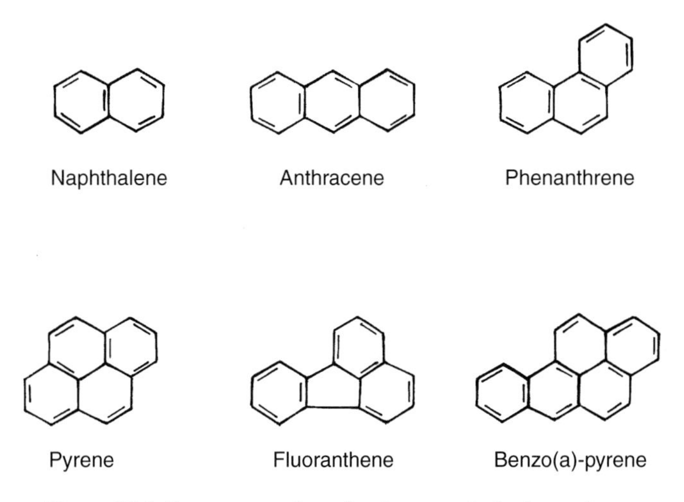

Figure 10.2 Some examples of polyaromatic hydrocarbons.

Soots are carbonaceous particles, hydrogenated and spherical to varying degrees, which can increase in size up to a diameter of about 20 nm, at which point they can group with other spheres to form long, coiled chains which are difficult to oxidise. Their surface contains radical sites which can provide suitable locations for the creation of stable bonds. For this reason, it is not surprising that soots can often hold molecules or polyaromatic radicals on their surface by adsorption, particularly during the cooling of smokes (where the term "smoke" describes a suspension of dust particles in a gas). However, the reader should not assume that this means that soots and polyaromatics are inevitably associated with one another, indeed polyaromatics can be detected where no smoke is produced. Nevertheless, there are correlations and certain polyaromatics, particularly those with an aliphatic side chain, undoubtedly play a key role in the mechanisms involved in soot formation.

Soots are not just pollutants. Without soots a candle would not emit any visible light and in a boiler's furnace they ensure good radiative and heat transfer. They only become dangerous, especially to the lungs, if they are not destroyed before leaving the flame.

Not surprisingly, the amount of soot produced varies according to the type of fuel used: wood (coniferous or deciduous), coal, oil-based products, or natural gas. The amount also depends on the combustion conditions, and can be large for a Diesel engine whilst a well-tuned spark-ignition engine can produce none. Indeed, as might be expected, it also increases with the equivalence ratio of the fuel mixture. To express soot production quantitatively, a critical ratio n_C/n_O may be defined as being the number of C atoms to O atoms for the mixture above which the flame becomes distinctly yellow in colour as a result of the presence of soot. This ratio obviously depends on the type of fuel, and is not the same for a diffusion flame compared with a premixed flame. In the latter case it is equal to 0.47 for propane and 0.83 for acetylene (although in this compound the n_C/n_H ratio = 1 is the highest in the aliphatic series), 0.57 for benzene and only 0.25 for ethanol. This point merits some discussion. Consider a fuel-rich mixture, i.e. where there is insufficient oxygen for all the C to give CO_2 and all the H to give H_2O, at thermodynamic equilibrium at flame temperatures. When this mixture burns, the products should include CO in addition to CO_2, and H_2 in addition to H_2O, but no unburnt C, at least not so long as there is enough oxygen for C to give CO. The aggregation of small particles into larger ones is, however, a process which reduces the entropy of the system and hence no soot should be found so long as the ratio n_C/n_O does not exceed unity, which is far from being the case. This means that thermodynamic equilibrium is not reached, and consequently that soot production is under what is termed "kinetic control", i.e. it depends on the kinetic mechanism.

Whilst a great deal of work is required before this mechanism is fully understood, we do have a general idea of what happens. Apparently, the steps leading to the formation of aromatics and hence to polyaromatics lie on the reaction path which, through thermal dehydrogenation, leads to alkanes such as CH_4 and C_2H_6, to alkyl radicals such as $CH_3^{\bullet}$, and $C_2H_5^{\bullet}$ then to alkenes such as C_2H_4, and then to alkynes such as C_2H_2. Subsequently, via dehydrogenation, C_2H_2 could lead to the $C_2H^{\bullet}$ radical then:

$$C_2H^{\bullet} + C_2H_2 \rightarrow C_4H_2 + H^{\bullet} \qquad (10.15)$$

If C_2H and diacetylene, C_4H_2, then react together, as indicated in Figure 10.3, the mechanism would then be continued as also suggested by this figure. The sequence could then be repeated, namely the addition of C_2H_2, then cyclisation and so on.

Soot can be eliminated by various types of oxidation, notably by reaction (10.5) shown above. Furthermore, this reaction can be promoted by using an additive, such as barite, which favours OH radical production.

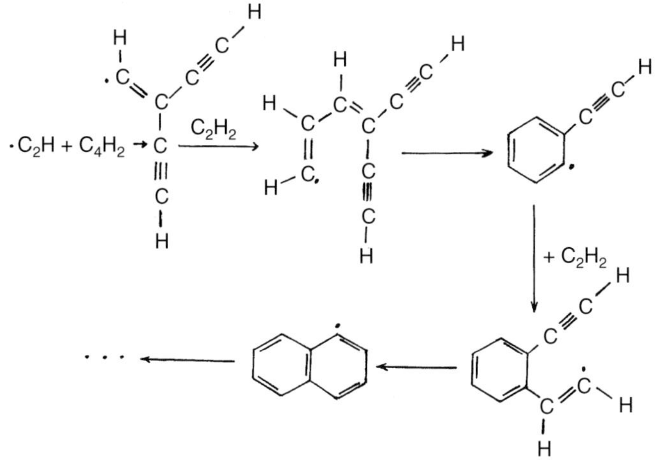

Figure 10.3 Mechanism suggested for the explanation of soot formation.

10.4 POLLUTION BY NITROGEN OXIDES (NO_X)

10.4.1 General points

The notation NO_X originally described NO and NO_2 in general, but is now taken to include all the gas-phase oxides of nitrogen: NO, NO_2, N_2O_3, N_2O and even HNO_3. This definition will be used here.

At high temperatures, N_2 and O_2 can react with each other to give NO, through the equilibrium involving $N_2/O_2/NO$. The amount of NO present in the equilibrium mixture increases if the temperature increases, (see worked example 1.7 in chapter 1), although the system does not regenerate N_2 and O_2 on returning to ambient temperature (see section 1.2.3 in chapter 1). Hence if air is used as the oxidant, then this NO is an immediate cause of pollution. However, NO can also be produced by the oxidation of combined nitrogen which is often present in the fuel itself in the form of amines, pyrroles, pyridines, porphyrins, etc. The NO can then combine with O_2 or react with peroxy radicals RO_2, where R is an alkyl radical, according to the reaction:

$$NO + RO_2^{\bullet} \rightarrow NO_2 + RO^{\bullet} \qquad (10.16)$$

which yields NO_2. Subsequently, the following reaction can occur:

$$NO + NO_2 \rightarrow N_2O_3 \tag{10.17}$$

with equilibrium possibly attained. Finally, through combining with atomic oxygen or otherwise, N_2 can give N_2O. The final result is a blend of NO, NO_2, N_2O_3 and N_2O, globally referred to as NO_X. The two most dangerous of these oxides for the respiratory tract are NO and NO_2. The former becomes dangerous above a mixing ratio of 25 ppm, and the later at just 5 ppm. Furthermore, NO_2 plays a major role in the mechanism leading to the formation of compounds such as peroxylacetyl nitrate (PAN), $CH_3COO_2NO_2$, present in polluted air over large cities such as Los Angeles. PAN and related compounds can cause severe irritation of the eyes and respiratory tract, whereas N_2O plays an active role in producing holes in the ozone layer. NO and NO_2 can react with water to produce acidic solutions containing nitrous and especially nitric acids. This process may contribute to the problem of "acid rain".

For more details on this subject, the reader might like to refer to "Les oxydes d'azote (hors carburation), nuisances, formation, réduction, mesures. réglementation", *Revue Générale de Thermique*, (published by the French Society of Heat Science), and in particular to the paper by De Soete[6].

10.4.2 Gas-phase mechanisms leading to NO_X emission

Studies have identified three mechanisms which may be interconnected to a certain extent.

- The first, the principle of which has been known since 1946, is that studied by Zeldovitch and includes the two steps:

$$O^{\bullet} + N_2 \rightarrow NO + N^{\bullet} \tag{10.18}$$

$$N^{\bullet} + O_2 \rightarrow NO + O^{\bullet} \tag{10.19}$$

where the atoms of the N radical are generally considered to obey the SSA (see chapter 2). The first governs the global rate, and has a high activation energy of around 315 $kJ{\cdot}mol^{-1}$. Its rate, and hence that of the reaction sequence (10.18) and (10.19), only therefore becomes important at high temperature. Thus the term "thermal" is often applied to the Zeldovitch mechanism. Still as a result of the first step, NO formation is also favoured by a long reaction time and predominates in mixtures which are lean or approaching stoichiometric. For stoichiometric or rich mixtures the following reaction can also occur:

$$N^{\bullet} + OH^{\bullet} \rightarrow NO + H^{\bullet} \tag{10.20}$$

6. De Soete G.G. (1989) Mécanismes de formation et de destruction des oxydes d'azote dans la combustion. *Revue Générale de Thermique,* 28, 331, 353.

For lean mixtures, but at lower temperatures, a mechanism may exist which passes via N_2O:

$$O^{\bullet} + N_2 + M \rightarrow N_2O + M \qquad (10.21)$$

$$O^{\bullet} + N_2O \rightarrow NO + NO \qquad (10.22)$$

$$H^{\bullet} + N_2O \rightarrow NH^{\bullet} + NO \qquad (10.23)$$

- The second mechanism is linked to the presence of nitrogen chemically-combined in the fuel, which is particularly the case for coal and its derived fuels. The mechanism is not well understood, but has been shown to be faster than the Zeldovitch mechanism, since the latter is based on practically unreactive molecular nitrogen. It is normally assumed that these fairly large molecules, which contain the chemically-combined nitrogen, are decomposed by heat into smaller molecules, such as HCN and NH_3, which in turn react.

- The third mechanism involves what is known as "prompt NO" formation, and was first proposed by Fenimore in 1971. The author had in fact found that in hydrocarbon-rich flames at one bar or more, the measured NO concentration profiles for the combustion zone extrapolated to the mouth of the burner did not appear to tend to zero, at least for his apparatus. However, using the same apparatus, the NO concentration did tend to zero for CO—air and H_2—air flames. A mechanism had therefore to be imagined which required the presence of a carbon species (other than CO). It was finally concluded that $CH^{\bullet}$ was at the root of the problem. This radical is in fact present (as are many other carbon-based radicals such as $CH_3^{\bullet}$ and $CH_2O^{\bullet}$, see Fig. 10.4), right from the start of the combustion and before the flame front arrives, but much less so after and, of course, much more so in rich than in poor mixtures. It is generally accepted that the first step in the mechanism leading to prompt NO formation is:

$$CH^{\bullet} + N_2 \rightarrow HCN + N^{\bullet} \qquad (10.24)$$

Reaction (10.24) is endothermic but much less so than reaction (10.18), consuming roughly 10 kJ·mol^{-1} compared with 320 kJ·mol^{-1} respectively, and hence it occurs much more readily. After this step the mechanism is complex and has not as yet been fully explained. Plots such as that given in Figure 10.5, obtained by G. De Soete, clearly show that the HCN concentration passes through a maximum prior to the rise in NO, and that this rise is very steep at the beginning before becoming slower. It seems perfectly logical to think that the steep slope, which is the sign of a high rate, relates to prompt NO whilst the less steep slope corresponds to thermal NO. The former could not be observed with the Meker-type burner used by Fenimore.

In the case of the oxidation of methane in well-stirred reactors (see chapter 2) and for short contact times (around 2 ms), if the equivalence ratio rises above 1.2 then almost all of the NO produced by the chemically-fixed nitrogen in the fuel, i.e. everything excluding the NO produced from molecular N_2, is due to prompt NO formation. Conversely, longer reaction times favour thermal NO

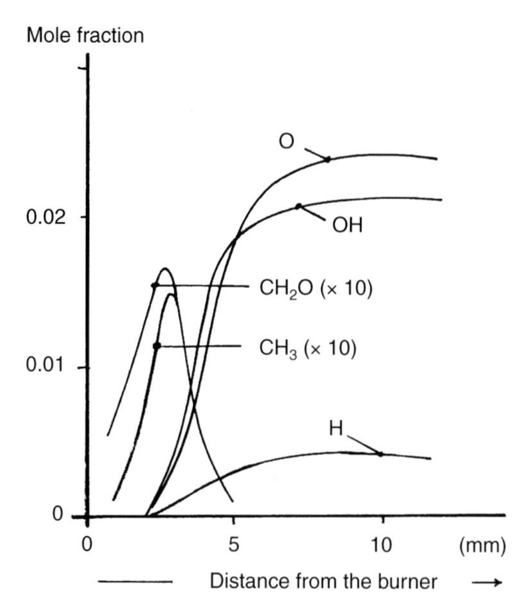

Figure 10.4 Profiles of O, OH, H, CH_2O and CH_3 in a premixed CH_4—O_2 flame at low pressure (After J. Peeters and G. Mahnen (1972) *14th Symposium (International) on combustion*, p. 133, The Combustion Institute. Already cited with reference to Figure 2.3).

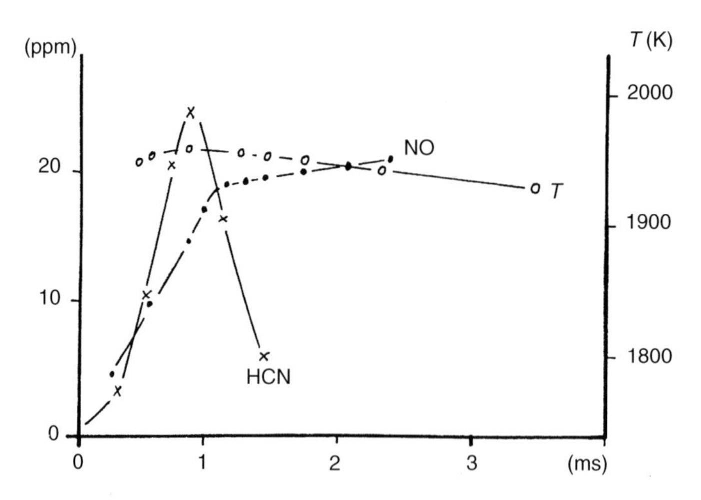

Figure 10.5 Profiles of NO, HCN and temperature measured in a flat, premixed, rich flame of composition:

$10.3\ C_2H_4 + 23.9\ O_2 + 65.8\ N_2$. Flow rate = 131 $cm^3 \cdot s^{-1}$

(After De Soete G.G. (1989) *Revue Générale de Thermique*, p. 353).

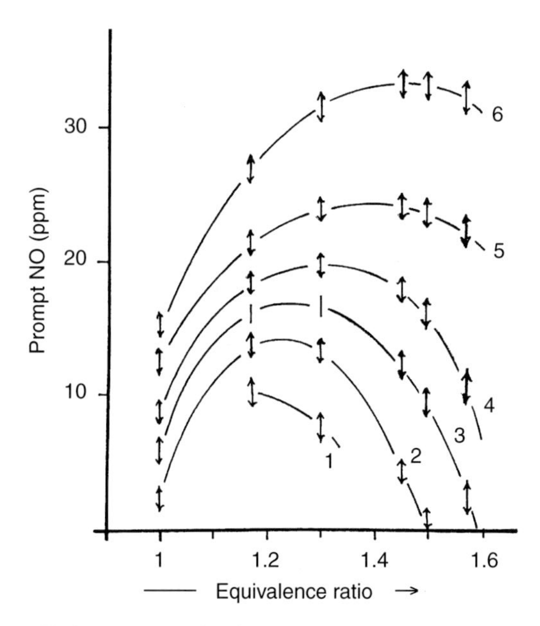

Figure 10.6 Total production of prompt NO as a function of equivalence ratio, at different temperatures. Flames of composition $C_2H_4/O_2/N_2$ with 62% N_2. At 1700 K (curve 1), 1800 (curve 2), 1900 (curve 3), 2000 (curve 4), 2100 (curve 5) and 2200 (curve 6) (After De Soete G.G. (1989) *Revue Générale de Thermique,* p. 353).

formation. Figure 10.6 produced by G. De Soete and relating to premixed $C_2H_4—O_2—N_2$ flames indicates that the total production of prompt NO increases if the temperature is raised and, for a given temperature, passes through a maximum as the equivalence ratio increases.

Finally, the nitrous oxide N_2O is either consumed or produced; its principal formation reactions being:

$$NH^{\bullet} + NO \rightarrow N_2O + H^{\bullet} \tag{10.25}$$

and:

$$NCO^{\bullet} + NO \rightarrow N_2O + CO \tag{10.26}$$

although the quantities produced remain low. N_2O actually reacts with radicals, particularly $H^{\bullet}$:

$$N_2O + H^{\bullet} \rightarrow N_2 + OH^{\bullet} \tag{10.27}$$

leading to its disappearance. Moreover, if apparatus is used in which N_2O production occurs in a region which is poor in H radicals, then more N_2O is found in the burnt gases.

10.4.3 Heterogeneous production and reduction of NO and N_2O

The combustion of a solid fuel from which the volatile matter has been removed by heating, as is the case for coke, must be governed by a heterogeneous mechanism. Since the first step is the chemisorption of O_2 at a surface carbon site (–C) of unfilled valency, then the subsequent desorption releases CO and CO_2. However, if the solid's surface also contains sites, represented by (–CN), where C is chemically bonded to a nitrogen, then the following reaction can occur:

$$O_2\,(\text{gas}) + (-\text{C}) + (-\text{CN}) \rightarrow \text{on the surface, OC} + \text{CNO} \qquad (10.28)$$

which through desorption leads to:

$$\text{surface CNO} \rightarrow \text{NO (gas)} + (-\text{C})\ (\text{regenerated}) \qquad (10.29)$$

and:

$$(-\text{CN}) + \text{CNO} \rightarrow N_2O + 2\,(-\text{C})\ (\text{regenerated}) \qquad (10.30)$$

The presence of hydrogen complicates this sequence and results in the emission of HCN and NH_3.

N_2O and NO_2 can be formed from other heterogeneous mechanisms. Indeed, for several reasons, notably if $CaCO_3$ is added to reduce emissions of SO_2 and SO_3, sulphates, sulphites and solid calcium sulphides may all be produced inside combustion zones. The oxidation state of the sulphur is the same in SO_2 as in the sulphites, namely + 4, and equally in SO_3 as in the sulphates, namely + 6. Thus in the same way in the gas phase:

$$2NO + SO_2 \rightarrow N_2O + SO_3 \qquad (10.31)$$

$$NO + SO_3 \rightarrow NO_2 + SO_2 \qquad (10.32)$$

a gas-solid mechanism of the following type can thus be envisaged:

$$2NO + CaSO_{n-1} \rightarrow N_2O + CaSO_n \qquad (10.33)$$

10.4.4 Controlling NO_X emissions

10.4.4.1 In the gas phase

The technique of "stratified" combustion is often employed. This technique covers all forms of combustion in which either the fuel and/or air are supplied from several burners each with varying equivalence ratios or, for a single burner, if the spatial distribution of the fuel and air, or of the air alone, is different.

Two-stage stratified combustion is used to minimise NO_X emissions arising from nitrogen bound in the fuel. The first stage operates with a mixture whose equivalence ratio is between 1.3 and 1.5. The total combined nitrogen, not including molecular nitrogen N_2, is minimised under these conditions. Below an equivalence ratio of 1.3, a large amount of NO is again found and above 1.5 a large amount of HCN results. The second stage uses a lean mixture to complete the combustion of the fuel and at a low enough temperature to prevent the emission of thermal NO. Simultaneously, during the second stage, virtually all of the combined nitrogen released in the first stage, which had not already given NO, does so during the second. Clearly the success of the technique depends on the control of a number of parameters, such as the equivalence ratio and residence time, and of the choice of their values. Numerical simulations are therefore of great help in guiding feasibility studies to a useful conclusion.

A second method (the De-NO_X process, patented by Exxon) treats the burnt gases in a post-combustion process. In this way, by injecting NH_3 into the exhaust gases of fixed burners, a sequence of reactions can be initiated which leads from NH_3 to $NH_2^{\bullet}$ to NO and, finally, to N_2. However, this injection must not deviate from a narrow temperature band (from 1000 to 1500 K). Experiments and numerical simulations show that the two main reactions are:

$$NH_2^{\bullet} + NO \rightarrow N_2H^{\bullet} \text{ (or perhaps } N_2 + H^{\bullet}) + OH^{\bullet} \quad (10.34)$$

and:

$$NH_2^{\bullet} + NO \rightarrow N_2 + H_2O \quad (10.35)$$

The RAPRENO_X process ("RAPid REduction of NO_X") takes advantage of the fact that above 700 K the compound HNCO reacts rapidly with NO mole per mole. In the resulting mixture, CO, CO_2 and H_2O are identified simultaneously with the removal of NO and HNCO, with neither HCN nor NH_3 detected. Under these conditions, all the nitrogen must be converted into molecular nitrogen by the end of the process. A chemical balance confirms that one mole of NO is consumed for each mole of HNCO consumed.

10.4.4.2 Heterogeneous control

Above about 700°C, NO can be reduced to nitrogen over coke or graphite. With (–C) again denoting a surface carbon of unfilled valency, it is assumed that :

$$NO + 2(-C) \rightarrow -CO + (-NC) \quad (10.36)$$

$$2(-NC) \rightarrow N_2 + 2(-C) \quad (10.37)$$

with the second reaction occurring rapidly. The desorption of CO then liberates a site and, simultaneously and faster still, chemisorbed CO can react with gas-phase CO to produce gaseous CO_2 and regenerate another site:

$$\text{CO adsorbed} \rightarrow \text{CO gas} + (-C) \quad (10.38)$$

$$\text{CO adsorbed} + \text{CO gas} \rightarrow CO_2 + (-C) \quad (10.39)$$

By adding gas-phase CO, the temperature above which this mechanism can occur is reduced by around 400°C, undoubtedly because the presence of CO favours reaction (10.39), which increases the number of –C sites.

By adding H_2, which is also a reducing agent, or if the fuel contains hydrogen, an analogous effect may be produced but with, in addition, the emission of NH_3, which in turn can be reduced either to N_2 or to HCN, which upon reaction with NO can also give N_2. Figure 10.7, produced by G. De Soete, shows how the amount of NO is reduced and N_2 increased, with NH_3 and HCN passing through a maximum and gradually disappearing beyond 900°C.

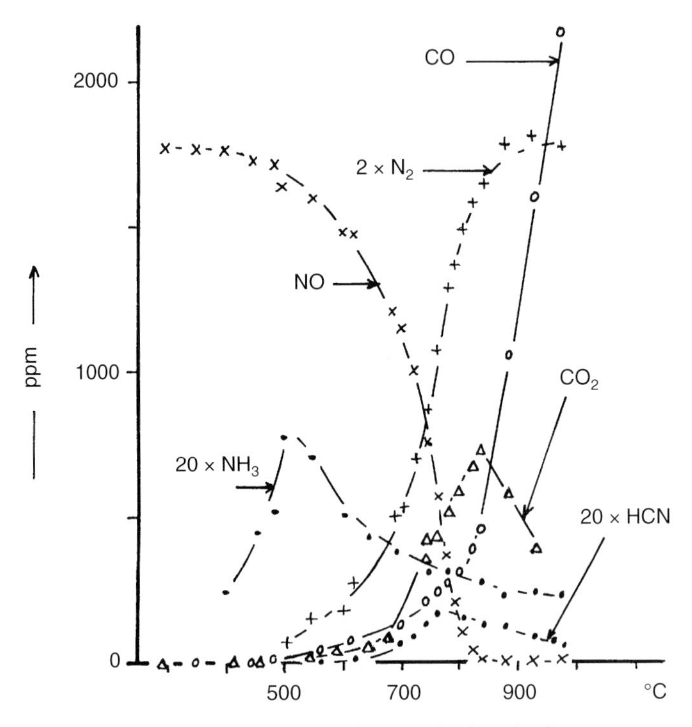

Figure 10.7 Formation of NH_3 and HCN during the heterogeneous reduction of NO over coke obtained from Rietspruit coal (South African) (After De Soete G.G. (1989) *Revue Générale de Thermique*, p. 353).

The catalytic reduction of NO over precious metals (catalytic converter) or over metal oxides is also used, producing molecular nitrogen but also small quantities of N_2O, depending on the efficiency of the reduction process.

N_2O can also react with surface carbon, in accordance with the following mechanism:

$$N_2O + (-C) \rightarrow N_2 + CO \text{ adsorbed} \quad (10.40)$$

$$N_2O + CO \text{ adsorbed} \rightarrow CO_2 + N_2 + (-C) \text{ regenerated} \quad (10.41)$$

It is not therefore surprising that the addition of CO has no effect as a promoter on this occasion. In fact, through reaction (10.41), N_2O "cleans" the surface in the same way as CO did by reaction (10.39).

10.5 POLLUTION BY SULPHUR OXIDES (SO_X)

10.5.1 General points

As with nitrogen, chemically-combined sulphur may be found in fossil fuels; indeed coal contains from 1 to 10% by mass. The combustion of this combined sulphur leads mainly to sulphur dioxide, SO_2, and to a lesser extent sulphur trioxide, SO_3. This said, practically all sulphur compounds, burnt or not, are dangerous.

Under normal conditions of temperature and pressure, SO_2 is a gas, which is noxious at mixing ratios of above 5 ppm in air. SO_3 is a solid but melts above 17°C and boils at 45°C. At a high enough temperature the following equilibrium can therefore exist:

$$SO_2 + 0.5\,O_2 \rightleftharpoons SO_3 \qquad (10.42)$$

This reaction is catalysed by metal oxides, especially those of transition metals. In air, SO_2 can give SO_3 according to a mechanism which is still poorly understood. SO_X is often used to describe blends of SO_2 and SO_3. On reacting with water, SO_2 gives H_2SO_3, sulphurous acid, while SO_3 gives sulphuric acid, H_2SO_4. These two can also contribute, with NO_X, to the production of "acid rain". All these compounds are dangerous to man, vegetation, animal life, and even stonework.

10.5.2 Mechanisms leading to SO_2 and SO_3 in flames

These mechanisms have not been greatly studied since the emissions of SO_2 and SO_3 seem unavoidable. However, there is interest in combined pollution effects involving sulphur-containing and nitrogen-containing compounds. In flames, SO_2 catalyses the recombination reactions of the main radical chain propagators, such as $O^\bullet$, $H^\bullet$, $OH^\bullet$, etc. and referred to below as $X^\bullet$ and $Y^\bullet$, through reactions of the type:

$$X^\bullet + SO_2 + M \rightarrow XSO_2 + M \qquad (10.43)$$

$$Y^\bullet + XSO_2 \rightarrow XY + SO_2 \qquad (10.44)$$

Now, in the Zeldovitch mechanism, O is required in the first stage, i.e. the stage in which the chemical bond between the N atoms in the N_2 molecule is broken, namely reaction (10.18) above. The situation is more complex for nitrogen combined in the fuel. For the moment it would appear that the simultaneous presence of combined sulphur and nitrogen increases the emissions produced by combined nitrogen in hydrocarbon flames, hence of HCN, NH_3 and NO.

10.5.3 Controlling SO_X emission

Since little can be done at the present time, then in anticipation of future developments, a deliberate choice is made to only burn low-sulphur fuels in high-risk geographical areas or during critical periods, notably in large industrialised centres when meteorological conditions do not favour large-scale mixing, hence diffusion, of gaseous pollutants. SO_2 can also be scrubbed using solid minerals such as dolomite, however, the process is only effective to a certain extent, poorly understood, and is hence poorly implemented and costly.

INDEX

This index refers to the subject defined by the word listed as well as to its immediate context. Each word or expression is followed by page numbers, with the principal references indicated in bold type.

Acid rain : **358**.

Activated complex : 79, **80**.

Activation (*see also* **Arrhenius**)
- energy, E : **77**, 79, 82, 85.
- temperature, $T_a = E/R$: **77**.

Active centre : **84**, 87, 127.

Adiabatic
- combustion temperature, : 33, **40**, 43, 243, 245, 246, 300, 302, 304.
- transformation : **32**, 33, 40.

Advancement (or extent) of reaction : **37**, 42, 63.

Anchoring (of a flame) (*see* **Attachment**).

Anti-knock : **135**.

Arrhenius' law : **77**, 78-80, 88, 89, 139, 142, 144, 227.

Attachment (of a flame) : **184**, 185, 186, **233**.

Autoturbulence : **159**.

Balance equations
- basic : **104**, **106**, **117**, 120, 121.
- energy : **105**, **107**, 137, 141, 163, 261, 290.
- mass : **107**, 144, 162, 168, 177, 247, 249, 253, 290.
- momentum : **107**, 162, 290.

Balance stoichiometric reaction, balance reaction : **32**, 61.

Barodiffusion : **99**.

Bath gas : **78**, 136.

Bimolecular reaction : **63**, 79.

Boltzmann constant : **44**, 79, 151.

Bomb calorimeter : **40**.

Boussinesq and Reynolds' approach : **121**.

Buoyancy : 2, **173**, 246, 262.

Burke-Schumann flame : **109**, 111, 174.

Burner
- all types : **18-22**, 153.
- Bunsen : 45, 157, **158**, 181, **182**, 192, 229, 263, 267.
- coal-fired : **22**.
- diffusion flame : **184**.
- flat flame : **67**.
- fuel oil : 21, **22**.
- gas : **19**, **21**, 196, 197, 214.
- Meker : **359**.
- pilot flame : **186**.

Burning velocity (*see* **Velocity**).

Calorific value
- gross (GCV) : **46**, 56.
- net (NCV) : **46**, 57.

Carrier gas : **69**.

Catalytic converter : **364**.

Chain reaction
- all types : **127**.
- branching : **85**, **87**, 136, 146, 147, 170.
- degenerate branching : **87**, 147.
- initiation : **84**.
- link : **84**.
- propagation : **84**.
- straight-chain : **85**, 87.
- termination : **84**, 85, 147.

Chapman-Jouguet (C.J.) condition : 275, **277**, 278, 279, 284, 287, 293, 294, 298.

Chemical equilibrium : **39**, 41-43, 59, 165, 166, 177, 178, 223, 245, 251, 350, 356, 357.

Chemical equilibrium (hypothesis of approximate) (*see* **Infinitely rapid chemical reactions**).

Chemical potential : **38**.
Chemiluminescence : **129**.
Closed system : **31**, 63, 66, 125.
CO (combustion of) : **86**.
CO, CO_2 (pollution by) : **351**, **352**.
Coal (combustion of) : 22, **31**, 35, 127.
Coaldust : **127**.
Compressibility coefficient : **270**.
Conserved scalar method : **251**.
Convection : **2**, 93, 95, 144, **166**, 173, 233.
Critical radius for ignition : 304, **305**, 307, 309.

d^2 rule : **327**, 339.
Damköhler number : **228**, 331.
Deflagration : 23, 153, **154**, **164**, 276, 277.
Deflagration-detonation transition : **272**, 316.
Detonability limits : **284**.
Detonation states (*see also* **Subsonic flow** and **Supersonic flow**)
strong : **277**.
weak : 24, **277**.
Detonation structures
one-dimensional (or straight or planar) detonation : **164**, **269**, **275**, 278, 281, 287, 316.
three-dimensional structures : **284**.
Detonator : **32**.
Diesel : **9**, 33, 353, 356.
Diffusion (of mass)
molecular : **2**, 64, 87, 93, 95-97, **99**, 151.
turbulent : **217**.
Diffusion (of momentum) : 102, 106, 119.
Diffusivity
thermal, laminar : **101**, 161.
thermal, turbulent : **122**.
Drop, droplet combustion : **173**, 342, 344.

Eddy : **200**, 201.
Eddy break-up (or EBU) : **223**, 252, 346.
Elementary chemical reaction : 31, **62**, 63, 72, 73, 77, 79, 85.
Enthalpy
of formation : **35**, 56.
of reaction : **36**.
stagnation : **105**.
thermodynamic function : **34**, 38, 163, 169.
Entropy : **38**, 280, 292, 356.
Equilibrium constant : **40**, 42, 59.
Equivalence ratio : 43, **47**.
Ergodic assumption : **199, 216**.
Eulerian approach : **346**.
Explosion : 33, 86, 87, **125**-**127**, 129, 139, 140, 143-145, 148.
Explosion limits : 86, 87, **129**, 131-133, **134,** 148.
Explosive : **32**.
Extent of reaction (*see* **Advancement**).

Fanno curve : **291**, 292, 294.
Favre mean : **119**, **200,** 223, 230, 246, 261.
Fick's law : **97**, **99,** 110, 122.
Final (maximum) temperature (*see also* **Adiabatic combustion temperature**) : 52-55.
Fire : **23**, 126.
Firedamp : **127**.
Fire point : **134**.
Flame (explosion, self-ignition)
cool flame : 131, **132**, 144.
decomposition flame : **32**, 130, 131.
long-delay flame : 131, **132**.
normal flame : **132**.
Flame propagation
counterflow : **235**, 245.
flat : 65, **67**, 157, 360.
laminar diffusion : 109, 153, **173**, 174, 176.
laminar premixed : 66, 153, **155,** 158, 162, 168, 180, 186.
turbulent diffusion : **233**, 237, 245.
turbulent premixed : 195, 196, **206**, **207**, 216, 221, 229.
Flame shape
jet flame : 109, **173**, 175, 176, 180, **183**, 232, 249, 254-256, 258, **342**.
oblique flame : 12, **195**.
one-dimensional : **64**.
pilot flame : **186**.
triple flame : **184**, 185, 233, 238, 330, 336.
Flame, flamelet structures in turbulent premixed flows
thickened : **207**, 208-212, 214, 215.

wrinkled : 196, **207**, 208, 211, 213, 236.
wrinkled-thickened : **207**, 208-213.

Flame, flamelet structures in turbulent diffusion flows : **234**, 237, 238.

Flame front : 154-156, **157**, 165, 175.

Flame holder, flame holding (*see* **Stabilisation** in a recirculation zone).

Flame sizes
height, length : **113**, 116, 122, **175**, 176, 178, 198, 246.
thickness : **111**, 116, 122, 157, 158, 165, 167, 168, 304.

Flammability limits : **171**, 172, 284.

Flash point : **134**.

Fluid particle : **109**, **116**, 117, 119, 254, 345, 346.

Flow
mean turbulent : **117**.
laminar : **153**, 158, 173, 174.
locally-homogeneous : **346**.
plug-flow : **69**.
turbulent : 7, 69, **116**, 173, 195, 285.

Flow velocity (*see* **Velocity**).

Fourier's law : **100**, 101, 166, 304.

Frank-Kamenetskii : **137**, 149, 301.

Free energy : **38**.

Free enthalpy : **38**.

Free radical (*see also* **Active centre**) : 72, 78, 82, 83, **84**.

Frequency factor (*see* **Pre-exponential factor**).

Friction tensor : 102, **103**.

Fundamental (or normal) burning velocity (*see* **Velocity**).

Fuel lean, fuel rich : 42, **47**, **48**.

Gibbs function (*see* **Free enthalpy**).

Glow (*see also* **Chemiluminescence**) : **86**.

Grebel line : **49**.

Greenhouse effect : **351**.

Group combustion : **334**, 335, 337.

Heat conduction (heat diffusion) : **100**, 101, 137, 138, 156, 197.

Heterogeneous : **362**, 363.

Homogeneous : **95**, 203, 346.

Hugoniot, Hugoniot-Rankine : **163**, 274, 283, 290, 293.

Hydrocarbons (combustion of) : **87**.

Hydrogen (combustion of) : **86**, 91.

Ignition
by auxiliary flame : **181**, 183.
by energy deposition : 299-301, **308**, 309-312.
by hot zone, hot spot : **302**, 303.
by spark : **7**, 8, 131,172, 181.
self-ignition : 33, **126** (*see also* **Explosion**).

Ignition, explosion temperature : **139**, 167.

Ignition with consecutive pressure perturbations : **312**.

Incomplete combustion : 41, **48**, 50, 51, 350, 352.

Incomplete reaction : **41**.

Induction delay (or period) : **87**, 129, 131, 148, 316.

Infinitely rapid chemical reactions hypothesis : 95, 109, **113**, 174, 188, 233, 245-248, 251, 256, 322.

Inhibition, inhibitor : **170**.

Internal energy : **34**, 163.

Irreversible processes (thermodynamics of) : **98**.

Isotherm, isothermal : **146**, **164**, **180**, 210.

Isotropic : **201**, 202, 203.

k-ϵ turbulence model : **204**, 261.

Kinetic energy of turbulence : **200**, 261.

Kinetic mechanism : **62**, 85, 89, 90.

Knock, knocking (or pinking) : **131**, 132, 135, 354.

Kolmogorov (*see* **Scales of turbulence**).

Lagrangian approach : **203**, 347, 348.

Lead (pollution by) : **353**.

Lewis number : **102**, 169.

Liquid fuel : 13, 16, 25, 134, 319, **320**.

Lobe : **131**.

Local chemical equilibrium (hypothesis) (*see also* **Infinitely rapid chemical reactions**) : **180**, 183.

Locally homogeneous flow approach (LHF) : **346**.
Longwell reactor (*see* **WSTR**).
Low-sulphur fuel : **366**.

Mach
number : **277,** 280, 281, 293.
reflection : **285,** 286.
stem : **285, 286**.
Magnussen model : 252.
Mass action (law of) : **40**.
Mass fraction : **96**.
Mean free path : **100**, 151.
Mean residence time : **188**.
Mist (*see* **Drop**).
Mixture fraction : **243**, 249-251.
Mole fraction : **99**, 192.
Molecular collisions : 44, **79**.
Molecularity : **62**, 77.
Mollier diagram : **290**, 292, 296.

Navier-Stokes equations : **119**, 261.
Negative temperature dependency : **144**, 145.
Newton's law : **102**.
NO_x (pollution by) : 59, **357**, 358.

Octane number : **135**.
Open system : **31**, 64, 66.
Ostwald diagrams : **48**, 60.
Oxidant : **31**.
Oxidation number : **32**.
Ozone layer : **358**.

PAH, polyaromatic hydrocarbons (pollution by) : **355**.
PAN (pollution by) : **353**, 354, 358.
Peclet number : **116**, 161,176, 190, **331**.
Peninsular region : **148**.
Percolation combustion : **333**, 335, 337.
Pinking (*see* **knocking**).
Plug flow reactor : **69**.
Plume, flame envelope : **46**.
Pockets (combustion in) : **334**, 335, 337.
Potential energy : **80**, 81-83.
Prandtl number : **103**, 122.
Pre-exponential factor : **77**.
Probability density function (PDF, P) : 200, 211, **212**, 213-215, 226, 243, 248, 250, 254, 255.
Profiles of temperatures and concentrations : **127**, **144**, 146, 160, 161, 180.
Progress variable : **223**, 226, 227, 229.
Promoter : **170**.
Prompt NO : **359**, 361.
Propagating centre (*see* **active centre**).
Propellant (mono, bi ...) : 15, 17, 18, 27, **32**, 42, 183, 287, 319.
Pyrolysis : **89**, 90.

Quenching
chemical : **43-44**.
diameter : **161**, 190.

Radiation : 3, 4, 23, **246**, 320.
Ramjet : **10**, 195, 289.
Rayleigh (equation, line, curve) : **162**, 164, **275**, 276, 278, 282, 290-294.
Reaction coordinate : **80**, 82.
Reaction degree (*see* **Advancement of reaction**).
Reaction order (partial, overall) : **75**, 76.
Reaction rate : 63, 64, 73, 75, 77, 79, 89, 91, 137, 167, 168, 187, 191.
Recirculation zone : 12, **183**, 185, 194, 195.
Reynolds
approach : **117**.
equations : **119**.
number : **117**, **203**.
tensor : **119**, 200.
Rocket : **15**, 16, 17, **32**, 42.

Scales of turbulence
integral length : 200, **201**, 235.
integral time : **203**.
Kolmogorov length : **202**, 203, 207, **209**, 215, 232, 236.
Kolmogorov time : **209**.
spectrum of turbulence length scales : **201**, 202.
Schmidt number : **103**, 110, 122, 185.
Self-ignition (*see* **Explosion**).
Shock : 72, **269**, 273, 275, 276, 278, 280.

Shock tube : **71**.
Simulation : **74**, **363**.
Single chemical reaction (*see* **Infinitely rapid chemical reaction**).
Slip stream, slip line : **285**, **286**.
Slow reaction (oxidation) : **86,** 128, 130.
Smoke : **355**.
Solid fuel : **23,** 31, 362.
Solid particle, soots : 355, **356**, 357.
Soret effect : **99**.
SO_x (pollution by) : **365**.
Spark-ignition engine : **7**, 33, 135, 197, 353, 356.
Speed of sound (of shock) : **270**, 271, 272, 274, 277, 279, 288, 289.
Spray (*see* **Drop**).
SSA, **Steady-state approximation** : **73**, 89, 148, 358.
Stabilisation (laminar flames)
blow-off : 181, **182**.
flash-back : 181, **182**.
of a flat flame : **67**.
of diffusion jet flames : **183**, 184.
in a recirculation zone : **185**, 189, 194.
on a Bunsen burner : **158**, 181, **182**.
Stabilisation (turbulent flames)
in a recirculation zone : **10**, 11.
lifted flame : **233**.
Stagnation (or total) enthalpy : **105**, 108.
State function : **34**, 38.
Steady-state : **64**, 66, 108, 110, 138, 140, 168.
Step (*see* **Elementary chemical reaction**).
Stoichiometry : **32**, 34, 57, 110, 168.
Stratified combustion : **362**.
Strong detonation : **277**.
Subsonic flow : **277**, 292.
Supersonic flow : 24, **277**, 280, 289, 292.
Swirl : **21**, 233.

Temperature (concept of) : **44**.
Temperature fluctuations : **211**.
Termolecular reaction : **63**, 83.
Thermal conductivity : **100**, 138, 151, 166.
Thermal equilibrium : 44, **45**.
Thermal theory of explosions : **137**.
Thermodiffusion (*see* **Soret effect**).
Third body : **83**, 147.
Toor (model of) : **247** (*see also* **Infinitely rapid chemical reactions hypothesis**).
Transition state : **80**.
Turbojet : **13**, **14**.
Turboprop : **13**.
Turbulence dissipation rate : **204**.
Turbulence models
k-ε : **204**.
premixed turbulent flames : **221**, 222.
non-premixed turbulent flames : **245**.
Turbulence production : **117**, **196**, 261, 285.

Unburnt (*see* **Incomplete combustion**).
Unimolecular reaction : **63**.

Velocity
flow : **154**, 162, 174, 185, 273, 277.
fundamental (or normal) burning : **154**, 155, 168, 170, **217**.
propagation, spatial burning : 64, **154**, **155**.
Very fast chemical reactions hypothesis : (*see* **Infinitely rapid chemical reactions hypothesis**).
Viscosity
dynamic : **102**.
kinematic : **103**, 117, 122, 185, 203, 205.

Wall effect : 69, **138**, 141, 147, 148, 161, 181, 272.
Weak detonation : **277**.
WSTR, Well stirred (or continuously-stirred) tank reactor : **69**, 186, 214, 359.

Zeldovitch mechanism : **358**, 366.
Zeldovitch variable : 177.

ACHEVÉ D'IMPRIMER

EN SEPTEMBRE 1998

PAR L'IMPRIMERIE NOUVELLE

45800 SAINT-JEAN-DE-BRAYE – 36357

N° d'éditeur : 991

Dépôt légal : septembre 1998

IMPRIMÉ EN FRANCE